Cell and Tissue Culture in Forestry

FORESTRY SCIENCES

Baas P, ed: New Perspectives in Wood Anatomy. 1982. ISBN 90-247-2526-7

Prins CFL, ed: Production, Marketing and Use of Finger-Jointed Sawnwood. 1982. ISBN 90-247-2569-0

Oldeman RAA, et al., eds: Tropical Hardwood Utilization: Practice and Prospects. 1982. ISBN 90-247-2581-X

Den Ouden P and Boom BK: Manual of Cultivated Conifers: Hardy in Cold and Warm-Temperate Zone. 1982. ISBN 90-247-2148-2

Bonga JM and Durzan DJ, eds: Tissue Culture in Forestry. 1982. ISBN 90-247-2660-3

Satoo T and Magwick HAI: Forest Biomass. 1982. ISBN 90-247-2710-3

Van Nao T, ed: Forest Fire Prevention and Control. 1982. ISBN 90-247-3050-3

Douglas J: A Re-appraisal of Forestry Development in Developing Countries. 1983. ISBN 90-247-2830-4

Gordon JC and Wheeler CT, eds: Biological Nitrogen Fixation in Forest Ecosystems: Foundations and Applications. 1983. ISBN 90-247-2849-5

Németh MV: The Virus-Mycoplasma and Rikettsia Disease of Fruit Trees. ISBN 90-247-2868-1

Duryea ML and Landis TD, eds: Forest Nursery Manual: Production of Bareroot Seedlings. 1984. ISBN 90-247-2913-0

Hummel FC, ed: Forest Policy: A Contribution to Resource Development. 1984. ISBN 90-247-2883-5

Manion PD, ed: Scleroderris Canker of Conifers. 1984. ISBN 90-247-2912-2

Duryea ML and Brown GN, eds: Seedling Physiology and Reforestation Success. 1984. ISBN 90-247-2949-1

Staaf KAG and Wiksten NA: Tree Harvesting Techniques. 1984. ISBN 90-247-2994-7

Boyd JD: Biophysical Control of Microfibril Orientation in Plant Cell Walls. 1985. ISBN 90-247-3101-1

Findlay WPK, ed: Preservation of Timber in the Tropics. 1985. ISBN 90-247-3112-7

Samset I: Winch and Cable Systems. 1985. ISBN 90-247-3205-0

Leary RA: Interaction Theory in Forest Ecology and Management. 1985. ISBN 90-247-3220-4

Gessel SP: Forest Site and Productivity. 1986. ISBN 90-247-3284-0

Hennessey TC, Dougherty PM, Kossuth SV and Johnson JD, eds: Stress Physiology and Forest Productivity. 1986. ISBN 90-247-3359-6

Shepherd KR: Plantation Silviculture. 1986. ISBN 90-247-3379-0

Sohlberg S and Sokolov VE, eds: Practical Application of Remote Sensing in Forestry. 1986. ISBN 90-247-3392-8

Bonga JM and Durzan DJ, eds: Cell and Tissue Culture in Forestry. Volume 1: General Principles and Biotechnology. 1987. ISBN 90-247-3430-4

Bonga JM and Durzan DJ, eds: Cell and Tissue Culture in Forestry. Volume 2: Specific Principles and Methods: Growth and Developments. 1987. ISBN 90-247-3431-2

Bonga JM and Durzan DJ, eds: Cell and Tissue Culture in Forestry. Volume 3: Case Histories: Gymnosperms, Angiosperms and Palms. 1987. ISBN 90-247-3432-0
(Set Vols. 1–3: ISBN 90-247-3433-9)

Cell and Tissue Culture in Forestry

Volume 3
Case Histories: Gymnosperms, Angiosperms and Palms

edited by

J.M. BONGA and DON J. DURZAN
Canadian Forestry Service / University of California

1987 **MARTINUS NIJHOFF PUBLISHERS**
a member of the KLUWER ACADEMIC PUBLISHERS GROUP
DORDRECHT / BOSTON / LANCASTER

Distributors

for the United States and Canada: Kluwer Academic Publishers, P.O. Box 358, Accord Station, Hingham, MA 02018-0358, USA
for the UK and Ireland: Kluwer Academic Publishers, MTP Press Limited, Falcon House, Queen Square, Lancaster LA1 1RN, UK
for all other countries: Kluwer Academic Publishers Group, Distribution Center, P.O. Box 322, 3300 AH Dordrecht, The Netherlands

Library of Congress Cataloging in Publication Data

Cell and tissue culture in forestry.

(Forestry sciences)
Rev. ed. of: Tissue culture in forestry. 1982.
Includes index.
Contents: v. 1. General principles and biotechnology -- v. 2. Specific principles and methods -- v. 3. Case histories--gymnosperms, angiosperms, and palms.
1. Plant cell culture--Collected works. 2. Plant tissue culture--Collected works. 3. Trees--Propagation--In vitro--Collected works. 4. Forests and forestry--collected works. I. Bonga, J. M. II. Durzan, D. J. III. Title: Tissue culture in forestry. IV. Series.
SD403.5.C45 1987 634.9'56 86-23767
ISBN 90-247-3433-9 (set)

ISBN 90-247-3432-0

PRINTED IN THE NETHERLANDS

VOLUME 3

TABLE OF CONTENTS

1. INTRODUCTION
1.1. Distribution
1.2. Economic importance

CONTRIBUTORS

ABOEL-NIL MM, Weyerhaeuser Company, Tacoma, Wash., 98477, USA

AGRAWAL DC, Biochemistry Division, National Chemical Laboratory, Pune - 411008, India

AHUJA MR, Institute of Forest Genetics, Sieker Landstrasse 2, D-2070 Grosshansdorf, Germany

BALL EA, University of California, Santa Cruz, Redwood Institute, Applied Sciences Building, Santa Cruz, Calif. 95064, USA

BIGOT C, Ministere de l' Agriculture, Ecole Nationale Superieure d'Horticulture, Ecole Nationale Superieure du Paysage, 4 Rue Hardy, 78009 Versailles Cedex, France

BORNMAN CH, Cell and Tissue Culture Laboratory, Hilleshog AG, Box 302, S-261 23 Landskrona, Sweden

BOULAY M, AFOCEL, Laboratoire de Physiologie, Domaine de l' Etancon, 77370 Nangis, France

CHALUPA V, Forestry and Game Management Research Institute, Jiloviste-Strnady, 255 01 Praha 5, Zbraslav-Strnady, Czechoslovakia

DEBERGH P, Rijksuniversiteit Gent, Faculteit van de Landbouwwetenschappen, Laboratorium voor Tuinbouwplantenteelt, Coupure Links, B-9000 Gent, Belgium

DRIVER JA, Plant Research Laboratory, Driver Nursery, 2737 North Avenue, Modesta, Calif. 95351, USA

DURZAN DJ, University of California, Davis, Department of Pomology, 1035 Wickson Hall, Davis, Calif. 95616, USA

ENGELMANN F, Ministère de l' Agriculture, Ecole Nationale Supérieure d'Horticulture, Ecole Nationale Supérieure du Paysage, 4 Rue Hardy, 78009 Versailles Cedex, France

GUPTA PK, University of California, Davis, Department of Pomology, 1035 Wickson Hall, Davis, Calif. 95616, USA

HORGAN K, Forest Research Institute, Private Bag, Rotorua, New Zealand

ISIKAWA H, Siviculture Division, Forestry and Forest Products Research Institute, P.O. Box 16, Tsukuba Norin Kenkyu Danchi, Kukizaki-machi, Ibaraki-ken 305, Japan

JELASKA S, University of Zagreb, Department of Botany, Rooseveltov trg 6/III, 41001 Zagreb, P.O. Box 933, Yugoslavia

JOHN A, Forestry Commission, Northern Research Station, Roslin, Midlothian, EH25 9SY, England

JORDAN M, Universidad Catolica de Chile, Laboratoria de Botanica, Facultad de ciencias Biologicas, Casilla 114-D, Santiago, Chile

KANN RP, State University of New York at Stony Brook, Department of Biochemistry, Stony Brook, N.Y. 11794-5215, USA

KENDURKAR SV, Biochemistry Division, National Chemical Laboratory, Pune - 411008, India

KHUSPE SS, Biochemistry Division, National Chemical Laboratory, Pune - 411008, India

KRIKORIAN A, State University of New York at Stony Brook, Department of Biochemistry, Stony Brook, N.Y. 11794-5215, USA

KULKARNI VM, Biochemistry Division, National Chemical Laboratory, Pune - 411008, India

MAENE L, Rijksuniversiteit Gent, Faculteit van de Lanbouwwetenschappen, Laboratorium voor Tuinbouwplantenteelt, Coupure Links, B-9000 Gent, Belgium

MASCARENHAS AF, Biochemistry Division, National Chemical Laboratory, Pune - 411008, India
MCCOWN BH, University of Wisconsin-Madison, Department of Horticulture, 1575 Linden Drive, Madison, WI 53706, USA
MCCOWN DD, Knight Hollow Nursery, Inc., 2433 University Avenue, Madison, WI 53705, USA
MCGRANAHAN G, University of California, Davis, Department of Pomology, 1035 Wickson Hall, Davis, Calif. 95616, USA
MEHRA-PALTA A, International Paper Company, P.O. Box 797, Tuxedo Park, New York, N.Y. 10987, USA
MEHTA UJ, Biochemistry Division, National Chemical Laboratory, Pune - 411008, India
NAGMANI R, The Institute of Paper Chemistry, P.O. Box 1039, Appleton, Wis. 54912, USA
NAIR S, Biochemistry Division, National Chemical Laboratory, Pune - 411008, India
OHYAMA K, National Institute of Agrobiological Resources, Department of Cell Biology, Yatabe, Tsukuba, Ibaraki, 305 Japan
OKA S, National Institute of Agrobiological Resources, Department of Cell Biology, Yatabe, Tsukuba, Ibaraki, 305 Japan
PARANJOTHY K, Biology Division, Palm Oil Research Institute of Malaysia, No.6, Persiaran Institusi, Bandar Baru Bangi, Selangor, Malaysia, P.O. Box 10620, Kuala Lumpur, Malaysia
SHAW D, University of California, Davis, Department of Pomology, 1035 Wickson Hall, Davis, Calif. 95616, USA
SOMMER HE, University of Georgia, School of Forest Resources, Athens, Georgia 30602, USA
THOMPSON DG, International Paper Company, P.O. Box 797, Tuxedo Park, New York, N.Y., 10987, USA
TISSERAT B, US Department of Agriculture, Agricultural Research Service, Western Region, Fruit and Vegetable Chemistry Laboratory, 263 South Chester Avenue, Pasadena, Calif. 91106, USA
TULECKE W, Antioch College, Yellow Springs, Ohio 45387, USA
VENKETESWARAN S, University of Houston, Central Campus, Department of Biology, Houston, Tex. 77004, USA
VILLALOBOS VM, Colegio de Postgraduos, Institucion de Ensenanza e Investigacion en Ciencias Agricolas, Chapingo, Mexico
WEBB DT, Agrogen Biotechnologies Inc., Agrogen Building, 520 W. 6th Avenue, Vancouver, B.C., V5Z 4H5, Canada
WEBB KJ, Welsh Plant Breeding Station, Plas Gogerddan, Aberystwyth, SY23 3EB, Dyfed, England
WETZSTEIN HY, University of Georgia, School of Forest Resources, Athens, Georgia 30602, USA

1. INTRODUCTION

This is the third in a series of three new volumes dealing with cell and tissue culture of forest trees, a research area that has developed very rapidly since our 1982 Edition.

The first volume deals with general aspects of nutrient media and physical elements of the culture environment, and contains chapters on the genetics and physiology of clonal propagation. In addition, because of the increasing importance of the subjects, it contains several chapters discussing genetic variation, genetic engineering, and ultrastructure.

The second volume deals with specific topics, in particular in the field of growth and development. Protoplast culture, haploid and triploid culture, embryo rescue, and somatic embryogenesis are important elements in attempts to obtain genetically improved sexual and clonal stock. Progress in these areas is covered by several chapters. Various aspects of morphogenesis and organogenesis, including positional effects, are also discussed. Now that more species are being micropropagated every year, it is increasingly important to pay attention to problems with the finished products, e.g., plantlets, artificial seeds. Therefore, chapters on pathology, mycorrhizae, senescence, cold storage, germplasm preservation, and nursery practice are included.

In the third volume, case histories are presented for a wide variety of gymnosperm and angiosperm (including palm) species. These species represent a wide range of geographic areas. In some cases, micropropagation efforts have recently been started and large-scale field-testing has not yet been carried out. For others, the initial large-scale field-testing is completed and micropropagation is practiced on a commercial scale.

In summary, the three volumes cover a wide range of research activities in the field of cell and tissue culture of forest trees, attempt to identify future trends, and where possible, identify the appropriate biotechnologies.

2. PICEA ABIES

CH BORNMAN

1. INTRODUCTION

Picea abies (L.) Karst. [formerly *P. excelsa* (Lam.) Link], known as Common, European or Norway spruce, is one of the most important timber and ornamental conifers in mid- and southeastern Europe, northern Europe, and Siberia (60), playing a vital role in the international trade balance of at least two Nordic countries, Sweden and Finland. Forty-five percent of the growing stock of Sweden's forests, which occupy the northern temperate coniferous belt and the southern temperate mixed forest belt, consist of *P. abies*, the balance being made up mainly of *Pinus sylvestris*, but also of *Pinus contorta* (introduced from northwestern America) and *Betula* spp.

Productive forestry in Sweden comprises about 60 percent of the total land area and the volume of wood, that is total trunk volume including bark, logged annually is 70 million m^3. Since the volume of annual felling is just about equivalent to that of annual growth, the future expansion of this country's forest industry is expected to be closely linked to attempts aimed at shortening the present 70 (southern Sweden)- to 140 (northern Sweden)-year rotation cycles, as well as to improvement in vigour, fibre quality and enhanced tolerance to low temperature stress, pathogens and insect attack. Precisely these features demand serious consideration of the clonal option, an option that is potentially attainable through vegetative macro- and micro-propagation.

Most economically important agronomic, vegetable and forest tree crops are seed propagated and hitherto there has been little incentive for vegetative propagation. However, it is expected that at least a part of the genetic variation not normally transmitted intact through sexual reproduction can be fixed by interbreeding of homozygous trees or by vegetative clonal propagation.

The objective with this chapter is not to review the existing extensive literature on *Picea abies*, but rather to assemble what we know about the response of this species to *in vitro* culture technology and to evaluate realistically the goal of mass clonal micropropagation. After all, ultimately it is upon the performance of the tissue-cultured spruce tree in terms of competitive price, superior quality, increased growth and enhanced stress tolerance that the technology of *in vitro* culture will stand or fall. Like nearly all members of the Pinaceae, *P. abies* has displayed recalcitrance to manipulation *in vitro*, but despite its relative intractability it remains one of the most interesting conifers for *in vitro* culture studies for the following reasons:

(1) detailed anatomical, morphological and physiological information is available on its leaf, embryonic shoot and apical meristem (14, 19, 27, 37, 38, 40, 46-48);

(2) the number of reports on the culture of its tissues *in vitro* has accumulated rapidly over the past decade (6-12, 15-18, 28, 31-34, 39, 48, 49, 53, 57, 61-67);

(3) cotyledons, leaves of flushing buds and leaf primordia of resting buds, axillary buds, megagametophytes and immature embryos are among the variety of organs and tissues that are being exploited;

(4) it was the first coniferous species in which somatic embryogenesis was unequivocally demonstrated (29-30); and

(5) it is the member of the Pinaceae from which the largest number of plants is propagated commercially through the rooting of cloned cuttings (ca 3-5 million annually) (2).

2. TERMINOLOGY

A number of terms will be used and to avoid ambiguity they are defined as follows:

adventitious - arising in an abnormal position and implying cellular dedifferentiation.

brachyblast - used synonymously with needle bundle, fascicle, short shoot, dwarf shoot, dwarf spur; specifically refers to the suppressed bud-like structure of meristematic, quiescent tissue extended by secondary, linear leaves (pseudophylls) borne in fascicles and surrounded at the base by sheaths of bud scales; in the genus Pinus.

bud - compact, undeveloped, short-stemmed shoot with immature leaves; terminal, axillary, lateral, adventitious, preventitious, epicormic.

clone - a population of cells derived from a single cell by mitoses; a group of plants that have arisen by vegetative reproduction from a single parent, having identical genetic material.

cotyledon - seed leaf, cotyledonary needle, embryonic leaf.

embryo - zygotically-produced sporophytic plant, before germination in seed plants; the sequence of embryo development is referred to as embryogeny.

embryoid - somatically-produced sporophytic plant, embryo-like in structure; the sequence of embryoid development is referred to as embryoidy, adventive embryogeny, somatic embryogeny.

endosperm - nutritive tissue formed within the angiosperm embryosac, from the central cell containing the primary endosperm nucleus.

epicormic bud - preformed, suppressed bud, formed in continuity with the apical meristem; distinguished from an adventitious bud by the presence of a vascular trace to the primary xylem (13).

maturation - a concept broadly interrelated with juvenility, rejuvenation, adolescence, ageing, maturity and senescence; specifically connected with decline; can be considered in chronological, ontogenetical and physiological terms.

megagametophyte - female gametophyte in heterosporous plants such as the gymnosperms; embryosac within the angiosperm ovule.

meristemoid - an active locus of growth; a nodule of undifferentiated tissue from which new cells and/or adventitious structures arise; used originally to denote differentiated tissues such as root hairs, stomata, procambial strands.

preventitious bud - rudimentary and/or suppressed endogenous axillary bud primordium present in an apparently "empty" leaf axil (25).

primary leaf - scale-like and deciduous, bearing in its axil the secondary linear leaves (see brachyblast).

primordium - a cell or organ in its earliest stage of differentiation; post-meristemoidal stage.

secondary leaf - see brachyblast.

somaclonal variation - genetic variability generated by plant cell and protoplast culture (clones derived somatically, somaclones); also protoclonal variation, protoclones; calliclones.

sphaeroblast - nodule of wood, separated from the central woody cylinder, gives rise to shoots with juvenile characters (particularly in apples, pears) (3); trace bud. (Has also been used to describe an embryo or embryoid whose cotyledons are not separated, giving rise to a shoot with a globular mass.)

Parallel terms: the following terms (suggested by Dr. W. J. Libby) are used to distinguish, and establish equivalence between, the processes of deriving plants from seeds, from cuttings _in vivo_ and from explants _in vitro_ -

1. seed ⟶ germinant ⟶ seedling
2. cutting ⟶ rooted cutting ⟶ steckling
3. explant ⟶ plantlet ⟶ plantling

3. APPLICATION OF _IN VITRO_ CULTURE TECHNOLOGY TO _PICEA ABIES_

Figure 1 illustrates the spectrum of present day _in vitro_ culture techniques that are used in economic crop plant improvement and against which the potential applications to _P. abies_ will be considered. Protoplast, cell, tissue and organ culture techniques are broadly grouped into the 10 applications shown to the left of Fig. 1. In the centre of this figure the ideal progression from protoplast to whole plant is shown, but although protoplasts and cells can be isolated, or even callus regenerated from some tissues of mature trees, in reality it is only from the tissues of embryonic to early juvenile plants (e.g. 14-day- or 0.03-year-old, Fig. 1) that sustained organogenesis is achievable at present. Extending this limit by two orders of magnitude to 3-year-old plants or by three to 30-year-old trees would constitute a significant advance in Norway spruce research. To the right of Fig. 1 is shown the role of the ovule and its component (e.g. immature embryonic) and adjacent (e.g. nucellar) tissues in the production of somatic embryos as well as in the potential production of parthenogenetic haploid embryos and apomictic seed.

4. BASIC RESEARCH

It is upon long-term basic research in at least two key areas that the successful outcome of applications 2 to 10 (Fig. 1) will eventually depend. These areas are the regulation of development and the control of maturation.

4.1. Regulation of development

In order to gain a better understanding of the regulation of development, that is of growth, differentiation and morphogenesis, increasing use is being made of *in vitro* culture techniques. However, in tissue culture the normal synchronized sequences of events that constitute an organism's vegetative and reproductive life cycle are usually dramatically disturbed. Then, in efforts to restore the interrupted developmental processes and in turn to achieve the desired response, many tissue culturalists rely almost solely on manipulation of plant growth regulators, ignoring other important contributing factors such as genotype, physical and chemical environments, and stage of development and/or pretreatment of the donor plants (8, 10, 12, 20).

If it is assumed that morphogenesis in tissues cultured *in vitro* is best achieved by the appropriate manipulation of growth regulator levels in the culture medium, the investigator should at least be aware that in some explant systems, development seemingly cannot be controlled simply via the quantitative interaction of exogenously applied auxin and cytokinin. Two of many examples will be used to illustrate this aspect. In *P. abies*, it has been observed (8) that the concentration of exogenous auxin which, in conjunction with cytokinin, promotes shoot bud initiation midway along a leaf excised from a flushing bud simultaneously inhibits initiation at the leaf's base. Omission of auxin is necessary for optimal induction of competent shoot buds and subsequent development of shoots. Apparently, the level of endogenous auxin decreases progressively from the leaf's base to its apex, so that the concentration of exogenously applied auxin which promotes morphogenesis from differentiated cells part-way along the leaf's axis, suppresses it in mestimatic cells situated in the cushion of the leaf's base. Wernicke *et al.* (68) observed a similar phenomenon in young leaves of *Triticum aestivum*. Low concentrations of 2,4-dichlorophenoxyacetic acid resulted in adventitious root formation from basal highly immature parts of the young leaves, whereas twofold higher levels had to be applied to stimulate growth from more mature parts toward the apex of the leaf although simultaneously suppressing it in the younger parts.

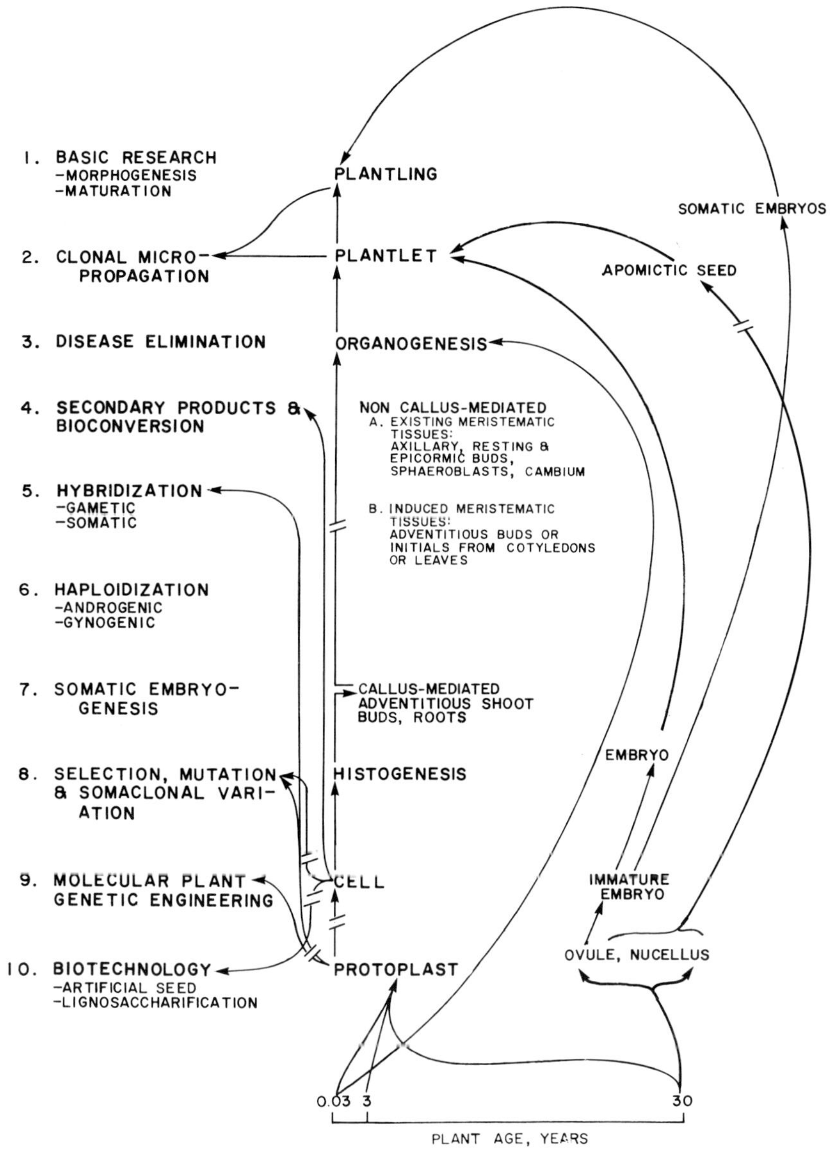

FIGURE 1. Actual and potential applications of *in vitro* technology to *Picea abies*. Activities of *in vitro* culture technology in the clonal improvement and propagation of *Picea abies* can be grouped into 10 applications. The existing

as well as the potential uses of protoplasts, cells, calli and regenerated plants are indicated by interconnecting arrows.

In the centre of Fig. 1 the ideal progression from protoplast to plantling is shown. Although protoplasts can be isolated from appropriate tissues (e.g. from needles of resting or flushing buds) of plants of any age (0.03, 3 and 30 years are shown), only cells in the second division stage (and therefore neither callus nor plantlings) can be regenerated at present. Sustained organogenesis is possible only from embryonic tissues (0.03 year- or two-week-old cotyledons). To the right is shown the role of the immature zygotic embryo in the induction of somatic embryos, and the potential role of the ovule and its component tissues in the production of apomictic seed. The ovule, also, is a potential source for the production of parthenogenetic haploid embryos.

Exogenously applied auxins alone are often effective for the initiation and maintenance of tissues in culture, especially members of the Poaceae. Similarly, as will be shown later, it would appear that cytokinins alone are effective in the induction of adventitious structures in conifer tissues cultured _in vitro_. This does not, of course, preclude interaction between exogenously applied plant growth regulators and the endogenous hormones. For a critical understanding of developmental regulation, information is required not only on endogenous plant growth regulator levels, but also on how these levels change in specific explant sites during specific differentiation responses. With very few exceptions in tissue culture generally and conifer tissue culture specifically, this information is not available. However, Fosket (26) believes that having such information will not necessarily insure adequate explanations of developmental phenomena. Nevertheless, even in the absence of analytical data on changing endogenous plant growth regulator levels, observations on the rate of macromolecular synthesis, carbon metabolism, uptake kinetics (9, 34, 41, 57-59, 61) as well as on cell and tissue sites and cytological response (32, 71) do allow certain deductions to be made with regard to growth regulator-induced control of development in coniferous tissues.

4.1.1. _Role of cytokinin_. As far as it is affected by the exogenous application of plant growth regulators, development in _Picea abies_ tissues cultured _in vitro_ appears primarily to be influenced by cytokinin. Work over a number of years on 7-day-old _Pinus radiata_ cotyledons in the laboratory of T. A. Thorpe, Calgary, Canada (1, 5, 41, 56, 58, 59, 71), which probably is applicable to other conifers, has led to the conclusion that (1) cytokinin should be present in an agarified nutrient medium during the first 72 h of culture, but that changes in rate and distribution of synthesis of macromolecules, developmental pathways and mitotic indices are induced as early as the first 24 h; (2) cytokinin is directly involved in adventitious bud induction; (3) cytokinin and photomorphogenic light are required for development of meristematic tissue and shoots, although initially light can be withheld for up to 10 days; and (3) cytokinin acts primarily on cells of the cotyledon's stomatal complexes that are in surface contact with the medium.

In our laboratories in Landskrona and Lund, Sweden, research (6-12, 31-34, 61) on 14-day-old cotyledons, leaves of flushing buds and leaf primordia of embryonic shoots of *Picea abies*, over the past 7 years has shown that (1) hormone gradients exist within a single leaf and that the levels of cytokinin applied exogenously (together with those of endogenous auxins) control qualitatively the type of adventitious structures induced and quantitatively the numbers of adventitious shoot buds and, that additionally, these cytokinin levels also determine the shoot buds' positional development; (2) in terms of morphogenesis, cytokinin and auxin each display both a positive, that is, organogenic and a negative, that is, histogenic effect since individually each may induce either adventitious shoot buds (cytokinin) or adventitious roots (auxin), whereas increasing the level of cytokinin in a root-induction programme or that of auxin in an adventitious bud-induction programme results in the predominant development of callus; (3) account must be taken of the physiological condition of the cotyledon and leaf explants and the degree of anatomical specialization of their epidermal, hypodermal and mesophyll tissues since these display different levels of development within the foliar organ from its apex to its base; and (4) the differentiating subsidiary cells of the stomatal complexes of cotyledons and leaves and differentiating cells of non-determined presumptive leaf primordia appear to be the target cells for cytokinin uptake, becoming primary promeristemoids that are particularly receptive to applied cytokinin.

Uptake of cytokinin by explanted *Cyanara* tissues was shown by Debergh *et al*. (21) to be affected by the matrix potential of the agar nutrient medium. Bornman and Vogelmann (9) studied the effect of gel strength on BA-induced adventitious bud formation and vitrification in *P. abies* and by regression analysis showed a highly significant inverse correlation between ^{14}C-BA accumulation and degree of both agar and non-agar gel stiffness. Although significantly greater numbers of buds per explant were induced at low to medium levels of rigidity (2.5-10 g agar l^{-1}, 1-5 g Gelrite l^{-1}), this advantage was nullified at low levels as a result of the high incidence of vitrification. Vitrification could be avoided by treatment of explants with a 125 µM BA pulse for 2 h, followed by subculturing on a hormone-free nutrient medium gelled with 10 g agar l^{-1}.

Vogelmann _et al_. (61) also used _P. abies_ primary explants to carry out a time course analysis of different concentrations of ^{14}C-BA and found uptake to be linear for the initial 60 min, with the amount of cytokinin taken up reaching a concentration about one-third of that of the medium. Concentration dependence experiments showed that BA uptake was directly proportional to the external concentration and consistent with a passive mode of uptake. After pulsing _P. abies_ embryos with BA, van Staden _et al_. (57) found that they accumulated little BA but that instead BA-nucleotides appeared as major metabolites. They concluded that the effect of BA was mediated via biologically active nucleotides.

4.1.2. _Culture medium_. Although culture media have become chemically better defined, they still contain variables, including for example, type and brand of agar, species of plant growth regulator, methods of preparation, etc. Many media differ only in degree, some combining the macronutrients of one with micronutrients of another. Many different media have been used for _P. abies_ (compare 8, 15, 29, 31, 39, 49, 53, 55, 62, 67) and this species seems to have been broadly tolerant to them all. However, still no maintenance media for growth and yield, or production media for differentiation and development, have been developed. Culture vessels of inadequate design are still being used: vessels that do not allow the basic physiological requirements of evapotranspiration and gas exchange to be fulfilled, resulting in inadequate diffusion of ions and transport of materials.

4.2. _Control of maturation_

The fundamental question posed by Romberger (50) remains: "Does ageing have its basis in events that occur at cellular levels in the meristems themselves, or is it imposed upon the meristems by the remainder of the organism?" The ability to control maturation experimentally is important for two reasons. (1) Attainment of true physiological rejuvenation is basic to the whole process of multicopy production of superior genotypes. At present in _P. abies_ micropropagation is reproducibly achieved only from embryonic, immature and early juvenile tissues and, as is to be expected from an open-pollinated species, this is causing considerable experimental variation (8). (2) Once attained, it is important, for the continued production of stecklings, that juvenility be maintained in the cuttings archive or cutting orchard. In a breeding population, on the other hand, acceleration of maturation may be desirable. This is often induced by artificial promotion of growth and of flowering.

5. CLONAL MICROPROPAGATION

The rationale for developing a system for the mass clonal micropropagation of P. abies at least as far as Sweden is concerned, should be viewed within the general framework of vegetative macropropagation since the factors affecting the latter are almost certainly bound to influence the former. Vegetative macropropagation of Norway spruce stecklings (i.e., rooted cuttings), which is becoming a highly developed technology (2, 23, 24, 35, 36, 52), embraces two aims, namely large-scale multiplication of selected genotypes and production of large numbers of plants from scarce and costly seed that originates from controlled seed orchard pollination (2). However, there are three factors that contribute a measure of uncertainty to both clonal macro- and micropropagation. These are non-genetic effects, forestry legislation and purchaser resistance.

5.1. Non-genetic effects

As regards large-scale multiplication, performance of cuttings, stecklings and transplanted cloned plants is influenced by three factors: (1) cyclophysis, or transmitted qualities that are related to the process of maturation and which involve expression of differences in characteristics such as growth rate, stem form, branching, root quality, time of flushing and age at flowering (42, 43, 51); (2) topophysis, or the positional influence of the ortet, which is often expressed in the exhibition of plagiotropic growth by the steckling some time after rooting, with duration of the reversion phase to orthotropic growth depending on the ramet's physiological age and usually associated with depression of growth (42, 51); and (3) periphysis, or environment-related effects, which sometimes persists long enough to confuse interpretation of cyclo- and topophysical effects.

Production of seed from Norway spruce seed orchards is intermittent and very low, whereas the annual seed consumption of this species in the Nordic countries is very high (about 10^4 kg). Mass micropropagation of plants from scarce and highly expensive seed produced from controlled crossings of desired genotypes is therefore seen as a potentially valuable aid in a vegetative multiplication program. Two advantages relate to the juvenile phase of seedlings derived from such material: the cyclo- and topophysical effects mentioned above are avoided and the lag phase in terms of numbers experienced during the early period of steckling production may be reduced by making use of tissue culture. However, the disadvantages of seedling explant tissues remain, namely their inherent variability and heterogeneity, and the lack of selection methods that allow early selection of potentially useful variability.

5.2. Constraints on clonal forestry imposed by legislation

In order to avoid the potential dangers inherent in monoclonal afforestation, constraining limitations have been placed by legislation on clonal forestry in Sweden. Nine seed orchard or breeding areas have been designated for *P. abies*, with clonal forestry being allowed in only five of these between 1982 and 1989. Five million vegetatively propagated trees can be planted per area, that is, a nation-wide total of 45 million, or 22.5% of the total annual market for spruce planting stock. Legislation further determines the minimum number of (1) clones to be used, (2) trialling sites, (3) years of trialling, (4) unrelated clones to be used, and (5) clonal parents to be employed in a clonal mixture. Also, specified are the maximum number of (1) ramets allowable per clone and (2) ramets per seed in the intended bulking.

5.3. Purchaser resistance

In Sweden there are three purchasing groups: private forest estate owners who purchase 50% of the total of 200 million spruce seedlings planted annually, companies (30%), and the State (20%). The purchaser's sourcing decisions are based on economic grounds and he must (1) be convinced that vegetatively propagated plantlings and stecklings will be qualitatively and quantitatively superior to seedlings, (2) because of the value of the tree in the future, be driven to purchase now at the discounted net present value rather than the minimum present cost (which means high starting-up costs), and (3) accept the uncertainty associated with a rotation cycle of not less than about 70 years.

Large forestry companies also tend to make strategic as well as economic decisions. They prefer to control seedling production themselves, having stocks of the appropriate seedlings close at hand at the right planting time.

Stecklings, with a cost ratio of stecklings to seedlings of 2.5:1, are considered by them to be very expensive.

The current cost ratio of plantling to steckling is ca 10:1, making the *in vitro* grown plantlet 25 times more expensive than a seedling. In addition, two major interrelated problems face the laboratory aiming to commercialize the production *in vitro* of *P. abies* plantlings: one is biological, the other is numerical. Lack of basic knowledge of the biology of the spruce tree is at present hampering mass clonal micropropagation, where numbers corresponding to at least 20 to 25% of the annual demand in Sweden for Norway spruce seedlings (i.e., 40 to 50 million), are regarded as an economically desirable commercial goal. At present the contributions of stecklings and plantlings to this demand are 1 and 0.0001%, respectively!

Presently, three pathways are available for the clonal micropropagation of *P. abies*. These are non-callus-mediated organogenesis via *existing* meristematic tissues, non callus-mediated organogenesis via *induced* meristematic tissues, and via somatic embryogenesis. (In the genus *Pinus*, brachyblasts constitute an additional potential pathway.)

6. ORGANOGENESIS

For the induction of adventitious buds and the release from suppression of axillary and preventitious buds, a cytokinin (usually benzyladenine, BA) may be applied either by incorporation in an agarified medium, or as a high concentration, short duration (about 2 h) pulse to the explants, or as a very high concentration, intermittent spray to greenhouse-grown seedlings (69). In the case of solid nutrient media account should be taken of the long term effects of the gelling agent, especially its matrix potential, on the uptake by explanted tissues of cytokinin (9, 21, 22). Auxins (napthaleneacetic acid (NAA) and indolebutyric acid (IBA)) may also be applied as very high concentration, medium duration pulses (12 h) for root initiation *in vivo* of the shoots developed *in vitro* from either *de novo* cytokinin-induced buds or cytokinin-stimulated bud development.

Positioning of the explants has been found to affect induction of adventitious buds both in terms of distribution and number. For example, a primary explant of a 14-day-old spruce seedling consists of the cotyledonary node with ca 7 cotyledonary needles, an emerging plumule and a hypocotyl stub. This explant can be placed upright in an agar-based medium with its cotyledons erect or flattened against the medium (see 6.2.1.). Alternatively, either before or after pulse treatment of the primary explant, the cotyledons can be removed and, as secondary explants, placed on agarified media either containing or lacking plant growth regulators.

6.1. Callus-mediated organogenesis

Generally, in tissue culture, the hormonal control of organogenesis is preceded by a histogenic phase involving the proliferation of callus from explant tissues, cells or protoplasts. However, whereas callus can be proliferated and even maintained (8, 15, 55), the inability to initiate adventitious shoot buds from this callus and accordingly the inability to regenerate plants has been one of the major limitations of tissue culture in the Pinaceae, not least in *P. abies*. In fact, *P. abies* is by no means unique, as sustained morphogenic capacity still is the exception rather than the rule among all major economic crop plants. At present, callus-mediated plant regeneration (Fig. 1) plays no role in the clonal micropropagation of *P. abies*, a factor which in spite of other alternatives as will be shown below, has a most serious consequence: it prevents the full utilization of protoplast and cell culture, which in a whole-plant regeneration pathway have to pass a callus phase, thus in large part making applications 8 and 9 (Fig. 1) inaccessible to the Norway spruce tree breeder.

6.2. Non callus-mediated organogenesis from induced meristematic tissues

6.2.1. Adventitious buds from cotyledons. Non callus-mediated adventitious bud induction may be obtained from cotyledons of 14-day-old germinants (the range varies from 5-21 days in certain Pinaceae: see also Thorpe and Biondi, 56) in the absence of auxin but in the presence of cytokinin (8). Cotyledons may be subcultured on agar-solidified media incorporating about 5 μM BA or, as secondary explants on hormone-free media, following administration of BA as a high concentration (125 μM), short-duration (2-3 h) pulse to seedling primary explants. After initiation of adventitious buds, shoots are excised and elongated with the aid of far-red irradiation for one or two weeks and then rooted *in vivo* following a long duration (12 h) pulse with 625 μM IBA. Pulse treatments, as compared to BA-containing agar media, result in the induction of greater numbers of adventitious buds over three- to four-week shorter culture periods. The inclusion of exogenous auxin, even at nanomolar levels, promotes histo- rather than morphogenesis. Propagation ratios of regenerated plants to seeds of 10:1, or a mean of 1.4 plants per cotyledon, have been obtained by using pulse treatments (Table 1).

Primary explants may also be positioned in such a way that the individual cotyledons are in surface contact with the agar over their whole 12-15 mm length, a method that works well for Pseudotsuga menziesii (see cover of Science, 11 February 1983). However, although a greater number of adventitious structures are initiated, only about 30 to 40% of these can be excised and subcultured, since the larger proportion of such structures are incompletely developed pseudobuds, fasciated buds or leaflike protuberances (phylloids).

Figure 2 is a composite representation of the variety of adventitious structures that have been observed on 14-day-old cotyledons cultured separately on media containing either a cytokinin or an auxin, or a combination of both of these hormones. In the absence of applied cytokinin, auxin (e.g. IBA 5-15 μM) invariably induces roots, the initiation of which is traceable to meristemoids that arise in newly-formed parenchyma around the vascular cylinder at the cotyledon's base (in Pinus sylvestris roots are also initiated acropetally along the cotyledon). Callusing nearly always accompanies root initiation. In the absence of applied auxin, cytokinin (BA) incorporated in the medium or administered as a pulse results in the initiation of adventitious buds, as well as of budlike structures identified as pseudobulbils, pseudobuds, phymas, phylloids and nodules (12). However, the proportion of the budlike structures, especially of pseudobuds and phylloids, was observed to increase with an increase in concentration of applied cytokinin or with a decrease in the matrix potential of the gel (resulting not only in increased uptake of cytokinin but also in hyperhydration or vitrification of tissues. The term pseudobulbil has been used to denote adventitious structures that are not visually identifiable as adventitious buds, but which eventually may either give rise to shoot buds or become necrotic and cease further development. Pseudobuds are extended, usually vitrified, bud-like structures, but often have damaged or retarded apical meristems and short, distended leaflike appendages. Phymas consist of pustular nodules of undifferentiated cells, and are up to 3 mm and more in diameter. Phylloids are leaf-like structures with unusually large parenchyma cells, either lacking or with only weakly-developed vascular tissue.

6.2.2. Adventitious buds from leaves. The response of leaves from flushing buds has been found to depend on mainly four factors (31, 32):

Table 1. Summary of results of experiments on the induction of adventitious buds in two-week-old seedling primary explants and stimulation and development of axillary and preventitious buds in 5- to 7-week old seedlings of Picea abies following application of benzyladenine in nutrient agar, or as a pulse, or as a spray to run-off. Results represent the means of 7 experiments conducted in 1983/84. For comparable data on Pinus sylvestris and Pinus nigra see references 10 and 4, respectively.

	Benzyladenine		
	5 μM In nutrient agar (Primary explant)	125 μM As a 2-h pulse (Primary explant)	1250 μM As 5 foliar sprays (1 spray every 48 h) (Seedling)
Age at start, weeks	2	3	5-7
Adventitious buds:			
induced	37	56	11
excised	14	23	3
Axillary buds:			
excised	1	4	24
# weeks	25	22	30
Ratio of plantlings:seed	7:1	10:1	18:1

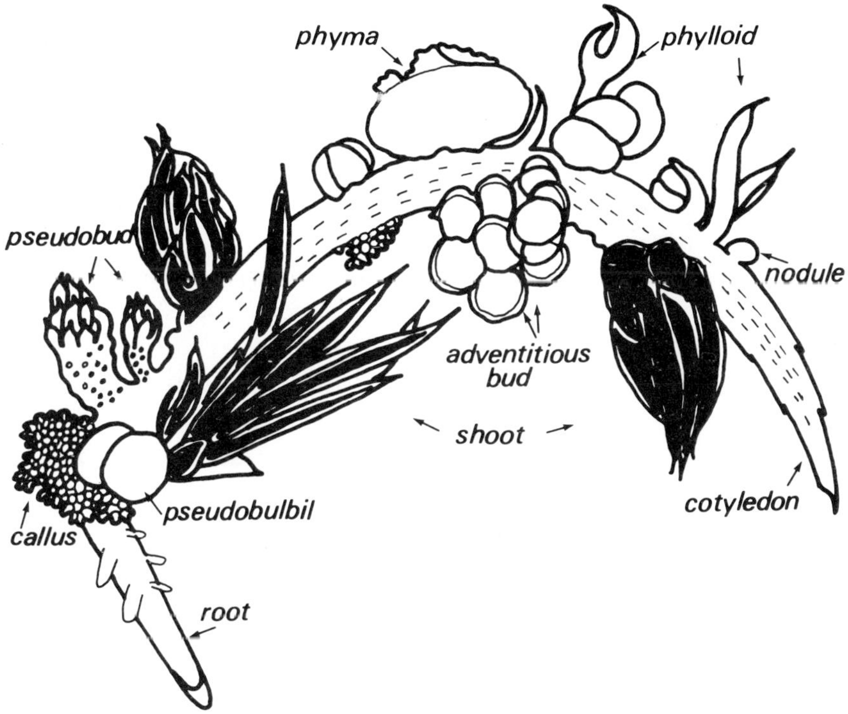

FIGURE 2. Composite representation of the variety of adventitious structures that may be induced on a 14-day-old cotyledon of P. abies cultured on a medium containing either a cytokinin and/or an auxin (12). See 6.2.1. for explanation.

(1) the method of explanting, (2) the age of the leaf, (3) the position on the leaf's axis, and (4) the ratio of cytokinin to auxin. Careful excision of the leaf explant ensures that the basal cushion of presumptive meristematic tissue (preventitious bud tissue) proximal to the abscission zone is included with the explant. The number of adventitious bud primordia (Fig. 3) that can be induced decreases with leaf age (i.e., length) as does their distribution from leaf base to leaf apex (i.e., in an acropetal direction distal to the abscission zone). Based on observations of the morphogenic effects of different ratios of BA to NAA, from 10.000:1 to 1:1, it was concluded that the level of endogenous auxin decreases progressively from the leaf's base to its tip, so that a particular concentration of exogenously applied auxin which promotes shoot-bud initiation midway along the leaf, simultaneously may inhibit it at the leaf's base (see 4.1.).

6.3. Non callus-mediated organogenesis from existing meristematic tissue

6.3.1. Axillary and preventitious buds. Non callus-mediated accelerated development of axillary buds is obtained when BA is administered as a pulse, in a hydroponic solution or as a foliar spray to 5- to 7-week-old glasshouse-grown Norway spruce seedlings, before their explantation to an auxin- and cytokinin-free medium *in vitro* (Table 1). Although adventitious buds may also be produced, they are less vigorous than those initiated on young cotyledons presumably because of growth suppression by the rapidly developing axillary buds. Propagation ratios of pulse-treated explants or of explants from cytokinin-sprayed donor material were 1.4 and 2.6 times higher than those of conventionally cultured tissues. When applied via the roots BA caused the necrosis of some axillary buds and excessive swelling and anthocyanin formation in the tissues of others with a resulting decrease in rate of plantling regeneration.

6.3.2. Resting buds and embryonic shoots. Multiplication in vitro from terminal and axillary meristems is quantitatively less productive than from non-meristematic tissues because of the time-consuming nature of the isolation procedure. Chalupa and Durzan (16) and von Arnold and Eriksson (63) have reported induction of adventitious buds on the resting buds of *Picea* spp., but plantlets were not regenerated. By including the crown, a band of thick-walled collenchymatous cells at the base of the embryonic shoot which is interpreted as an anatomical stricture with a transducing function (17), we (33) have been able to increase to ca five the number of adventitious shoot buds initiated

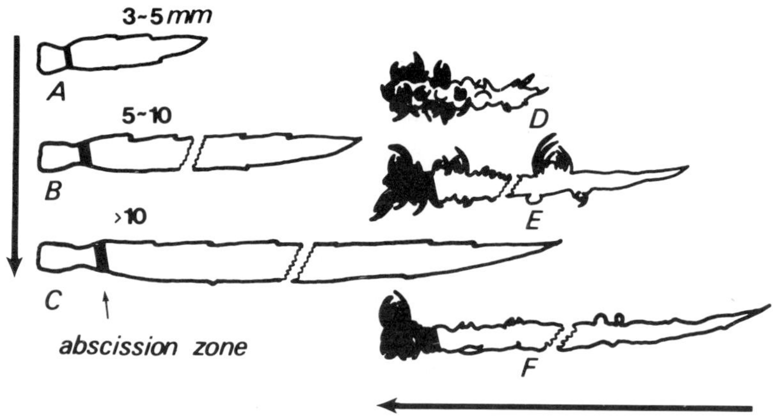

FIGURE 3. Induction of adventitious buds on leaves from flushing buds of *P. abies* depends *inter alia* on leaf age (varying in length from 3 to more than 10 mm). Arrows indicate that response in number of buds decreases with leaf age and that bud distribution increases from the leaf's morphological apex to its base (12).

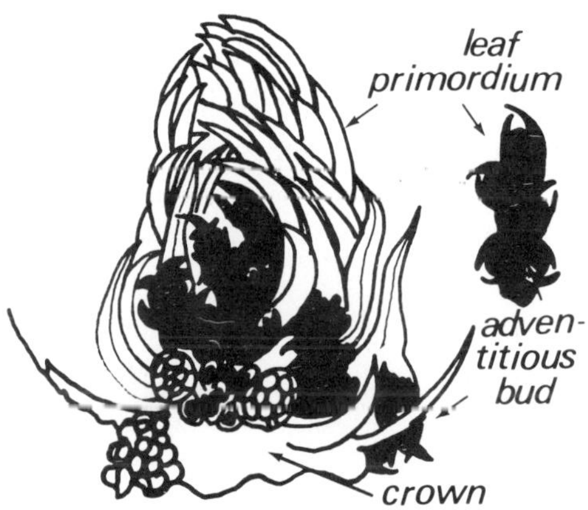

FIGURE 4. Embryonic shoot (resting bud *plus* crown) of *P. abies* with adventitious shoot buds initiated from the needle primoria (12).

from the needle primordia of embryonic shoots pulse-treated with BA (Fig. 4) and are now able to regenerate plantlets from such induced adventitious buds of 18- to 20-year-old clones of *Picea abies*.

Transport studies (34) have confirmed that the crown in its collenchymatous state acts as a selective barrier for the transport of radiolabelled indoleacetic acid and phosphate. Redetermination of the resting bud's or embryonic shoot's leaf primordia into shoot (bud) primordia is taken as an indication that the leaf primordia were not yet irreversibly determined as leaves. According to Romberger (46), apical growth in the conifer's leaf primordium begins to be replaced by intercalary growth when its length is less than ca 400 μM. However, intercalary growth is probably not a reliable parameter of determination since we found that initiation of bud meristemoids also occurred in primordia less than 400 μm long; it is probable that differentiation of procambial tissues in the leaf primordium is a better parameter. With the scanty evidence at hand, it appears that adventitious bud initiation is determined *de novo* by the cytokinin-controlled induction of meristemoids rather than via a reversal of the determined state.

6.3.3. *Epicormic buds and sphaeroblasts*. Two other types of buds deserve mention: epicormic buds and sphaeroblasts or trace buds. Epicormic buds are preformed but suppressed buds, formed in continuity with the apical meristem and capable of remaining viable indefinitely (13). Sphaeroblasts (3) are adventitious structures with circumferential cambia present as woody spheroids, which can form new apical meristems and grow out as adventitious and juvenile shoots. Epicormic buds occur among some conifers (13), but trace buds presumably are uncommon; as far as I know they have not been reported in Norway spruce. Because of their early initiation and long periods of subsequent quiescence, both of these types of buds presumably have retained their juvenility and are therefore of interest in attempts to rejuvenate mature trees (see 9.4.).

6.4. *Seedlings to plantlings*

As indicated earlier, seedling tissues can be used in different strategies in micromultiplication. Figure 5 is a schematic outline of the production pathways used for *P. abies*, *Pinus sylvestris* and *Pinus nigra*. For the initiation of adventitious buds from non-meristematic tissues, 14-day-old seedling explants (Fig. 5A) are pulsed with 125 μM BA for 2 h and then subcultured on hormone-free nutrient agar. After ca 6 to 7 weeks adventitious

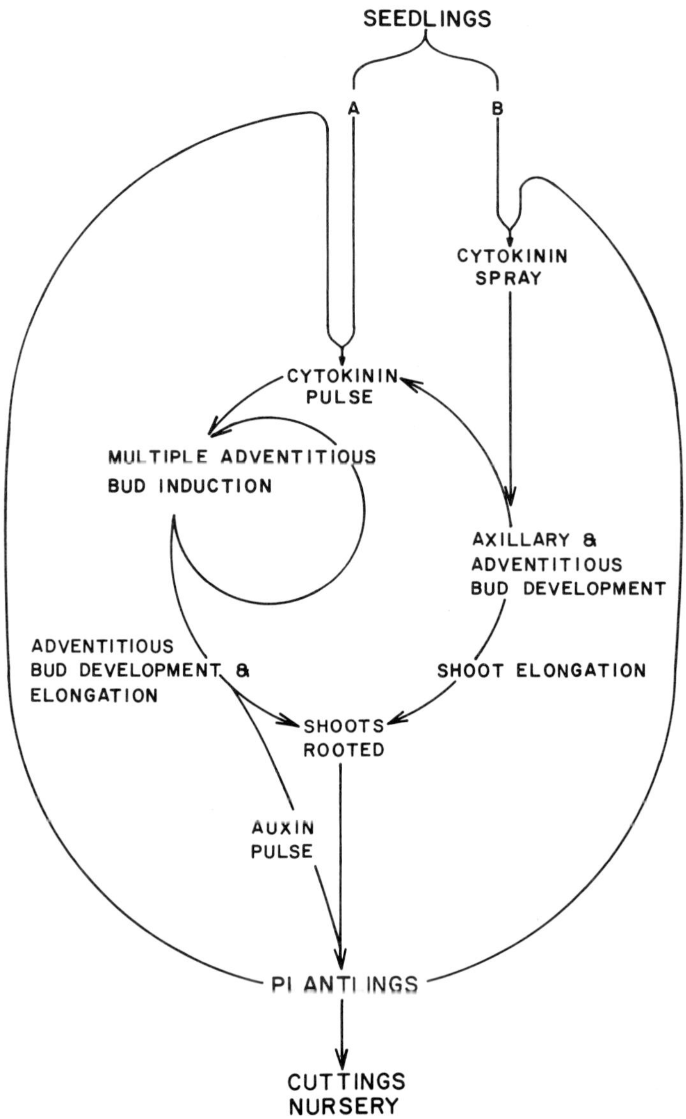

FIGURE 5. Outline of the production pathway of *P. abies* plantlings from 2- to 7-week-old seedlings derived from controlled pollination. A, 2-week-old and B, 5- to 7-week-old seedlings.

shoot buds are excised and maintained under conditions that promote development and elongation, while the original explant is transplanted to fresh medium for a second (and often third and fourth) cycle of bud production. Following shoot elongation (10, 12), an auxin (IBA or NAA) pulse is applied (625 µM, 12 h) and the shoots are rooted *in vivo* in a mistbed.

Five- to seven-week-old glasshouse-grown seedlings (Fig. 5B) can also be given about 5 foliar sprays (one every other day) with 1250 µM BA (10, 69) to stimulate the development of axillary as well as preventitious buds. Following removal of the buds for shoot elongation and rooting, the seedlings's debudded shoot is pulse-treated with BA and cut into explant segments for multiple adventitious bud induction.

Since each seed is treated as a clone, plantlings produced *in vitro* are usually recycled a number of times through the production pathway in order to establish a population of clones before introduction to the cuttings nursery for further clonal multiplication and observation.

7. SOMATIC EMBRYOGENESIS

As has been the case with many initially embryoid-recalcitrant mono- and dicotyledonous species, somatic embryos have now been induced in *P. abies* from immature zygotic embryos (29, 30, see also 10). The reproducibility of this system has yet to be demonstrated since at least one serious problem will have to be overcome, namely the tendency of the embryogenic culture to decline progressively (decrease in number of embryoids and increase in embryoidal aberration) with each passage or subculture. Therefore, when necessary, embroidy must be able to be induced repeatedly from source material. This demands precise knowledge of the stage of development at which the ovule-immature embryo complex is most sensitive to induction. Pollination of *P. abies* occurs in mid-May and fertilization in mid-June. It takes from 25 to 30 years before mega- and microstrobili are produced and when sufficient maturation is reached the production of only ovulate strobili prevails for some years, with as many as 7 years between cycles of strobilus formation.

8. BIOTECHNOLOGY

Although biotechnology in the past has been regarded as practically synonymous with the application of microbiological organisms, processes or systems to the manufacturing and service industries, present day usage of the term includes the component activities of higher plant cell and molecular biology. As regards the applications of *in vitro* culture technology to forest trees shown in Fig. 1, that of biotechnology, *sensu stricto*, would involve the successful process development of any or part of applications 2 to 9 in Fig. 1.

Under biotechnology as it relates to *Picea abies* (Fig. 1) I have singled out two processes as being particularly challenging. One is the encapsulation of somatic embryos to produce cloned artificial seed, a process that among others will depend on whether Norway spruce plantlings can be raised successfully from the embryoids. The other is the complete saccharification of lignocellulosic materials for the production of sucrose, ethanol and other energy-rich products which, according to Smith (54), represents the most difficult of all biotechnological problems.

9. OTHER APPLICATIONS

Reference to some of the other applications of *in vitro* culture technology in clonal forestry has been made before (10, 11, 70), but some comment on somaclonal variation, molecular plant genetic engineering, apomictic seed and natural vegetative propagation may be appropriate.

9.1. Somaclonal variation and genetic instability

It is known that somaclonal variation can come about as a result of (1) structural and/or numerical chromosome changes (deletions, duplications, inversions, translocations, aneuploidy, polyploidy), (2) changes in nuclear chromosomal DNA (point mutations, somatic recombination, gene amplification), and (3) changes in organellar DNA. It is believed that the major part of this variation originates during protoplast, cell and tissue culture as a result of the stresses and changes to which DNA replication, mitosis and cell division are subjected. Whether or not such Norway spruce variants will be of practical value will depend on how successfully plants of this species can be regenerated through *in vitro* culture. In addition it will take long testing times of 15 to 30 years to assess the value and stability of the variation.

The initial concern that genetic instability may be a characteristic of callus-mediated or even non-callus-mediated adventitious bud induction in non-meristematic tissues (45) was not corroborated for *Picea abies* (28). Chromosome stability in plants, in terms of type and frequency of mitotic abnormality such as lagging chromosomes, anaphase and telophase bridges, micronuclei, aneuploids and polyploids has been studied by Papes *et al*. (44) and Berljak *et al*. (4). In six-week-old *Pinus nigra* seedlings pulse-treated with 125 μM BA, it was found (4) that adventitious buds 10 and 22 days after pulse treatment showed a higher frequency of aberrancy than axillary buds, but both types of buds were able to undergo dramatic repair between day 10 and day 22. Regenerated plants appeared to be genetically stable.

9.2. *Molecular plant genetic engineering*

The application of molecular plant genetic engineering to *P. abies* will take many years to realize. Growth, vigour, photosynthetic capacity and cold stress tolerance represent complex polygenic traits, but even if monogenic traits such as certain types of herbicide resistance could be taken advantage of, appropriate vector systems will have to be found. Only two of the six genera in the Pinaceae, namely *Abies* and *Pseudotsuga*, appear to be susceptible to *Agrobacterium tumefaciens*.

9.3. *Apomictic seed*

In apomixis seeds are formed entirely from diploid ovular tissue, not following pollination and fertilization of the egg. Apomictic seedlings are not only juvenile, like sexual seedlings, but more important they are identical genetically to the genotype from which they are derived. In some tropical trees (*Citrus*, *Mangifera*), some Asian and North American *Malus* and in some *Juglans* and *Eucalyptus* hybrids spontaneous induction of apomixis has been found. The question is whether in *P. abies* apomixis (Fig. 1) may be induced by interspecific crossing or by some other technique such as induction of alternative cycles of rejuvenation and maturation by repeated grafting.

9.4. *Natural vegetative propagation*

Norway spruce trees in the Arctic Circle regions of Sweden, Finland and Norway flower irregularly and set seed poorly. As a result of the short growing season and depth of snow, seedlings that germinate have difficulty in becoming established. An interesting feature is that many of these trees are propagated vegetatively in a manner reminiscent of artificial trench layering. Over the course of years the lower branches of such trees are weighed down by snow in a bed of organic debris and moss. The branches strike root some distance from the trunk and following a change from plagiotropic to orthotropic growth, the branch end distal to the region of root initiation begins to assume tree symmetry.

It is quite common to find a mature tree or stump surrounded by a circle of vigorous young trees, spaced in a pattern correponding to the original phyllotaxy of the lowermost branches of the vegetative parent. Sometimes, although rooting has occurred, branches remain plagiotropic.

This phenomenon is now under investigation, as are attempts to mimic it experimentally.

10. SUMMARY AND CONCLUSIONS

Can sustained whole-plant regeneration of *Picea abies* presently be achieved *in vitro* from protoplasts, cells, callus, meristems, cotyledons and somatic embryos? With the exception of regeneration from cotyledons, the answer is "no". And from tissues that are pre-embryonic (nucellar, gametophytic, ovular), embryonic, juvenile, adolescent and mature? With the exception of regeneration from embryonic tissues the answer again is "no". At present, tissue culture-produced plantlings from seedling tissues are costly, have low propagation ratios (up to 18:1, plantlings:seed), are of limited genetic value unless produced from controlled pollination, and are subject to the same legal and silvicultural restraints applicable to stecklings.

Since those of us who work with *Picea abies* should account for the public and private monies that fund our research, we do injustice when promoting the view that *in vitro* culture technology is about to revolutionize clonal forestry. Although much progress has been made with a species that has served as a conifer model, one cannot escape the fact that tissue culture is but a tool and what really hampers progress is not so much a lack of tissue culture know-how as a lack of understanding of the basic biology of the gymnosperm. If scientists using cell and tissue culture techniques were to develop greater understanding of cellular and whole-plant biology, we may discover that we can use *in vitro* culture technology more effectively. Nevertheless, most - if not all- of the tissue culture methodologies of, for example, callus, cell and protoplast culture, somatic hybridization and haploidization probably will be firmly confined to the research laboratory for at least the rest of this century.

11. ACKNOWLEDGEMENTS

Ceskoslovenská Akademie Věd is gratefully acknowledged for permission to reproduce Figs 2-4, as are my colleagues Drs BG Bentzer and NO Bosemark for discussions and viewpoints concerning the application of *in vitro* culture technology to Norway spruce. Some of the material was adapted from reference 10.

12. REFERENCES

1. AITKEN J, KJ HORGAN, TA THORPE 1981 Influence of explant selection on the shoot-forming capacity of juvenile tissue of *Pinus radiata*. Can J For Res 11:112-117
2. BENTZER BG, CH BORNMAN 1987 Mass clonal micro- and macropropagation of cuttings with special reference to *Picea abies*. *In* TD Davis, N Sankhla, eds, *Adventitious Root Formation on Cuttings*. Dioscorides Press, Portland, OR (in press)
3. BALDINI E, B MOSSE 1956 Observations on the origin and development of sphaeroblasts in the apple. J Hort Sci 31:156-162
4. BERLJAK J, CH BORNMAN, S JELASKA, D PAPES, M VIDAKOVIC Stimulation of preventitious axillary bud development *in vitro* in *Pinus nigra* (in press)
5. BIONDI S, TA THORPE 1982 Growth regulator effects, metabolite changes and respiration during shoot initiation in cultured cotyledon explants of *Pinus radiata*. Boz Gaz 143:20-25
6. BORNMAN CH 1981 *In vitro* regeneration potential of the conifer phyllomorph. *In* G Eriksson, K Lundkvist, eds, Symp on Clonal Forestry. Dep For Genet, Swedish Univ Agric Sci Res Notes 32:43-56
7. BORNMAN CH, E JANSSON 1981 Regeneration of plants from the conifer with special reference to *Picea abies* and *Pinus sylvestris*. *In In vitro Culture of Forest Tree Species*. Proc Int Workshop IUFRO, sect S2 01 5, Fontainebleau, France, pp 41-53
8. BORNMAN CH 1983 Possibilities and constraints in the regeneration of trees from cotyledonary needles of *Picea abies in vitro*. Physiol Plant 57:5-16
9. BORNMAN CH, TC VOGELMANN 1984 Effect of rigidity of gel medium on benzyladenine-induced adventitious bud formation and vitrification *in vitro* in *Picea abies*. Physiol Plant 61:505-512
10. BORNMAN CH 1984 Application of *in vitro* culture technology in clonal forestry. *In* Proc Int Symp Recent Adv for Biotechnol, Traverse City, MI, pp 178-194
11. BORNMAN CH 1985 Regeneration *in vitro* of economically important crop plants in the Nordic countries. *In* CH Bornman, WK Heneen, CJ Jensen, A Lundquist, eds, Proc. 1st Nordic Cell and Tissue Culture Symp on Research, Breeding and Production of Crop Plants. Hereditas 3:7-13
12. BORNMAN CH 1985 Hormonal control of growth and differentiation in conifer tissues *in vitro*. Biol Plant 27:249-256
13. BRYAN JA 1981 Epicormic branching in Rocky Mountain Douglas-fir. Can J For Res 11:190-199
14. CAMPBELL AH, AE VINES 1938 The effect of *Lodphodermellina macrospora* (Hartig) Techon on leaf abscission in *Picea excelsa* Link. New Phytol 37:358-368
15. CHALUPA V, DJ DURZAN 1973 Growth of Norway spruce (*Picea abies* L. Karst.) tissue and cell cultures. Commun Inst For Cech 8:111-125
16. CHALUPA V, DJ DURZAN 1973 Growth and development of resting buds of conifers *in vitro*. Can J For Res 3:196-208

17. CHALUPA V 1977 Development of isolated Norway spruce and Douglas fir buds *in vitro*. Commun Inst For Cech 10:71-78
18. CHALUPA V 1977 Organogenesis in Norway spruce and Douglas fir tissue cultures. Commun Inst For Cech 10:79-87
19. CHAMBERLAIN CJ 1935 Gymnosperms. Structure and Evolution, Univ of Chicago Press, Chicago
20. CHRISTERSSON L 1978 The influence of photoperiod and temperature on the development of frost hardiness in seedlings of *Pinus sylvestris* and *Picea abies*. Physiol Plant 44:288-294
21. DEBERGH P, Y HARBAOUI, R LEMEUR 1981 Mass propagation of globe artichoke (*Cyanara scolymus*: evaluation of different hypotheses to overcome vitrification with special reference to water potential. Physiol Plant 53:181-187
22. DEBERGH P 1983 Effect of agar brand and concentration on the tissue culture medium. Physiol Plant 59:270-276
23. DIETRICHSON J, C KIERULF 1982 Selection of eight-year-old Norway spruce (*Picea abies* L. Karst.) plants in a progeny trial and mass production by cuttings. Medd Nor Inst Skogforsk 38:1-28
24. DORMLING I, H KELLERSTAM 1981 Rooting and rejuvenation in propagating old Norway spruce by cuttings. *In* G Eriksson, K Lundkvist, eds, Symp on Clonal Forestry. Dep For Genet, Swedish Univ Agric Sci Res Notes 32:65-72
25. FINK S 1984 Some cases of delayed or induced development of axillary buds from persisting detached meristems in conifers. Am J Bot 71:44-51
26. FOSKET DE 1979 Hormonal control of morphogenesis in cultured tissues *In* F Skoog, ed, Plant Growth Substances, Springer-Verlag New York NY, pp 362-369
27. FOSTER AS, EM GIFFORD, JR 1974 Comparative Morphology of Vascular Plants. WH Freeman & Co, San Francisco
28. HAKMAN I, S VON ARNOLD, A BENGTSSON 1984 Cytofluorometric measurements of nuclear DNA in adventitious buds and shoots of *Picea abies* regenerated *in vitro*. Physiol Plant 60:321-325
29. HAKMAN I, LC FOWKE, S VON ARNOLD, T ERIKSSON 1985 The development of somatic embryos in tissue cultures initiated from immature embryos of *Picea abies* (Norway spruce). Plant Sci 38:53-59
30. HAKMAN I, S VON ARNOLD 1985 Plantlet regeneration through somatic embryogenesis in *Picea abies* (Norway spruce). J Plant Physiol 121:149-158
31. JANSSON E, CH BORNMAN 1980 *In vitro* phyllomorphic regeneration of shoot buds and shoots in *Picea abies*. Physiol Plant 49:105-111
32. JANSSON E, CH BORNMAN 1981 *In vitro* initiation of adventitious structures in relation to the abscission zone in needle explants of *Picea abies*: anatomical considerations. Physiol Plant 53:191-197
33. JANSSON E, CH BORNMAN 1983 Morphogenesis in dormant embryonic shoots of *Picea abies*: influence of the crown and cold treatment. Physiol Plant 59:1-8
34. JANSSON E, P JENSEN, CH BORNMAN 1983 Effect of the crown on uptake and transport in embryonic shoots of *Picea abies*. Physiol Plant 59:521-527
35. KLEINSCHMIT J, W MULLER, I SCHMIDT, J PACZ 1973 Entwicklung der Stecklingsvermehrung von Fichte (*Picea abies* L. Karst.) zur Praxisreife. Silvae Genet 22:4-15
36. KLEINSCHMIT J, J SCHMIDT 1977 Experiences with *Picea abies* cuttings propagation in Germany and problems connected with large scale application. Silvae Genet 26:197-203

37. LIESE W, D DUJESIEFKEN 1986 Anatomie des Fichtenholzes. In H Schmidt-Vogt, ed, Die Fichte, Vol II, Verlag Paul Parey, Hamburg, Berlin, pp 373-389
38. MARCO HF 1939 The anatomy of spruce needles. J Agric Res 58:357-368
39. MOMOT TS 1977 Organogenesis of European spruce (Picea abies) L. Karst.) and pine (Pinus sylvestris L.) in isolated culture. Citologia i Genetika 4:497-498
40. NEGER FW 1911 Zur Mechanik des Nadelfalls der Fichte. Naturw Z Forst Landw 9:214-223
41. OBATA-SASAMOTO H, WM VILLALOBOS, TA THORPE 1984 ^{14}C-metabolism in cultured cotyledon explants of radiata pine. Physiol Plant 61:490-496
42. OLESEN PO 1978 On cyclophysis and topophysis. Silvae Genet 27:173-178
43. OLESEN PO 1982 The effect of cyclophysis on tracheid width and basic density in Norway spruce. For Tree Improv 15:1-80
44. PAPEŠ D, V GARAJ-VRHOVAC, S JELASKA, B KOLEVSKA-PLETIKAPIĆ 1983 Chromosome behaviour in cultured cell populations of higher plants. In PE Brandham, MD Bennet, eds Kew Chromosome Conf II: Proc 2nd Chromosome Conf, Royal Bot Gardens, Kew, England. Allen and Unwin, London, pp 155-163
45. PATEL KR, GP BERLYN 1981 Genetic instability of multiple buds of Pinus coulteri regenerated from tissue culture. Can J For Res 12:93-101
46. ROMBERGER JA 1963 Meristem, Growth and Development in Woody Plants. USDA For Serv Tech Bull 1293
47. ROMBERGER JA 1966 Developmental biology and the spruce tree. J Wash Acad Sci 56:69-81
48. ROMBERGER JA, RJ VARNELL, CA TABOR 1970 Culture of apical meristems and embryonic shoots of Picea abies - approach and techniques. USDA For Serv Tech Bull 1409:1-30
49. ROMBERGER JA, CA TABOR 1971 The Picea abies shoot apical meristem in culture I. Agar and autoclaving effects. Am J Bot 58:131-140
50. ROMBERGER JA 1976 An appraisal of prospects for research on juvenility in woody perennials. Acta Hortic 56:301-317
51. ROULUND H 1975 The effects of the cyclophysis and topophysis on the rooting and behaviour of Norway spruce cuttings. Acta Hortic 54:39-50
52. ROULUND H 1979 Stem form of cuttings related to age and position of scions (Picea abies L. Karst.). For Tree Improv 13:1-16
53. SIMOLA L, J HONKANEN 1983 Organogenesis and fine structure in megagametophyte callus lines of Picea abies. Physiol Plant 59:551-561
54. SMITH JE 1981 Biotechnology. Edward Arnold, London
55. STEINHART CE, LC STANDIFER, JR, F SKOOG 1961 Nutrient requirements for in vitro growth of spruce tissue. Am J Bot 48:465-472
56. THORPE TA, S BIONDI 1984 Conifers. In WR Sharp, DA Evans, PV Ammirato, Y Yamada, eds, Handbook of Plant Cell Culture, Vol II, MacMillan, New York, pp 435-470
57. VAN STADEN J, C FORSYTH, L BERGMAN, S VON ARNOLD 1986 Metabolism of benzyladenine by excised embryos of Picea abies. Physiol Plant 66:427-434
58. VILLALOBOS M, MJ OLIVER, EC YEUNG, TA THORPE 1984 Cytokinin-induced switch in development in excised cotyledons of radiata pine cultured in vitro. Physiol Plant 61:483-489
59. VILLALOBOS VM, DWM LEUNG, TA THORPE 1984 Light-cytokinin interaction in shoot formation in cultured cotyledon explants of radiata pine. Physiol Plant 61:513-517

60. VOGELLEHNER D 1977 Phylogenie, Taxonomie und Verbreitung der Gattung *Picea*. *In* H Schmidt-Vogt, ed, Die Fichte Vol I, Verlag Paul Parey, Hamburg, Berlin, pp 1-647

61. VOGELMANN TC, CH BORNMAN, P NISSEN 1984 Uptake of benzyladenine in explants of *Picea abies* and *Pinus sylvestris*. Physiol Plant 61:513-517

62. VON ARNOLD S, T ERIKSSON 1978 Induction of adventitious buds on embryos of Norway spruce grown *in vitro*. Physiol Plant 44:283-287

63. VON ARNOLD S, T ERIKSSON 1979 Bud induction on isolated needles of Norway spruce (*Picea abies* L. Karst.) grown *in vitro*. Plant Sci Lett 15:363-372

64. VON ARNOLD S 1982 Factors influencing formation, development and rooting of adventitious shoots from embryos of *Picea abies* (L.) Karst. Plant Sci Lett 27:275-287

65. VON ARNOLD S, T ERIKSSON 1985 Initial stages in the course of adventitious bud formation on embryos of *Picea abies*. Physiol Plant 64:41-47

66. VON ARNOLD S, T ERIKSSON 1986 Norway spruce (*Picea abies* L.). *In* YPS Bajaj, ed, Biotechnology in Agriculture and Forestry, Vol I, Trees I, Springer-Verlag, Heidelberg, pp 291-319

67. WEBB KJ, HE STREET 1977 Morphogenesis *in vitro* of *Pinus* and *Picea*. Acta Hortic 78:259-269

68. WERNICKE W, J GORST, L MILKEVITS 1986 The ambiguous role of 2,4-dichlorophenoxyacetic acid in wheat tissue culture. Physiol Plant (in press)

69. WHITEHILL SJ, WW SCHWABE 1975 Vegetative propagation of *Pinus sylvestris*. Physiol Plant 35:66-71

70. WINTON LL, RS STETTER 1974 Utilization of haploidy in tree breeding. *In* KJ Kasha, ed, Haploids in Higher Plants, Univ Guelph Proc Int Symp, Guelph, pp 259-274

71. YEUNG EC, J AITKEN, S BIONDI, TA THORPE 1981 Shoot histogenesis in cotyledon of radiata pine. Bot Gaz 142:494-501

3. SITKA SPRUCE (*Picea sitchensis* (Bong.) Carr.)

A JOHN and K J WEBB

1. INTRODUCTION

Sitka spruce has been widely planted as an exotic species in Western Europe for over 150 years since its introduction into the United Kingdom from Western North America by David Douglas in 1831. It is the most commonly planted conifer in the United Kingdom and at present accounts for 47% of all soft-wood plantations (2). It is also planted extensively in Norway, Denmark, Ireland, other parts of Europe and in Iceland and New Zealand. The productivity of commercial forests would benefit from the use of genetically improved seed for seedling production but the yield of such seed from established seed orchards is, as yet, small. Attention has therefore turned to the vegetative propagation of improved plant material to meet the demand for forest planting, at least until seed production from the orchards is adequate. Research was started in 1973 to determine whether *in vitro* methods could produce the large number of clonal plants required either from the relatively small quantities of improved seed that are available or from superior, mature trees. Three main approaches have been considered: organogenesis on callus derived from juvenile tissue, morphogenesis directly from juvenile and mature explants and the micropropagation of juvenile seedlings.

2. ORGANOGENESIS ON CALLUS DERIVED FROM JUVENILE TISSUE

Juvenile tissue is generally more responsive in culture than mature tissue. Accordingly, Webb and Street (15,16) concentrated on developing techniques for organogenesis from embryos and seedlings.

FIGURE 1. Callus with bud primordia (p) formed on excised embryos of Sitka spruce after 95 days with a 13.5 hr light period at 20°C and a 10.5 hr dark period at 7°C. Some of the primordia have developed into buds (b) (Bar - 1 mm).

2.1. Embryos

Mature embryos of Sitka spruce form adventitious buds under appropriate stimuli (15,16). In a series of experiments embryos were excised from surface-sterilized seeds and incubated on various media under a range of conditions. Preliminary tests had shown that embryos suffered from high mortality and a consequent low rate of callus and shoot regeneration. These problems could be overcome by placing the embryos in the dark immediately after excision.

The *de novo* production of shoots generally occurred on the callus on the embryos and rarely directly on the embryo (Fig. 1). On a modified Murashige and Skoog (MS) medium (called CI medium (1)) containing indole-3-acetic acid (IAA), indole-3-butyric acid (IBA), N^6-(Δ^2-isopentyl adenine) (2iP) and N^6 benzyl-aminopurine (BAP) at 5 x 10^{-6} M each between 33 and 69% of the embryos produced callus with adventititous buds. The first signs of shoot regeneration were apparent after 28 days in culture. The shoot primordia developed into typical buds with a regular phyllotactic arrangement, but leaf primrodia were also produced which appeared to be spatially separated from any bud. New primordia could be produced from the proliferating callus for up to 95 days. In some instances, the cultures continued to proliferate both callus and buds after transfer to fresh medium.

Various factors, including the plant genotype, affect the regeneration of shoots from the embryos. The influence of the environment and the constituents of the medium were investigated with a view to improving both the percentage of the explants that produced adventitious buds, and the number of buds formed per embryo. However, no significant differences were observed between embryos cultured in different physical environments, except that a 12 day dark pretreatment of the embryos prior to their incubation under high light intensities was beneficial (15,16).

Major effects were achieved by growth regulator changes in the medium. Auxin was required to give the best regeneration, although a high concentration of BAP alone did induce regeneration under some conditions. Alterations in the concentration and type of cytokinin strongly influenced shoot formation (15,16). The auxins of the CI medium were used throughout and the effects of the cytokinins zeatin, 2iP, BAP and kinetin alone were compared with the complete CI medium (containing both 2iP and BAP). Zeatin tended to stimulate outgrowth of roots from the original root meristem of the embryos rather than adventitious bud formation. Solitary buds were occasionally produced at the tips of the cotyledons. BAP and kinetin however induced buds in about 45% of the embryos. BAP induced numerous stunted, dark-green buds and isolated leaf primordia. Kinetin induced a small number of paler buds, which started to extend whilst still on the initiation medium. Up to 40 shoots, 1 cm in length, were produced per embryo. Callus which formed in the presence of either 2iP or kinetin alone did not proliferate after transfer to new medium. Some of the clonal shoots were successfully rooted producing plants that were morphologically similar (15).

2.2. Seedlings and two-year-old shoots

Adventitious buds also occurred, formed on seedling explants (15,16). Surface-sterilized seeds were germinated under sterile conditions and the 7-54 days old seedlings were cut into two segments: hypocotyl and hypocotyl plus cotyledons and shoot tip. When cultured on naphthalene-1-acetic acid (NAA) and BAP both at 10^{-5} and 10^{-7} M, a maximum of 55% of the hypocotyl segments produced callus. Some of these calluses were maintained through several passages. Buds could be formed from the calluses after transfer to a high cytokinin medium or when drops of a concentrated solution of BAP were applied directly to the callus, although these buds failed to elongate.

Adventitious buds were also induced on some hypocotyl segments when they were placed on the medium (BAP exceeding 10^{-6} M) in close proximity to the section containing the original shoot tip. Similarly, the small section of hypocotyl remaining attached to the shoot tip was capable of bud production. This suggests that a factor which promoted organogenesis was transported from the seedling apex to the hypocotyl below and passed from its base to the medium and thus to nearby hypocotyl segments which then acquired the capacity to form adventitious buds. The shoot tips extended when maintained on levels of BAP lower than 10^{-6} M with, occasionally, two or more shoots growing out.

Actively growing shoots were also excised from 2-year-old glasshouse grown seedlings and cultured. Callus and buds were produced from 65% of the cultures maintained on 10^{-7} M NAA and 10^{-6} BAP after 84 days. Further growth of the cultures occurred after transfer to fresh medium although no attempt was made to obtain elongation of the induced buds.

3. MICROPROPAGATION OF SHOOTS EXCISED FROM JUVENILE SEEDLINGS

In vitro methods may have the potential to produce the large number of plants required from the available small quantity of improved seed either directly or by producing stock plants for a large scale cuttings programme as described by Kleinschmit and Schmidt (10). John (4) has argued that although shoot formation on callus derived from mature embryos produces plants morphologically similar to normal seedlings, the method is unsuitable for large scale propagation because the numbers achieved are low. Therefore, other methods were investigated.

3.1. On hormone-free medium

Seeds were germinated in compost under non-sterile conditions and explants were taken after about one month when the cotyledons were open and the apical shoot was just beginning development (5). Shoots with about 2 cm of the hypocotyl left attached were excised from the root system, surface-sterilized and transferred to *in vitro* conditions. The explants developed similarly to seedlings grown in soil and produced axillary buds and shoots during their extension. A multiplication rate of approximately three axillary buds per explant per six week cycle was achieved on MS medium and this rate was not increased by the presence of plant growth regulators in the medium. John and Murray (9) found that the development of subcultured apical and axillary shoots differed. The two types of explant elongated at the same rate but the rate of axillary meristem production was different, i.e. apical and axillary shoots when subcultured produced axillary meristems at multiplication rates of 4.5 and 2.0 per explant respectively. In subsequent subcultures, shoots that had become a dominant apex in the previous culture produced axillary buds and shoots at a higher rate than those that had been an axillary bud or shoot. It was demonstrated further that axillary bud proliferation was not promoted by high temperatures (26°C), enhanced light levels, etiolation or plant growth regulators (6). Plant material was maintained in culture for up to two years through numerous subcultures without apparent ageing, loss of vigour or reduction in the rate of axillary bud and shoot proliferaton. However, there was a large clonal variation and the extrapolation of data has demonstrated that the annual multiplication rates of individual clones may vary from 15 to 5000 (7).

3.2. By vitrification

Vitrification has been observed in some Sitka spruce cultures (8). The phenomenon could occur after surface sterilization of young shoots or could arise spontaneously after the subculture of established cultures on hormone-free medium. Vitrification was characterized by a change in colour from blue-green to green, a wet appearance to the plant surface, a decrease in the amount and a change in the form of the needle surface wax and an enhancement of growth and development. When vitrification occurred after surface sterilization, it was found that vitrified cultures had a total of 9.2 buds per explant compared to 3.3 buds in normal unaffected cultures after 68 days. Fifty-nine percent of the vitrified cultures reverted to normal growth by the end of the incubation period. Cultures in which vitrification had arisen spontaneously were subcultured to yield vitrified and reverted cultures; normal cultures were established as a control. After 65 days, normal cultures had 2.6 buds per explant, vitrified cultures 6.1 buds and reverted cultures 4.5 buds, i.e., the enhanced growth of the vitrified cultures was still present, though to a lesser extent, even after they had apparently reverted to normal growth.

Vitrification occurs in about 20% of cultures after surface sterilization whereas spontaneous vitrification after subculture is relatively rare and occurs in about 1 to 2% of cultures. The phenomenon results in anatomical and morphological changes that render the cultures liable to desiccation unless they are maintained at 100% humidity and 20°C. Therefore, before the enhanced growth and development due to vitrification can be used for micropropagation, techniques are needed that will induce vitrification and also revert the vitrified plant material to normal growth before transfer from *in vitro* conditions. John (8) found that when normal cultures, on solid MS medium without growth substances, were submerged by adding sterile distilled water to the culture vessel, the new growth that developed was vitrified in all cases. After 97 days, unsubmerged cultures had a total of 2.9 buds per explant whereas cultures submerged in

FIGURE 2. Right:- Sitka spruce culture after 42 days submergence in sterile distilled water and 28 days in air after the water was poured away. There is prolific adventitious bud and shoot production at the base and along the length of the stem. The apex and many of the adventitious shoots have reverted to normal growth. Left:- Control, untreated culture. (Bar - 1 cm). Cultures were maintained at 20°C with a 16 hr light, 8 hr dark photoperiod and an irradiance of 10.5 Wm^{-2}.

1 cm of water had 18.1 buds and cultures submerged in 3 cm of water had 23.2 buds. The treated cultures reverted to normal growth only when the apices grew above the level of the water or when the water was poured out of the culture vessel. The much higher proliferation rates in the submerged cultures were due mainly to the development of adventitious buds in or near the axils of fully developed needles along the length of the vitrified stem (Fig. 2); the buds were not initiated in the apical meristem. A minimum submergence period of 42 days was required before high bud proliferation rates were achieved. As long as the adventitious buds produced were still very small they needed to be subcultured while still attached to a stem segment, but could be subcultured alone if the bud had developed into a short shoot with extending needles (John, unpublished data).

4. MORPHOGENESIS DIRECTLY ON JUVENILE AND MATURE EXPLANTS

Selby (13) has argued that traits such as vigour, crown form and timber properties can only be adequately assessed in mature individuals since correlations between juvenile and mature traits are poor (11). However, Gill (3) has suggested that reasonably reliable predictions of later performance can be made after eight to ten years growth from seed. Selby has studied morphogenesis in Sitka spruce with particular emphasis on non-juvenile tissues because of its practical significance for the cloning of superior individuals. Selby (13) collected buds just prior to flushing from the lower crown of juvenile (three-year-old) and mature (twenty-year-old) trees. The buds were surface sterilized, the bud scales were stripped away and the total primordial shoots were cultured initially on a bud induction medium containing IAA, IBA and BAP and subsequently on a hormone-free elongation medium (16). BAP concentrations between 10^{-6} M and 5 x 10^{-6} M were optimal for adventitious bud production after 8 weeks induction. Buds from 3-year-old trees produced twice as many adventitious buds as those from 20-year-old trees, i.e., 13 per explant compared to 6 (Fig. 3). Cold pretreatment of plant material at 4°C for 10 weeks, before incubation on the induction medium for periods of 8 to 12 weeks, resulted in the production of a relatively high number of adventitious buds (about 18 per explant) after a further seven weeks on the elongation medium. Adventitious bud production was not affected by the age of the tree. Neither the absence of auxin nor the addition of 2,4-dichlorophenoxyacetic acid (2,4-D) to the induction medium significantly affected adventitious bud production on buds from 3-year-old trees.

Thus Selby (13) has developed a method for the _in vitro_ culture of ortets that are in a post-dormant, unflushed condition. To be of use to the plant propagator, the technique needs to be extended either to induce morphogenesis on plant material in any physiological condition or to change the physiological condition of the plant material itself. Selby and Harvey (14) have demonstrated that morphogenesis could be

FIGURE 3. Adventitious bud production on an excised bud from a three-year-old tree after induction with 5×10^{-6} M BAP and 12 weeks on the bud elongation medium. Basal needles that fell away from the bud on subculture also formed adventitious buds. (Bar - 1 cm).

induced on buds throughout the winter resting period but rates were highest just prior to flushing; the introduction of already flushed buds into culture resulted in a low level in adventitious bud production. The rate of organogenesis on buds collected in November from 6-year-old ortets could be dramatically increased by preincubation on the induction medium without cytokinins for seven days before transfer to the bud induction medium proper. Selby and Harvey (14) also found that organogenesis on basal needles excised from buds collected and cultured *in vitro* in November was optimal, at about 2.4 adventitious buds per needle, if the buds were preincubated on cytokinin-free medium before transfer of the needles to the induction medium. The rate of adventitious bud production was similar on needles excised from 5-year-old buds collected in March but in this case preincubation of the buds reduced the subsequent rate of bud production.

5. THE ROOTING OF SHOOTS PRODUCED IN VITRO

5.1. In vitro

The rooting of micropropagules of Sitka spruce has proved difficult, regardless of the method of production. Webb (15) reported occasional rooting when calluses with adventitious buds were transferred from the initiation medium to a half-strength hormone-free medium. Excision of juvenile extended adventitious shoots and transfer to half-strength hormone-free medium resulted in about 5% rooting, with shoots initiated on 2iP and kinetin rooting at a higher frequency than those on BAP (16). Incubation of excised adventitious shoots on semi-solid media or on filter paper bridges above stationary liquid medium did not increase the level of rooting. However, Webb and Street found that the adventitious shoots could be rooted at levels of about 20% if they were removed from in vitro conditions and maintained under mist. Selby (12) rooted a few plantlets from meristem cultures and some adventitious shoots by prolonged incubation on a hormone-free medium. Selby (13) found that about 75% of buds taken from 3-year-old trees rooted when they were transferred intact from the bud induction medium to the bud elongation medium with the roots growing out from dense clumps of adventitious shoots. A few adventitious shoots excised from 3-year-old buds rooted whereas none rooted from more mature tissue. John (6) found that fewer than 1% of micropropagated juvenile axillary shoots produced roots spontaneously on agar although vitrified cultures rooted at a much higher rate than normal cultures. In vitro rooting was not promoted by a continuous treatment with a low concentration of auxin in the medium, by a transient treatment with a high auxin concentration or by variations in concentration of some constituents of the medium (John unpublished).

5.2. Under non-sterile conditions

The rooting of micropropagules has been much more successful under non-sterile conditions in a mist house (John unpublished). Approximately 38% of large juvenile axillary shoots (about 5 cm long) rooted if the callus that formed at the base of the shoot was removed before insertion into the propagation beds. Transfer of the axillary shoot cultures to fresh medium _in vitro_ for 30 days before insertion increased the level of rooting from 14 to 25% in cultures of small shoots (about 2.5 cm) compared with 91% in greenhouse grown seedling control cuttings. Environmental preconditioning of the micropropagules by reducing daylength and temperature to winter conditions over a period of 40 days, followed by a 20 day period of cold storage at 2°C in the dark, resulted in a subsequent rooting level of about 36%. It would appear that the major problem with rooting micropropagules under mist is not one of their inability to root but one of keeping them alive long enough for rooting to occur. The needles of micropropagated axillary shoots have less surface wax, a thinner cuticle, a thinner walled epidermis and less central vascular tissue than greenhouse grown seedlings and are therefore much more prone to rapid desiccation and death (8). Rooting levels of up to 70% have been achieved by inserting the micropropagated shoots into coarse sand in covered seed trays in the controlled environmental conditions of a growthroom (John unpublished).

The work on Sitka spruce has progresed to the stage where juvenile plant material can be micropropagated without an intervening and potentially damaging callus phase. Shoots can also be produced directly from tissues of mature trees. However, the routine rooting of shoots remains a problem in both of these systems. Future research will be directed towards refining these techniques and also towards the rejuvenation of mature trees.

6. ACKNOWLEDGEMENTS

We would like to thank Dr C Selby for the illustration in Fig. 3 and Dr C J A Samuel for critically reading the manuscript.

7. REFERENCES

1. CHENG TY 1975 Adventitious bud formation in culture of Douglas fir (*Pseudotsuga menziesii* (Mirb) Franco). Plant Sci Letters 5: 97-102
2. FORESTRY COMMISSION 1978 Planning and Economic Paper, 31
3. GILL JGS 1983 Comparisons of production costs and genetic benefits of transplants and rooted cuttings of *Picea sitchensis*. Forestry 56: 61-73
4. JOHN A 1979 Applications of plant tissue culture in forestry. Rep For Res, Edinburgh 1979: 54-55
5. JOHN A 1981 Micropropagation. Rep For Res, Edinburgh 1981: 31-32
6. JOHN A 1983 Micropropagation. Rep For Res, Edinburgh 1983: 28
7. JOHN A 1984 Micropropagation. Rep For Res, Edinburgh 1984: 31
8. JOHN A 1984 Vitrification in Sitka spruce cultures. Proc 41st Conf, East School Ser, Univ Nottingham (in press)
9. JOHN A, BW MURRAY 1981 Micropropagation of Sitka spruce. *In* Proc IUFRO Sect S2 01 5. Int Workshop "*In vitro*" Cultivation For Tree Species, Fontainebleau, France, pp65-70
10. KLEINSCHMIT J, J SCHMIDT 1977 Experiences with *Picea abies* cuttings propagation in Germany and problems connected with large scale application. Silvae Genet 26: 197-203
11. NAMKOONG G, RD BARNES, J BURLEY 1980 Screening for yield in forest tree breeding. Commonw For Rev 59: 61-68
12. SELBY C 1980 Tissue culture research in coniferous forest trees. For Res Comm Meet
13. SELBY C 1983 Adventitious bud formation in bud cultures of Sitka spruce. *In* AC Cassells and JA Kavanagh, eds, Plant Tissue Culture in Relation to Biotechnology, Royal Irish Academy, Dublin, pp111-119
14. SELBY C, BMR HARVEY 1985 The influence of natural and *in vitro* bud flushing on adventitious bud production in Sitka spruce bud and needle cultures. New Phytol 100: 549-562
15. WEBB KJ 1978 Growth and Morphogenesis of Tissue Cultures of *Pinus contorta* and *Picea sitchensis*. PhD Thesis, University of Leicester
16. WEBB KJ, HE STREET 1977 Morphogenesis *in vitro* of *Pinus* and *Picea*. Acta Hortic 78: 259-267

4. EUROPEAN PINES

S. JELASKA

1. INTRODUCTION

The three most important pines in Europe are *Pinus nigra* Arnold, *P. pinaster* Sol. and *P. sylvestris* L. All three species are two-needle pines that grow to a height of 25-30 m (maritime pine) or 30-40 m (Scots and black pine) (66).

Black pine (*P. nigra*) is a native of the Mediterranean and submediterranean areas of South Europe, North-West Africa and Asia Minor. Its natural distribution in these areas is discontinuous (16). In submediterranean forests and underbrush it generally occurs on arid, semiarid, stony and sandy terrains. Most of its races grow well on limestone and dolomitic soils and, therefore, are highly suitable for afforestation on Karstic-degraded terrain (66).

Maritime pine (*P. pinaster*) is native to the Pyreneesian peninsula, southern and southwestern France on the Atlantic coast, the northwestern coastal area of Italy, northern Africa from Morocco to Tunisia, and the islands of Sardinia and Corsica. It is planted vastly in the Atlantic area of France and in many Mediterranean localities. It grows on sandy, silicate soils and requires a highly humid atmosphere.

Scots pine (*P. sylvestris*) is the most widely distributed pine species. In Europe it is found from Scotland and Scandinavia to the Pyreenees, with discontinuous areas in the central part of Spain, south and middle Europe and in the north of Greece. In Asia it spreads from central Turkey and the Caucasus to north Manchuria and the Ohot Sea (16). Scots pine is a very variable species, growing on sandy and stony sites as well as on poor soils and peat-bogs in pure or mixed components. It grows at altitudes of up to 2000 m.

Vegetative propagation of these pines by means of rooted cuttings, especially if the material is taken from older trees, is very difficult. Grafting is carried out in the conventional way for mass vegetative propagation (66), but recently micropropagation has become a promising method for clonal multiplication of *Pinus* sp.

This chapter presents a review of methods of embryo culture, tissue culture and micropropagation of *Pinus nigra*, *P. pinaster* and *P. sylvestris*.

2. EMBRYO CULTURE

The technology for the culture of excised, zygotic embryos is applicable to many plant species and has become a routine tool for geneticists in many breeding programmes. Continued development of mature embryos is usually easily obtained on simple media containing mineral salts, sucrose and minor additions of growth regulators. However, the earlier the stage of development, the more difficult the culture of the embryo becomes. Achievement of this goal is, for many tree species, in itself the object of a research project.

Schmidt (56) observed the growth and development of *P. sylvestris*, *P. pinaster* and other embryos with time. He also observed changed colour in light and dark mostly in relation to chlorophyll formation. His work is probably among the first attempts to culture free embryos. Brunner (12) using a sterile nutrient solution without a carbohydrate source, cultured mature embryos of *P. pinaster*. This appears to be the first report of a serious attempt to culture excised conifer embryos on a sterile medium free of the female gametophyte. In some of his experiments macerated gamtetophyte was added to the medium as nutrient.

Embryo cultures of *P. nigra* and *P. sylvestris* were also studied by Thomas (60, 61, 63) (Table 1). He found that the development of isolated embryos of *P. sylvestris* cultured on Halperin´s medium (31) complemented with vitamins depended on their stage of differentiation. This medium permitted only a limited survival of undifferentiated embryos, but allowed normal growth of differentiated embryos (60, 61). However, *P. nigra* or *P. sylvestris* embryos excised from the ovule at an early stage of cleavage, but retaining their suspensor apparatus, developed normally when kept in close contact with a nurse culture on a conditioned medium (63).

3. INDUCTION AND ESTABLISHMENT OF CALLUS CULTURE

Callus tissue has been established in cultures of all three the pines (Table 2), and has been maintained over a prolonged period (17, 18, 19, 35, 37,

Table 1. Embryo culture.

Pinus	Stage of embryo at excision	Basal medium	Growth substances	Reaction	References
pinaster	?	Tollen	0	-	(12)
sylvestris *pinaster*	?	peptone, dextrose, albumin, galactose, asparagine		0	(56)
nigra	mature	White (69) 2% agar 2% glucose	malt extract 1%		(1, 2)
sylvestris	undifferentiated	Halperin (31) 2% sucrose 0.8% agar	casein hydrolysate 5 g/l *myo*-inositol 100 mg/l vitamins	1-8 days surviving without division	(60)
	differentiating			growth and division of apices	
	differentiated (4 mm long)	Halperin (31)		elongation from 1.5-3 cm	(61)
	15th day after fertilization	Halperin (31) 2% sucrose 0.8% agar	NAA 5.4 µM KIN 2.3 µM glutamine 50 mg/l *myo*-inositol 100 mg/l vitamins		
nigra		Halperin (31) conditioned by callus tissue		cell division and embryos developing	(63)

Table 2. Induction and establishment of callus culture.

Pinus	Explant source	Donor age	Basal medium	Growth substances	Establishment	References
pinaster	phloem, cambium xylem		Heller (34) 3% glucose	IAA, NAA, KIN 100 mg/l inositol	-	(44,45)
	hypocotyl	1 month	MS (49) 2% glucose 1.2% agar	22.6 µM 2.4-D 2.3 µM KIN 100 mg/l inositol 100 mg/l asparagine	+	(17,18)
	trunk of adult trees comprising phloem and cambium		MS (49) 2% glucose 1.2% agar	22.6 µM 2,4-D 2.3 µM KIN 100 mg/l inositol 100 mg/l asparagine	+	(19)
sylvestris	stem base phloem, cambium xylem	10 years	Heller (34)	IAA, NAA 10 nM-100 µM	-	(4)
	shoot	1 year	White (69) LS (47) 2% sucrose	22.6 µM 2,4-D 0.9 µM BA asparagine	+	(54)
	plumule cotyledon needle	14 days 20 weeks	Sommer *et al.*(59) 3% sucrose 0.6-0.7% agar	IAA, IBA, NAA (10-20 µM) 30 µM IAA + 5 µM KIN 3 mM inositol	-	(9)
nigra	stem base	10-12 yrs	Heller (34) 2% glucose 1% agar	NAA, IAA, 2,4-D (100 nM-100 µM)	-	(5)
	cortex of current	mature year's stem	Harvey (32)	5.4 µM NAA 2.3 KIN 0.1 mg urea	-	(33)
	hypocotyl	15 days	Halperin (31) 2% sucrose 0.8% agar	5.4 µM NAA 0.11 µM KIN 50 mg/l glutamine 100 mg/l inositol	+	(62)
	hypocotyl	8-10 days	MS (49) 2% sucrose 1-1.2% agar	22.6 µM 2,4-D 2.3 µM KIN 100 mg/l inositol 100 mg/l asparagine	+	(37,48, 40,41)
	receptive female strobilus	45 years	MS (49) 2% sucrose 0.8% agar	32.3 µM NAA 4.4 µM BA 100 mg/l inositol 100 mg/l asparagine	+	(35)
			BSTCM(27)			

38, 40, 41, 62). Table 2 shows that callus may be induced on explants from different parts of plants and also from plants of different ages (from 10-day-old seedlings to 45-year-old trees).

Establishment of callus, however, was only reported in cases when the medium was supplemented with high concentrations of an auxin such as 2,4-dichlorophenoxyacetic acid (2,4-D) or naphthaleneacetic acid (NAA), cytokinin (kinetin or benzyladenine), and vitamins and amino acids. Myo-inositol and aspargine were found to be important. Regeneration of organogenic structures was not observed in these cultures.

4. MICROPROPAGATION METHODS

Regeneration of pine plantlets *in vitro* from explants of juvenile tissues may be achieved in three different ways: the initiation of adventitious buds, the enhancement of axillary bud development, and the release of brachyblast meristems from suppression.

4.1. Adventitious buds

Non-callus-mediated adventitious bud induction is obtained from cotyledons of several day-old germinants in the three pine species (Table 3). The presence of the cytokinin N6-benzyladenine (BA) played an important role in regeneration, and it was used in all three the species.

Its application can be various. Thus buds were formed on *P. sylvestris* cotyledons of 14-day-old germinants in the presence of BA, either incorporated in the medium or applied as a high-concentration (125 µM), short-duration pulse (7, 9). The pulse treatment gave a higher average number of regenerated field-transplantable plants per seed than the non-pulsed treatment. Pulse-treatment gave best results when applied to 8-week-old seedlings (7).

Adventitious buds were induced in black pine embryos after 8.9 uM BA was added to a basal medium, consisting of half-strength Murashige and Skoog (MS) medium with 6.75 uM NAA and 2.5 or 4.9 uM IBA. Adventitious buds also appeared on 6 to 8-week-old seedling explants after being pulse-treated with 125 µM BA for 2h and then cultured on the same basal MS medium (Berljak *et al.* in preparation).

Adventitious budding was induced on two types (cotyledon and brachyblast) of *P. pinaster* organs (24). Cotyledons, derived from 8- to 10-day-old seedlings, showing morphogenic response to an appropriate mineral solution were used. The combination of 0.8 µM BA with 5 nM NAA was best in promoting organogenesis. Brachyblasts were derived from the cuttings of a 10-year-old tree, cultivated in the greenhouse, and regularly fertilized with a mineral nutrient solution. At the time of collection, these cuttings showed dome-shaped meristematic cell clusters on the top of their short-shoot buds. These buds, with their meristematic domes, were excised and cultured on the same medium that was used for the cotyledon explants. After 3 weeks in culture, 79% of the short shoots developed. Among these, 42% showed small excrescences at the basal surface of needles which subsequently developed into buds. Adventitious buds, generated from the black pine (42) as well as from _P. sylvestris_ brachyblasts (7), often developed into multiple shoots.

Tranvan (64) found that as the seedling aged an increase in cytokinin concentration became necessary for bud induction. Bornman (8) supposes that a hormone gradient exists within a single conifer cotyledon or leaf and that levels of endogenous auxin, in conjunction with those of cytokinin applied exogenously, control bud induction quantitatively and qualitatively and determine bud position.

With regard to adventitious bud induction in brachyblast explants of _P. sylvestris_, Bornman and Jansson (9) concluded that they can be induced by high concentrations of cytokinin (e.g. 5-10 µM each of BA and 6-dimethylallylaminopurine (2iP), or up to 20 µM BA. These buds were initiated around the brachyblast´s shoot apex in the presence and absence of auxin, regardless of whether coumarin was included in the medium.

4.1.1. _Histological origin of adventitious buds_. Longitudinal sections from cotyledon explants and short shoot tissues of _P. pinaster_ (24) showed that only the cells of the superficial layers of mesophyll tissue dedifferentiated, forming areas of primary meristematic cells from which bud primordia arose. In black pine, buds were induced from cell divisions in the epidermal or hypodermal layers of mature embryos (43). Vogelmann _et al_. (68) studied morphogenic responses in cotyledon explants of _P. sylvestris_ and _Picea abies_ obtained after treatment with different concentrations of benzyl (8-^{14}C) adenine applied in different ways (11). Their data indicate that there are strong diffusion barriers against BA in _P. sylvestris_ explants. Time course studies with the labelled BA showed that only about one-third of the BA in the medium was taken up by the explants. This is probably a result of the xeromorphic anatomy of the

Table 3. Adventitious bud induction

Pinus	Explant	Donor age	Basal medium	Growth substances and application	References
pinaster	plumule	2 months	Campbell and Durzan (13)	10 μM BA	(22, 26)
	cotyledon (1-cm long)	8-10 days	Cheng (15)	0.8 μM BA 5 nM NAA	(20, 24)
	brachyblast (7-cm needles)	10 years	1/2xMS (49) 1% agar 117 mM sucrose	10 μM BA 25 nM NAA	
nigra	embryo	0-2 days	1/2xMS macro-elements (49) 1xMS micro-elements 3% sucrose 0.8% agar	8.9 μM BA 6.7 μM NAA 2.5-4.9 μM IBA	(36, 42)
	seedling	6-8 weeks	MCM (6)	125 μM BA "pulse" - 2 h	(Berljak *et al.* in prep.)
sylvestris	embryo		Lin and Staba (46)	11.1 μM BA	(64)
	seedling		2% sucrose 0.8% agar	22.2 μM BA	
	seedling		MS (49) 3% sucrose	11.4 μM IAA 0.45-4.95 μM 2,4D	(48)
	plumule	14 days	Summer *et al.* (59)	BA + 2iP 5-10 μM each	(7)
	brachyblast	11 weeks	3% sucrose 0.6-0.7% agar	or 20 μM BA	(9)
	cotyledonary node with cotyledons and primordial shoot				
		2 weeks	MCM (6)	a) 5 μM BA	
		6-8 weeks	MCM (6)	b) 125 μM BA "pulse" - 2 h	(7)

cotyledons which possess a thick cuticle and a sub-epidermal cell layer and a low leaf surface to volume ratio. However, cotyledons do not have a hypodermis that could become sclerified after long periods of culture (6, 10). These observations and uptake results may partly explain different cellular origins of developing adventitious buds when BA is applied to cotyledons in different ways. When included in agar, meristemoids usually originated in subsidiary cells of the stomatal apparatus of cotyledons in surface contact with the agar, whereas after a pulse treatment they often originated in subepidermal tissues as well (6).

4.1.2. Peroxidases during the differentiation of adventitious buds. In addition to histogenic studies of adventitious buds in black pine embryo culture, the total peroxidase activity as well as the distribution of isoperoxidases were investigated (43). It was found that in the initial explants the peroxidase activity was low, then it increased slowly until the 12th day, while the protein content was decreasing at the same time. Between the 12th and the 19th days a considerable increase in peroxidase activity could be measured; maximal peroxidase activity accorded with minimal protein content. Between the 19th and 28th days of culturing the activity of the peroxidases increased. The electrophoretic pattern of the isoperoxidases showed a characteristic separation of two isoenzymes which occurred between the 12th and the 19th days, and disappeared on the 28th day. It was observed that the adventitious bud formation was accompanied by changes in peroxidase activity and in the isoenzyme pattern.

4.1.3. Chromosome behaviour in the primary culture of embryos and seedlings. Abnormalities in cell division in black pine embryos during the period from 0 to 38 days in culture were studied (50). All the investigated metaphases were diploid with 2n=24 (99.5%) and only 0.5% were tetraploid. All division abnormalities (anaphase and telephase bridges, micronuclei) occurred with a very low frequency.

Cytological analysis of adventitious and axillary buds, induced by a pulse-treatment with 125 uM BA of six-week-old *P. nigra* seedlings, showed that adventitious buds 10 and 22 days after pulsing have 7.8% and 3.7% aberrant cells, respectively, and axillary buds 2.7% and 1.4% over the same period of observation (Berljak *et al.* in preparation).

4.2. Axillary bud development

Vegetative primordia which may persist in leaf axils can develop buds capable of forming shoots *in vitro* (20). The development of axillary buds *in vitro* was achieved in all three the pines (Table 4).

Buds appeared in axils of cotyledons and juvenile leaves of explants of *P. pinaster*, cultured on a medium containing 10 µM BA (22). In this experiment the apical part of the hypocotyl with the apical meristem and cotyledons was excised from 20-day-old seedlings and cultured on a medium containing 10 µM BA and 0.025 µM NAA. After a 15-day-exposure to the hormones, well-organized axillary buds were formed (26). Comparing the effect of various mineral solutions, David *et al.* (26) noted a stimulatory effect by 15 mM of potassium in the medium. Initiation of buds by exogenous cytokinin was stimulated by the presence of cotyledons. Rancillac (51, 52), using similar explants of *P. pinaster*, showed that the addition of IAA, IBA or NAA did not stimulate the growth of cytokinin-induced axillary buds. On the contrary, auxins stimulated callus formation. Gibberellic acid, too, had a negative effect. On the other hand, a medium rich in minerals such as in MS medium (49) stimulated organogenesis in explants.

It was observed that a concentration of BA as high as 25 µM, and a duration of exposure to the hormone as long as 45 days was sometimes required for morphogenesis (21). A large variability in axillary bud-forming capacity was observed among explants, with the number of buds being induced by one cytokinin treatment varying from 4 to 37 per explant, the most common number being from 10 to 20 buds. Axillary bud induction was also obtained on 1-1.5 cm shoot tips from 20-day-old seedlings of *P. nigra*, as well as on the upper part of 1-year-old plants growing in the greenhouse (3, 36, 42).

Bornman (7) reported that in *P. sylvestris* accelerated development of axillary buds without callus was obtained when cytokinin was administrated as a pulse, in a hydroponic solution or as a foliar spray to 6- to 8-week-old glasshouse-grown seedlings, before excision and culture of epicotyls on an auxin- and cytokinin-free medium. Propagation ratios of 8-week-old donor explants pulse-treated or sprayed with BA were 2 to 5 times higher than those of 2-week-old tissues, which could be explained by the fact that a larger surface of tissues was subjected to the BA. Similar results were obtained in experiments on *P. nigra* seedlings (3) (Berljak *et al.* in preparation).

4.3. Brachyblasts

Release of the brachyblast meristem from suppression can be accomplished in explants of early juvenile material and from brachyblasts developed *in vitro*.

Table 4. Axillary bud/shoot formation and release of brachyblast bud suppression

Pinus	Explant	Donor age	Basal medium	Treatment BA µM	NAA nM	References
pinaster	seedling epicotyl	20 days	Campbell and Durzan (13)	10	25	(20,22,26)
	brachyblast	10-30 days	½ x MS (49)	10	25	(51,52)
nigra	shoot-tip *in vitro* brachyblast	20 days 1 year	½ x MS macro-elements (49) 1 x MS micro-elements	4.4	50	(3,36,42)
nigra	- " -	6-8 weeks	MCM (6)	125 "pulse"	0.0 2h	(Berljak *et al*. in prep.)
nigra	resting apical bud	2 years	½ x MS micro-elements (49) 1xMS micro-elements	4.4	50	(14)
sylvestris	seedling brachyblast	2 weeks 6-8 weeks	MCM (6)	5.0 in agar	0.0	(7,11,68)
	plumule explant	2 weeks 20 weeks		125 "pulse"	0.0 2h	

Cold acclimation, disbudding and a cytokinin spray treatment greatly enhanced uniformity of response. Regeneration from the brachyblast meristem follows three pathways: (a) release of bud suppression, induction of multiple adventitious buds *in vitro*, excision and elongation *in vitro*, and rooting of adventitious shoots *in vivo*; (b) release of bud suppression, shoot elongation *in vitro*, and subculturing or rooting *in vivo*; and (c) excision of the brachyblast before release of bud suppression, rooting *in vivo*, and shoot elongation *in vivo* (7).

4.4. Shoot elongation

To induce elongation of shoot buds, explants with either newly-formed or excised buds are usually transferred to hormone-free nutrient media, often with half the normal mineral strength or media of a different composition. Activated charcoal at a level between 0.3 and 2% is often added to the medium to enhance elongation (7, 26, 35). Shoots developed from buds after being cultured for some time on a nutrient medium without hormones. Bornman (6) irradiated cultures with far-red light for the first two weeks for additonal stimulation of shoot elongation.

To further improve the shoot development phase in cultures of buds of 21-day-old black pine seedlings three different basal media were examined (3): (a) half-strength MS macro-nutrients (49), (b) SH medium (55) and (c) GD medium (30). The level of sucrose was reduced to 2% and hormones were omitted. The best elongation and survival of buds was obtained with the following manipulation. After 4 to 6 weeks in primary culture, explants with induced axillary buds and brachyblasts were first placed for 3 weeks on half-strength MS macronutrients, and then for 3 weeks on GD medium. Finally, single shoots were isolated and transferred to fresh GD medium with or without the addition of 2% activated woody charcoal and maintained on this medium for 3 weeks.

This procedure made it possible to obtain well-developed and hardened 10-20 mm high shoots which could be submitted to the next *in vitro* procedure. It was found necessary to subculture explants every 21 to 28 days. During this period, second generation brachyblasts formed in the axils of primary leaves on the newly generated shoots.

4.5. Remultiplication

Special attention was paid to the possiblity of remultiplication of shoots produced *in vitro*. Well-developed shoots were exposed to cytokinin so as to induce additional axillary buds. Successive generations of shoots were obtained in all three the pines (7, 20, 41). Black pine cultures retain the capacity for producing multiple buds for longer than three years. Thus, permanent, although relatively slow multiplication is possible (3). Such multiplication is of interest not only for clonal propagation, but also as a potential means of gene pool preservation (67).

4.6. Root formation

4.6.1. *Adventitious rooting.* Root formation was induced in shoots of *P. pinaster* after soaking them in an IBA solution (4%) for 24 h, followed by transfer to a well-aerated, sterile mixture of perlite and peatmoss (26). Rooting occurred in 90% of the shoots from axillary buds of 20-day-old *P. pinaster* plants. Only 5% of the shoots from the apex of brachyblasts of 2- to 3-year-old trees rooted.

Using the same species, Rancillac (51) found that auxin concentration in agar medium affected the quality of the roots. NAA at 1 uM, induced many roots, at the base of the shoots, but also callus through which the establishment of vascular connections between roots and shoots often occurred. On the other hand, a concentration of 0.1 uM resulted in thin, normal roots, and no callus at the base of the shoot. A temperature of 20°C resulted in better root growth than one of 25°C. A nutrient medium low in minerals was best for root growth. Transfer to the greenhouse was possible after the roots had grown for 2-3 months on the sterile medium. To obtain adventitious *P. pinaster* shoots *in vitro* required a variety of rejuvenation treatments (29). These shoots formed roots (90%) within one month after exposure for 24 h, at low temperature, to an aqueous solution of 220 μM IBA with 26.9 uM NAA, and transfer to a sterile substrate of perlite and tourbe (3:1). Bornman and Jansson (9) obtained rooting in 35% of brachyblasts of *P. sylvestris*, treated with coumarin (10 μM).

Adventitious and axillary buds of *P. sylvestris* were excised, transferred to BA-free culture medium containing 1% activated charcoal (7) and irradiated with far-red light (6). These shoots elongated and were subsequently rooted in a mistbed. This rooting was most effective if the shoots were treated with a high-concentration, medium- to long-duration auxin pulse before transfer to the mistbed (6).

Several experiments on root formation in *P. nigra* shoot cultures have been carried out on shoots of juvenile-origin, generated through a cycle of shoot culture (3). However, these shoots, when submitted to different *in vitro*-manipulations with auxins and then transplanted, had a low root formation rate. Better results were achieved with axillary shoots obtained from cultures of 42-day-old seedlings (Berljak *et al.* in preparation). Shoot bases were treated with either "Floramon" powder (0.1% NAA) or with 5.4 μM NAA added to an agar substrate. After the powder treatment shoots were transplanted to peat: perlite or vermiculite under good greenhouse conditions. After 2-3 months 60-70% shoots rooted. Rancillac *et al.* (53) systematically studied the rooting ability of *P. pinaster* shoots derived from axillary buds *in vitro*. Their purpose was to experiment with material that was as homogeneous as possible. They investigated the influence of the substrate, the efficiency of the hormonal treatment on the intensity and the quality of rooting, and finally, the uniformity of results within and between clones. They showed that a mixture of peat and perlite (1/3, v/v) permitted good rooting and further development of the root system if the humidity was properly controlled. With the use of an agar medium difficulty in regulation of humidity was avoided; it supported root induction but resulted in limited growth. For hormonal treatment, explants submitted to a 12-day-exposure to NAA (1 μM) and then subcultured in an auxin-free medium, gave rooting percentages ranging from 80-100%. In terms of rooting quality (intensity, growth rate and branching ability), a 15- to 19-day-period of auxin treatment was more effective. Application of this optimal treatment to several clones revealed slight differences within and between clones (80 to 100% of rooting according to the clone). Regenerated plantlets were able to undergo autotrophic growth after transplantation to the greenhouse conditions.

4.6.2. Branching of short roots and mycorrhizae. Slankis (57, 58) studied the growth of 10-mm-long excised root tips, 10-mm-long, of *P. sylvestris* seedlings. Optimum growth was obtained with 4% glucose or 5.8% sucrose. Sucrose resulted in more branching.

The structure and mode of branching of the short roots of *P. pinaster* under semi-sterile and sterile conditions was studied (28). Using intact seedlings, excised buds or root tips, the short roots that developed in these were histologically similar to ectomycorrhizal roots except for the absence of the fungus. These short roots produced dichotomous branches in all the environmental conditions used. The formation of these mycorrhizogenic-like short roots seems to depend on genotype. It could be useful to select such short root forming genotypes to facilitate studies of mycorrhizal root formation in young trees.

4.6.3. Initiation of root primordia. The possiblity of root regeneration in cuttings of 10-day-old *P. nigra* seedlings was examined *in vitro* (39). Sucrose and IBA were both important in the regeneration process of root meristems. Sucrose (20 g l^{-1}) alone induced adventitious rhizogenesis in 50% of cuttings; higher percentages were obtained with sucrose in combination with IBA. In the most favourable combination, i.e. sucrose (20-40 g l^{-1}) and IBA (0.49-49.2 μM) the percentage of rooted cuttings was 100%. Root formation was not achieved on medium containing mineral components only.

The capacity of the shoot to synthesize the auxin synergists required for root initiation was tested in shoots of *P. pinaster* cloned *in vitro* for a 3-year-period (25). Shoots submitted to an auxin treatment (1 µM NAA, 18 days) demonstrated that the rooting ability *in vitro* persisted over 5 successive induction cycles of shooting (during a 9-month-period). Rooting requires a permanent synthesis of auxin synergists which activate the metabolism of cell dedifferentiation and root primordium initiation. Agar media permitted intense meristem initiation, but prevented active root elongation. After adding a mycorrhizal fungus, either *Pisolithus tinctorius* or *Hebeloma cylindrosporum*, to the agar medium the roots resumed growth and short lateral root formation was stimulated. These two phenomena, induced by fungal association, improved the quality of root systems and thus facilitated successful transplantation from test-tubes to the field. A two-step method of root system formation was suggested: primordium initiation in agar, including hormonal treatment; and root system development in perlite (or any aerated substrate) inoculated with a mycorrhizal fungus. Moisture control remains a problem over a long period.

Many questions in relation to the rooting of *in vitro*-produced shoots of two-needle pines still remain, but it appears that good greenhouse facilities with good humidity control are indispensable for good root development.

5. REGENERATION IN MATURE PLANTS

The subject of mature plant regeneration is extensively discussed elsewhere in this book. However, the data obtained with cultures of *P. pinaster* and *P. nigra* will be dealt with briefly.

Franclet *et al.* (29) have obtained regeneration from 11-year-old *P. pinaster* trees. Brachyblasts were excised from current shoots of intensively fertilized root cuttings, and were subsequently cultured *in-vitro*. These treatments had a rejuvenating effect, and plants with a needle morphology similar to that of young seedlings were regenerated from the brachyblasts as described in section 4.3., pathway (a). It was further observed that the ultimate behaviour is strongly influenced by environmental conditions applied after the *in vitro* phase. A rapid transfer to soil leads to reduced vigour, formation of long needles, bud dormancy, and a return to the physiological state of a mature tree. On the other hand, if after the *in vitro* phase plantlings are maintained under optimal nutritional and rooting conditions, it is possible to temporarily prolong the juvenile phase.

It was tried to establish multiple bud cultures of *P. nigra* donor plants which were older than one year (14). Resting vegetative buds of *P. nigra* and hybrid *P. nigrosylvis* Vid. were used as explants. The induction medium was the same as was used for axillary bud generation from apical part of hypocotyl explants of black pine seedlings as described in section 4.2. Only one clone of shoots was obtained, i.e. from brachyblasts from buds derived from a 2-year-old donor (14, 36).

6. MICROGRAFTING

Besides adventitious budding *in vitro*, rejuvenation of meristems could possibly be achieved by micrografting. Tranvan and David (65) studied the possibility of auto- and homografting *in vitro* in *P. pinaster* cultures. Apices, 5-7 mm in length, either collected from 2-3 month-old seedlings or produced *in vitro* from an 11-year-old tree by means of successive axillary budding cycles over a five-year-period, were successfully grafted on epicotyls of 2 to 3-month-old seedlings. The following steps in the procedure were important: immediately after decapitation of the seedlings (at the level of the first node of the epicotyl) a drop of sterile solution (1 g l^{-1}) of DIECA (sodium Diethyldithiocarbamate) was put on the cut surface both of the rootstock and the graft. The two parts of the graft were glued with 0.8% (Difco) agar mixed with a drop of glycerol. Grafted seedlings were first kept in the dark for 3-4 days and then one week under diffuse light. Subsequently they were transferred to normal light conditions (16-h photoperiod, 25 Wm^{-2}). Grafted plants were watered with a Knop solution and eventually transferred to the field.

7. CONCLUSION

The three species of European pine, namely *P. nigra*, *P. pinaster* and *P. sylvestris* have been shown to be amenable to the following in *in vitro* culture: induction of adventitious shoot buds, stimulation of axillary bud development, release of brachyblastic buds from suppression, and rooting of elongating shoots. *Pinus sylvestris* appears to be the most recalcitrant of the three species. Much basic research remains to be done with all three species before an efficient system of *in vitro* propagation can be developed.

Acknowledgements

Research on black pine tissue culture was sponsored by Grant No. YO-FS-90-JB-61 and the Science Research Council of SR Croatia.

8. REFERENCES

1. BARTELS H 1957a Kultur isolierter Koniferenembryonen I. Naturwissenschaften 44: 290-291
2. BARTELS H 1957b Kultur isolierter Koniferenembryonen II. Naturwissenschaften 44: 595-596
3. BERLJAK J 1984 Micropropagation of black pine (Pinus nigra Arn.) in vitro. M Sc Thesis Univ Zagreb pp 74 (in Croatian)
4. BOGDANOVIC M 1968 The influence of different factors on stem tissue cultures of Pinus silvestris L. Zemljiste i biljka 17: 135-145
5. BOGDANOVIC M, DJ JELENIC 1968 The effect of growth substances on the development of initial stem tissues culture of Pinus nigra Arn. Zemljiste i biljka 17: 305-313 (In Serbian)
6. BORNMAN CH 1983 Possibilities and constraints in the regeneration of trees from cotyledonary needles of Picea abies in vitro. Physiol Plant 57: 5-16
7. BORNMAN CH 1984 Application of in vitro culture technology in clonal forestry. In Proc Int Symp Recent Adv For Biotechnol, Traverse City, Michigan, June 10-14, pp 178-194
8. BORNMAN CH 1985 Hormonal control of growth and differentiation in conifer tissues in vitro. Biol Plant (In press)
9. BORNMAN CH, E JANSSON 1980 Organogenesis in cultured Pinus sylvestris tissue. Z Pflanzenphysiol 96: 1-6
10. BORNMAN CH, E JANSSON 1982 Regeneration of plants from the conifer leaf with special reference to Picea abies and Pinus sylvestris. In Proc IUFRO Sect S2 01 5. Int Workshop "In Vitro" Cultivation For Tree Species, Fontainebleau, France, pp 41-53
11. BORNMAN, CH, TC VOGELMANN 1984 Effect of rigidity of gel medium in benzyladenine-induced adventitious bud formation and vitrification in vitro in Picea abies. Physiol Plant 61: 505-512
12. BRUNNER G 1932 Beiträge zur Entwicklungsphysiologie der Kieferkeimlinge. Jahrb Wiss Bot 76: 407-440
13. CAMPBELL RA, DJ DURZAN 1976 Vegetative propagation of Picea glauca by tissue culture. Can J For Res 6: 240-243
14. CESAR V 1983 Organogenesis in isolated bud culture of European Black Pine (Pinus nigra Arn.) and hybrid P. nigrosylvis Vid. M Sc Thesis, Univ Zagreb pp 83 (In Croatian)
15. CHENG TY 1977 Factors effecting adventitious bud formation of cotyledon culture of Douglas fir. Plant Sci Lett 9: 179-187
16. CHRITCHFIELD WB, EL JR LITTLE 1966 Geographic distribution of the pines of the World. USDA For Serv, Misc Publ 991
17. DAVID A 1965 Influence de divers facteurs sur les cultures initiales de tissue de pin maritime. C R Acad Sci 260D: 2058-2060

18. DAVID A 1970 Obtention d'une souche de Pinus pinaster Sol à partir de fragments d'hypocotyles. C R Acad Sci 271D: 1866-1868
19. DAVID A 1971 Obtention d'une souche tissulaire de Pinus pinaster Sol à partir de tissus du tronc de l'arbre. C R Acad Sci 273D: 2236-2238
20. DAVID A 1979 Manifestations de diverses potentialities organogènes et micropropagation végétative chez le Pin maritime (Pinus pinaster Sol.). Ann Resc Sylvicoles AFOCEL, 57-75
21. DAVID A 1982 In vitro propagation of gymnosperms. In JM Bonga, DJ Durzan, eds, Tissue Culture in Forestry. Martinus Nijhoff Publ, The Hague, pp 72-108
22. DAVID A, H David 1977 Manifestations de diverses potentialités organogènes d'organes ou de fragments d'organs de Pin maritime (Pinus pinaster Sol) en culture in vitro. C R Acad Sci 284D: 627-630
23. DAVID A, H DAVID, M FAYE, K ISEMUKALI 1979 Culture "in vitro" et micropropagation du pin maritime. AFOCEL 12: 33-40
24. DAVID A, H DAVID, T MATEILLE 1982 In vitro adventitious budding on Pinus pinaster cotyledons and needles. Physiol Plant 56: 102-107
25. DAVID A, M FAYE, M RANCILLAC 1983 Influence of auxin and mycorrhizal fungi on the in vitro formation and growth of Pinus pinaster roots. Plant Soil 71: 501-505
26. DAVID H, K ISEMUKALI, A DAVID 1978 Obtention de plants de Pin maritime (Pinus pinaster Sol) à partir de brachyblastes ou d'apex caulinaires de très jeunes sujets cultivés in vitro. C R Acad Sci 287D: 245-248
27. DE FOSSARD RA 1976 Tissue Culture for Plant Propagators. Univ New England, Armidale, Australia
28. FAYE M, M Rancillac A DAVID 1981 Determinism of the mycorrhizogenic root formation in Pinus pinaster Sol. C R Acad Sci 290D: 927-930
29. FRANCLET A, A DAVID, H DAVID, M BOULAY 1980 Première mise en évidence morphologique d'un rajeunissement de méristèmes primaires caulinaires de Pin maritime âgé (Pinus pinaster Sol). C R Acad Sci 290D: 927-930
30. GRESSHOFF PM, CH DOY 1972 Development and differentiation of haploid Lycopersicon esculentum (tomato). Planta 107: 161-170
31. HALPERIN W 1966 Alternative morphogenetic events in cell suspensions. Am J Bot 53: 443-453
32. HARVEY AE 1967 Tissue culture of Pinus monticola on a chemically defined medium. Can J Bot 45: 1783-1787
33. HARVEY AE, JL GRASHAM 1969 Procedures and media for obtaining tissue cultures of 12 conifer species. Can J Bot 47: 547-549
34. HELLER R 1953 Recherches sur la nutrition minérale des tissues végétaux cultivés in vitro. Ann Sci Nat Bot Biol Veg 14: 1-223
35. JELASKA S, B KOLEVSKA-PLETIKAPIC, M VIDAKOVIC 1979 Effect of broad spectrum tissue culture media on Pinus nigra callus. Works Facul Agric Univ Sarajevo 31: 43
36. JELASKA S, B KOLEVSKA-PLETIKAPIC, M VIDAKOVIC 1982 Bud regeneration in Pinus nigra embryo and seedling tissue culture. In Proc IUFRO Sect S2 01 5. Int Workshop "In Vitro" Cultivation For Tree Species, Fontainebleau, France, pp 159-166

37. KOLEVSKA-PLETIKAPIC B 1974 Callus culture of Austrian pine (*Pinus nigra* Arn.) on defined medium. Acta Bot Croat 33: 69–72

38. KOLEVSKA-PLETIKAPIC B 1978 The effect of the nutrient medium composition on callus growth of European black pine (*Pinus nigra* Arn.) - Clone K. Acta Bot Croat 37: 75–82 (In Croatian)

39. KOLEVSKA-PLETIKAPIC B 1980 Root regeneration *in vitro* by seedling cuttings of *Pinus nigra* Arn. Acta Bot Croat 39: 65–70

40. KOLEVSKA-PLETIKAPIC B 1982a *In vitro* studies of callus growth of *Pinus nigra* Arn. 1. Effect of some anorganic and organic nitrogen sources. Acta Bot Croat 41: 41–48

41. KOLEVSKA-PLETIKAPIC B 1982b *In vitro* studies of callus growth of *Pinus nigra* Arn. 2. Growth regulators requirements and their effects. Acta Bot Croat 41: 49–55

42. KOLEVSKA-PLETIKAPIC B, S JELASKA, J BERLJAK, M VIDAKOVIC 1983 Bud and shoot formation in juvenile tissue culture of *Pinus nigra*. Silvae Genet 32: 115–119

43. KOLEVSKA-PLETIKAPIC B, M KRSNIK-RASOL 1984 Analysis of adventitious bud formation in *Pinus nigra* Arn. embryo culture. Acta Bot Croat 43: 49–57

44. LAVAUD JJ 1969 Action conjuguée de quelques mauxines et de kinétine ou du mésoinositol sur de Pin maritime cultivés *in vitro*. C R Acad Sci 269D: 834–837

45. LAVAUD JJ 1970 Phénomènes d'histogenèse se produisant au cours du dévelopment des cultures initiales de tissues de Pin maritime; influence de certains modificateurs de croissance. C R Acad Sci 270D: 1116–1119

46. LIN ML, J STABA 1961 Peppermint and spearmint tissue culture. I Callus formation in submerged culture. Lloydia 24: 139–145

47. LINSMAIER EM, F SKOOG 1965 Organic growth factor requirement of tobacco tissue culture. Physiol Plant 18: 100–127

48. MOMOT TS 1977 Organogenesis of European spruce (*Picea abies* (L.) Karst and pine (*Pinus silvestris* L.) on isolated culture. Citologia i Genetika 4: 497–498

49. MURASHIGE T, F SKOOG 1962 A revised medium for rapid growth and bioassays with tobacco tissue cultures. Physiol Plant 15: 473–497

50. PAPES D, V GARAJ-VRHOVAC, S JELASKA, B KOLEVSKA-PLETIKAPIC 1983 Chromosome behaviour in cultured cell populations of higher plants. Kew Chromosome Conference II. George Allen & Unwin: 155-163

51. RANCILLAC M 1979 Mise au point d'une méthode de multiplication végétative "in vitro" du pin maritime (*Pinus pinaster* Sol) pour la constitution de clones à partir de semences. AFOCEL 12: 41–48

52. RANCILLAC M 1981 Perspectives d'application des cultures d'organes *in vitro* à la multiplication végétative du Pin maritime, *Pinus pinaster* Sol. Ann Sci For 38: 55–69

53. RANCILLAC M, M FAYE, A DAVID 1982 *In vitro* rooting of clones shoots in *Pinus pinaster*. Physiol Plant 56: 97–101

54. ROGOZINSKA JH 1970 Culture of Scots pine callus and its nutritional requirements. Acta Soc Bot Pol 39: 151–160

55. SCHENK VR, AC HILDEBRANDT 1972 Medium and techniques for induction and growth of monocotyledonous and dicotyledonous plant cell cultures. Can J Bot 50: 199–204

56. SCHMIDT A 1924 Ueber die Chlorophyllbild in Koniferenembryo. Bot Arch 5: 260–282

57. SLANKIS V 1947 Influence of the sugar concentration on the growth of isolated pine roots. Nature 160: 645-646

58. SLANKIS V 1948 Verschiedene Zuckerarten als Kohlhydratquelle für isolierte Wurzeln von *Pinus silvestris*. Physiol Plant 1: 278-289

59. SOMMER HE, CL BROWN, PP KORMANIK 1975 Differentiation of plantlets in longleaf pine (*Pinus palustris* Mill) tissue cultured *in vitro*. Bot Gaz 136: 196-200

60. THOMAS MJ 1970a Premières recherches sur les besoins nutritifs des embryons isolés du *Pinus silvestris* L. Embryons indifférencies et en voie de différenciation. C R Acad Sci 270D: 1120-1123

61. THOMAS MJ 1970b Premières recherches sur les besoins nutritifs des embryons isolés du *Pinus silvestris* L. Embryons différencies. C R Acad Sci 270D: 2648-2651

62. THOMAS MJ 1972a Sur l'isolement d'une souche de tissus du *Pinus nigra* Arn. susceptible de fournir des cultures-nourrices. C R Acad Sci 274D: 852-855

63. THOMAS MJ 1972b Comportement des embryons de trois espèces de Pins (*Pinus mugo* Turra, *Pinus silvestris* L. et *Pinus nigra* Arn.), isolés au moment de leur clivage et cultivés *in vitro*, en présence de cultures-nourrices. C R Acad Sci 274D: 2655-2658

64. TRANVAN H 1979 *In vitro* adventitious bud formation on isolated seedlings of *Pinus silvestris* L. Biol Plant 21: 230-233

65. TRANVAN H, A DAVID 1984 Greffage *in vitro* du Pin maritime (*Pinus pinaster* Sol). Can J Bot (In press)

66. VIDAKOVIC M 1982 Conifers - Morphology and Variability. Liber, Zagreb (In Croatian)

67. VIDAKOVIC M, S JELASKA 1983 Preservation of the gene pool of forest tree species. Genetika 15: 369-375

68. VOGELMANN TC, CH BORNMAN, P NISSEN 1984 Uptake of benzyladenine in explants of *Picea abies* and *Pinus sylvestris*. Physiol Plant 61: 513-517

69. WHITE PR 1943 A handbook of Plant Culture. The Jacques Catell Press, Inc, Tempe, Ariz

5. TISSUE CULTURE OF EASTERN NORTH AMERICAN CONIFERS

Asha Mehra-PALTA and David G. THOMPSON

1. INTRODUCTION

The forests of eastern North America (defined in this chapter as eastern Canada, and the United States east of the Mississippi River) account for a significant part of the total forest resource of the continent. Approximately 60% out of the total one billion acres of commercial forest land lies in the eastern part of North America. Coniferous or softwood species account for about 80% of the total volume of sawtimber in the region. This amounts to about 20% and 37% of the total softwood sawtimber of the United States and Canada respectively.

The forests of eastern North America are divided into two general forest types. Most of eastern Canada lies in the Boreal Forest type, while part of southeastern Canada and all of the eastern U.S. lies in the Eastern Deciduous Forest type (71). Central North America with its very favorable climate and evenly distributed precipitation has one of the most diverse and complex plant communities in the world. Of the literally hundreds of forest tree species, the majority are hardwoods. Conifers are best adapted to the less favorable sites and are most numerous in the northern and southern edges of the continent. Thus, most of the economically important coniferous forest stands are found in the Boreal Forest of eastern Canada and the Northeastern Coniferous and Southeastern Pine Forests of the U.S.

The Boreal Forest, with its short growing season, cool temperatures, and very wet spring followed by a very dry summer and autumn is best suited for coniferous species. Over 70% of the total wood volume in this forest type is softwood. This region is populated with species of spruce, pine, fir, tamarack and cedar. The spruces are represented by black (*Picea mariana*) and white (*P. glauca*) spruce. The pines include jack (*Pinus banksiana*) and red (*P. resinosa*) pine. Balsam fir (*Abies balsamea*), tamarack or eastern larch (*Larix laricina*) and northern white cedar (*Thuja occidentalis*) are also economically important species.

The cool humid climate of the Northeastern Coniferous Forest also favors coniferous species with about 48% of the total volume being softwood. This forest type mixes with the Northeastern Hardwood Forest, but conifers predominate on sites that are either too wet or too dry for hardwoods. In addition to the species of conifers found in the Boreal Forest, this region contains red spruce (*Picea rubens*), eastern white pine (*Pinus strobus*), pitch pine (*P. rigida*), eastern hemlock (*Tsuga canadensis*), as well as, Atlantic white cedar (*Chamaecyparis thyoides*) and eastern red cedar (*Juniperus virginiana*). The Northeastern Coniferous Forest extends westward to the Lake states of the U.S. where white, red, and jack pine, black and white spruce, eastern red and northern white cedar, eastern hemlock, tamarack and balsam fir predominate.

The Southeastern Pine Forest is characterized by a subtropical climate which in the past has been frequented by fires that tended to favor pines over hardwoods. In this forest type, the softwoods account for about 50% of the wood volume. Eastern white and pitch pines, Atlantic white and eastern red cedar, red spruce and eastern hemlock which are found in the Northeastern Coniferous Forest are also found here. The uniquely southern pines include loblolly (*Pinus taeda*), longleaf (*P. palustris*), pond (*P. serotina*), sand (*P. clausa*), shortleaf (*P. echinata*), slash (*P. elliottii*), and Virginia (*P. virginiana*). Bald cypress (*Taxodium distichum*) is a conifer unique to flooded sites in the southeastern U.S.

Although the forests of eastern North America contain a wide range of conifer species, not all of them are of economic importance. As a result, there are no published reports of tissue and cell culture work with many of these species. This chapter will review the progress that has been made in the tissue and cell culture of conifers native to eastern North America. The propagation of trees from both embryonic and mature tissues as well as developments in callus cultures, haploid tissues and protoplasts will be discussed. Work with loblolly pine (*Pinus taeda*), an important pine in the southeastern U.S. and the focus of a significant amount of work, will be discussed in a later chapter devoted entirely to that species. We have, however, taken the liberty of discussing work on introduced species not covered in other chapters.

2. REVIEW OF LITERATURE

2.1. Embryonic tissues

Presently, the most successful tissue culture propagation technique for conifers is the induction of adventitious buds on cultured embryonic tissues. Table 1 lists the published reports on bud induction from embryo tissues of eastern North American conifers. Sommer and Brown (73) and Sommer et al. (76) were the first to report bud induction on embryos of longleaf, slash, and loblolly pines cultured *in vitro* on a Gresshoff and Doy (38) medium containing 4.4 to 22μM benzylaminopurine (BAP) and 10μM napthaleneacetic acid (NAA). They observed numerous protuberances along the surface of the cotyledons which later developed into buds. When transferred to a basal medium without exogenous growth regulators some of the buds elongated into shoots and a few could later be rooted. Buds arose from epidermal and subepidermal cells which divided, giving rise to meristematic centers with varying degrees of organization. Brown and Sommer (17) obtained bud induction from cotyledons of several pine species. They reported much genotypic variation in bud induction and rooting potential from embryos of seed from wild populations or half-sib families placed on identical nutrient media. Such variations within a species have been reported for several other conifers. Using intact embryos, Stine and Sommer (77) induced adventitious buds on longleaf and slash pine as well as embryos of their interspecific hybrid.

Minocha (58) reported that normal growth of excised eastern white pine embryos was inhibited when placed on a Murashige and Skoog (60) medium (MS) containing an auxin transport inhibitor triiodobenzoic acid (TIBA). However, cotyledons of some embryos produced adventitious shoots. An average of 12 shoots per embryo were obtained with 2μM TIBA and this number could be increased with the addition of 4.4 to 8.8μM BAP. Flinn and Webb (35) obtained shoots on eastern white pine embryos cultured on a Schenk and Hildebrandt (69) medium containing cytokinin. Contrary to the observations of Minocha (58), they found that cytokinin was essential for bud induction, and that neither TIBA nor auxin stimulated bud formation.

Campbell and Durzan (18) reported bud induction on hypocotyl segments of white spruce when placed on a nutrient medium (27) containing 8.8μM BAP. On this medium, up to 100 "scale-like organs" were formed per embryo. Some of these elongated into shoots when placed on a medium lacking growth regulators, and a few were rooted. In a study on black and white spruce, Rumary and Thorpe (68) induced adventitious buds on epicotyl segments cultured on a Schenk and Hildebrandt (69) medium supplemented with 4μM each of the cytokinins BAP and 2-isopentenyladenine (2iP). Subsequent shoot elongation occurred on a medium containing a conifer derived charcoal. The inclusion of charcoal to stimulate shoot elongation has also been reported for a number of other conifers (15, 23, 57). Although the role of charcoal is not clear, it has been reported to adsorb excess plant growth regulators and phenolic substances (37). Rooting of black and white spruce shoots was possible by dipping the shoots in a sterile rooting powder containing the auxin indolebutyric acid (IBA), and then placing them in sterile vermiculite containing half-strength mineral salts, sucrose and charcoal. Over 80% rooting was possible in both species. Using this procedure, about 40 plantlets could be produced per seed within 170 days (68).

Karnosky and Diner (47) reported on the propagation of European larch (_Larix decidua_, an exotic species planted widely in the northeastern and lake states) and jack pine by the cotyledon method. There was considerable variability between clones in both bud induction and rooting frequency. They found that the average number of adventitious buds produced per embryo was 24.6 and 3.6 for larch and pine respectively. In larch they observed a 47.7% rooting frequency (range was between 0 and 100%). Huang and Tsai (44) reported on the propagation of Scots pine (Pinus sylvestris) by the cotyledon method. This species was introduced from Europe for use in the production of Christmas trees.

In conclusion, the technique of inducing adventitious buds to form on embryonic tissues has become almost routine. It should be fairly easy to accomplish this with almost any species; however, bud elongation and rooting may prove to be more difficult in some species. The spruces (68) and the larches (47) appear to be more responsive than most other conifers. Technical limitations of this method, such as bud elongation and root induction, need to be resolved to maximize propagation rates. The genetic gain possible is limited to the quality of the untested embryos used to initiate the cultures. The method, however, is useful in multiplying specific crosses and genetically improved material from seed orchards. The cost to produce tissue culture plantlets has to be reduced considerably before these techniques become economically feasible (72).

2.2. Mature tissues

Many conifer species have been propagated from tissues of embryos or young seedlings (22), while propagation from mature tissues of adult trees has proven to be more difficult. The advantage of culturing mature tissues is that trees old enough to have demonstrated their superior genotype can be reproduced. Bonga (12) has proposed two reasons why plantlet regeneration from mature tissues has not been possible. He suggests that in mature trees there are either no organogeneticaly competent cells or those cells that are competent do not differentiate because neighboring cells impose some form of inhibition upon them. Both of these conditions may be temporary and may change over time. During the annual growth cycle, there are times when tissues programmed to follow a certain developmental pathway can be switched to a different developmental pattern (12). This may be achieved during primordia initiation or meiosis. When these developmental changes occur, it is possible that organogenetic competency in some cells may be temporarily switched on, or if the cells are already competent, they may be released from inhibition by neighboring cells (11, 65).

The *in vitro* growth of resting vegetative buds of white, black, and Norway spruce, Balsam fir and Douglas-fir was reported by Chalupa and Durzan (20). Buds grew rapidly on a Risser and White (66) medium containing inorganic nitrogen sources. The addition of organic nitrogen stimulated growth, although changes in shoot weight and axis length were not always significant. Sucrose levels above 2% (w/v) greatly improved the shoot weight and length with 6% producing optimal growth.

Shoot and embryo-like structures have been observed in cultures of various mature tissues of conifers (Table 2). Bonga (9) reported the formation of new apical buds on cultured dormant shoots of 15 to 20 year-old balsam fir trees. Soaking in water or various growth regulator solutions was necessary to induce the growth of the original bud, and the formation of adventitious buds and embryo-like structures. None of these structures grew into plantlets. Later, Bonga (10) observed that young vegetative shoots dissected from the early spring flush of balsam fir and white spruce formed shoot-like protuberances from the base of the young needles. These shoots when excised and placed on fresh nutrient medium grew very slowly. None of these shoots ever formed roots.

Adventitious shoot induction was reported by Bonga (13) on discs of female cones taken from one 25 year-old European larch tree. These shoots were mainly formed on the cut abaxial surface of the bracts. After the original shoots were separated, more shoots were formed from the callus, however, these shoots could not be rooted. Adventitious buds were also reported on callus obtained from vegetative shoots of the same tree (12). None of these adventitious buds elongated but a few roots were formed. Recently, Mohammed *et al*. (59) reported production of shoots from vegetative buds of 1 to 17 year-old white spruce trees collected just prior to bud elongation. Shoot formation was more prevalent from buds of 1 to 4 year-old trees as compared to those from 12 to 17 year-old trees.

Propagation from tissues of mature trees that have demonstrated their superiority over time allows for the capture of maximum genetic gain. Of the tissues used to date, female cones and vegetative shoot tips appear to be the most responsive for regeneration. From the results discussed above, the future looks promising, but more work is needed on developing these techniques into practical tools for tree improvement programs.

2.3. Callus and cell suspension cultures

The ability to regenerate plants from unorganized callus and cell cultures would make possible the propagation of trees of any age; the production of haploids and homozygous diploids; the induction, isolation, and selection of mutants; the production of somatic hybrids; and the modification of cells by "genetic engineering" techniques.

Undifferentiated callus cultures of conifers have been grown since the early days of plant tissue culture. Among the conifers of eastern North America, growth of jack pine callus was reported in the 1950's (55). This callus could be maintained on a simple mineral salt medium only if it contained unautoclaved malt extract or a seed extract of jack pine. Both normal and "tumor" tissues of white spruce were grown on a complex medium (64) which led to a series of studies on the cultural conditions (84), nutritional requirements (66, 83), site of callus initiation (81), and the cellular differentiation of spruce in culture (82). Later, Brown and Lawrence (16) used a slightly modified Murashige and Skoog (60) medium to grow callus of longleaf pine.

Callus of other eastern North American conifers has been used to study a variety of plant processes. These include: callus growth (21, 28), nitrogen metabolism (3, 26, 30, 31, 32), adenine catabolism (2), β-alanine biosynthesis (4), oxygen consumption (40), sterol composition (52), fatty acid composition (51), tannin accumulation (5, 19), tannin biosynthesis (63), phenylalanine ammonia-lyase activity (53), lignin biosynthesis (41, 43), heartwood formation (45), and chloroplast ontogeny (6, 70).

Cell suspension cultures of several conifers of eastern North America have been reported over the years. White spruce and jack pine have been studied in perhaps the most detail (19, 24, 25, 26, 27, 29, 32, 33, 34). Durzan et al.(27) grew cell suspension cultures of white spruce either under constant light and warm temperatures or under alternating conditions of warm days and cool nights. Under the alternating conditions free sugars, tannins and aldehyde levels were increased over the cells growing under constant conditions.

In spite of all these studies, plantlet regeneration from conifer callus or suspension cultures has not been conclusively demonstrated. Risser and White (66) reported that callus of white spruce showed some signs of cellular differentiation on primary explants, but after testing many compounds reported to induce shoot differentiation in other plants, they observed no shoot formation. In suspension cultures of white spruce Durzan et al. (27) observed vascularization of cell clumps growing under conditions of constant light and warm temperatures, but did not see a similar response in cultures grown under warm days and cool nights.

Minocha (58) grew callus of eastern white pine on over one hundred different auxin/cytokinin combinations with no evidence of shoot, root or embryo formation. Also, working with eastern white pine, Kaul (48) reported the formation of vegetative buds in callus cultures. Bud formation was inconsistent but occurred in up to 30% of the callus cultures (49). While these latter observations are far from conclusive, they suggest that conifer callus cultures may indeed retain the ability to regenerate plants but the conditions necessary to permit this have not yet been provided.

Recently, Hakman et al. (39) reported the formation of somatic embryos in callus cultures initiated from immature embryos of Norway spruce. The use of immature embryos to initiate callus cultures is important because these may be more responsive to our attempts to induce organogenesis/embryogenesis, as has been demonstrated with cereal and legume species. Cultures derived from these tissues deserve more attention (see chapter by Tulecke and Durzan).

2.4. Haploids

Haploids are important materials in the hands of geneticists because they can be used to produce homozygous diploids for use in breeding programs. This is particularly important in forest tree breeding because of the long generation time, and the number of breeding cycles necessary to produce homozygous diploids by inbreeding. In the gymnosperms there are two major sources of haploid cells, pollen and the female gametophyte. Each microspore in the microsporangium is genetically unique, therefore, a callus derived from more than one microspore will result in a callus with mixed genotypes. All cells of the megagametophyte, however, originate from one original megaspore and thus have a common genome (7).

Bonga (7) obtained a mixture of haploid and diploid callus from microsporophylls of red pine cultured on Brown and Lawrence (16) medium containing the auxin 2,4-dichlorophenoxyacetic acid. The formation of haploid callus was greatly affected by the developmental stage of the microspores at the time of culture. This is a problem in the culture of pollen because all the pollen grains may not be at the same stage of development. Bilaterally symmetrical structures were seen in the callus cultures but these did not develop further. Callus development from immature pollen of red pine was stimulated by centrifugation before culture (14). A haploid callus was generated from megagametophyte tissue of white spruce (8). Callus has also been obtained from the megagametophyte of longleaf pine and was maintained over several subcultures as a suspension culture (75). However, no plants were regenerated from this callus.

Nagmani and Bonga (61) observed that callus tissue induced from immature megagametophyte tissue of 25-year-old European larch trees would form somatic embryos, some of which developed into small plantlets. These structures seemed to mimic the stages of zygotic embryo development. The plantlets could be transferred to soil, but did not grow further. In light of these results with immature megagametophyte tissues of European larch, the culture of these tissues seems to be a fruitful area for further research.

2.5. Protoplasts

Protoplast fusion may provide a means to develop unique hybrids that cannot be obtained by conventional techniques. Protoplasts may also be useful in the uptake of foreign DNA into plant cells. The future success of somatic hybridization and genetic transformation depends to a large extent on our ability to regenerate callus from protoplasts, and then plants from callus.

Although protoplasts have been isolated and cultured for short periods of time in a number of conifers (Douglas-fir, maritime pine and Norway spruce), little work has been done with the conifers of eastern North America. Winton et al. (80) isolated protoplasts from shortleaf pine callus but provided no details on the results. Although Teasdale and Rugini (78) reported on the production, culture and formation of callus colonies from protoplasts of loblolly pine, attempts to repeat this work have not been successful. In the vast majority of conifers studied, it has been difficult to produce viable protoplasts capable of regenerating cell walls and dividing to form a callus mass capable of sustained growth. Such problems must be overcome to permit progress in this area.

3. FUTURE PROSPECTS

In the last 50 years, the southeastern U.S. has become a center for intensive forest management and is a major source of wood for the world. In contrast, the use of intensive forest management practices in the northeastern U.S. and eastern Canada have not followed a similar path despite their vast forest resource. Both the southeastern and northern parts of eastern North America are similar in many respects, with climate perhaps being the largest difference between the two regions. If climate and its effect on growth rate is indeed the major factor limiting the productivity of the northern region, one may ask how can this situation be remedied?

Over the last 30 years intensive forest management practices, including tree improvement programs, have been responsible for dramatic increases in forest productivity in the southeast. Similar, but perhaps not quite as dramatic, accomplishments would be possible with intensive forest management methods applied to northern species. The problem seems to be one of economics. Presently, the demand for wood in the north is not high enough to justify the expense of the establishment and intensive management of plantations of genetically improved material. This will occur when (and if) there is a shortage of wood in conjunction with an increased demand for forest products.

Rapid genetic improvement of a species can be achieved if truly superior individuals are selected and exactly reproduced by vegetative propagation. This would allow for rapid and immediate deployment of improved material without the delays and genetic variations associated with sexual reproduction (46, 79). At the same time, these superior individuals could be incorporated into breeding programs to develop new outstanding trees.

To date, the majority of work on tissue and cell culture of conifers has been concentrated on the multiplication of trees via embryonic tissues. This is because, by now, it is a fairly routine system that can be adapted to almost any conifer. This process has utility in multiplying small amounts of genetically improved material and selected control pollinated seeds. The method is of greatest use in species such as larch where seed crops are typically small and irregular (47). Presently, in some species this process is at a stage where, with technical refinements, it could be used as a tree improvement tool. It will have to compete with propagation by rooted cuttings of seedlings which will provide the same genetic gain, but at a lower cost (1),unless large scale propagation via somatic embryogenesis becomes possible.

The real benefit of tissue culture as a propagation method lies in the ability to multiply material old enough to have demonstrated its superior characteristics. A greater genetic gain could be captured if tissues from mature trees could be grown in culture,and made to act more juvenile (recapture the high rooting potential and growth rate of seedlings). In such a system it would theoretically be necessary to "rejuvenate" only one shoot and then propagate it by rooted cuttings. The little work that has been done in this area shows some promise, but more research is needed.

Regeneration of trees from callus and cell suspension cultures also has not received sufficient attention. Not only would this technique be useful for propagating both juvenile and mature trees, but it would make possible the production of homozygous diploid trees, screening and regeneration of trees with selected traits, production of somatic hybrids, and regeneration of trees from genetically modified cells. These techniques are discussed in detail in other chapters, and are possible only if we are able to regenerate complete plants from unorganized cells. To date, this remains to be a goal for the future, partly because we understand so little about the process of organ formation, and partly because there are so few people actively working on this problem. An empirical approach, together with an understanding of the biochemical processes involved in regeneration, will accelerate progress to achieve this goal. The recent reports on somatic embryogenesis from immature embryos (39) and from haploid tissues (61) look promising, and should speed development in this area.

Researchers in tissue and cell culture programs should continue to associate themselves with forest geneticists and tree improvement scientists, both to capitalize on their knowledge, and to gain insight into what problems are of greatest importance. Alternative uses of tissue and cell culture, which do not depend on the ability to regenerate complete plants from unorganized tissues, should be applied to solve practical problems. Examples are screening of families for the ability to osmotically adjust, an important strategy for drought tolerance (62), and screening for disease resistance *in vitro* (36). Tissue and cell culture methods can also be helpful to the physiologist and biochemist who may be attempting to understand some of the complexities of the intact plant. As an example, the environmental control of cold hardiness has been studied in cell cultures of Scots pine (42).

It has been 20 years since the first report of adventitious bud formation on cultured conifer embryos (50). Of all the conifers of eastern North America that have been grown and propagated _in vitro_, the most well developed is Loblolly pine. This species will provide a good model of how these laboratory methods can be developed and used in conjunction with tree improvement programs towards a practical end (56). If tissue culture methods become important tools in loblolly pine tree improvement programs, and if economics permit, these same methods could soon be applied to other commercially important conifer species in eastern North America.

Table 1. Summary of selected reports on Bud Induction from Embryonic Tissues

Species	Medium	Growth Regulators	Reference
Pines			
Longleaf	Gresshoff & Doy (38)	BAP 4.4-22µM	(73)
		NAA 10µM	(75)
	Sommer & Brown (75)	BAP 4.4-22µM NAA 0.5µM	(77)
Slash	Gresshoff & Doy (38)	BAP 4.4-22µM NAA 10µM	(17)
	Sommer & Brown (75)	BAP 4.4-22µM NAA 0.5µM	(77)
Shortleaf	Gresshoff & Doy (38)	BAP 4.4-22µM NAA 10µM	(17)
Virginia	Gresshoff & Doy (38)	BAP 4.4-22µM NAA 10µM	(17)
Eastern white	Murashige & Skoog (60)	BAP 4.4µM TIBA 2µM	(58)
	Shenk & Hildebrandt (69)	BAP	(35)
Scots	Gresshoff & Doy (38)	Zeatin 6.8µM NAA 0.05µM	(44)

Species	Medium	Growth Regulators	Reference
Jack	Gresshoff & Doy (38)	BAP 10μM	(18)
Spruces			
White	Durzan et al. (27)	BAP 10μM	(18)
	Shenk & Hildebrandt (69)	BAP 4μM 2iP 4μM	(68)
Black	Schenk & Hildebrandt (69)	BAP 4μM 2iP 4μM	(68)
Larch			
European	Gresshoff & Doy (38)	BAP 22μM	(47)

Table 2. Summary of selected reports on Bud Induction from Mature Tissues.

Species	Medium	Explant	Growth Regulator	Reference
Balsam fir	Romberger *et al*.(67)	dormant vegetative buds	IBA 5μM	(9)
	Romberger *et al*.(67)	vegetative buds	IBA 5μM	(10)
European larch	Litvay *et al*. (54)	female strobili, vegetative buds	BAP 8.8μM IBA 0.5μM	(12)
	Litvay *et al*. (54)	female strobili, vegetative buds	BAP 10μM	(13)

4. ACKNOWLEDGEMENTS

We wish to express our appreciation to Madeleine Reilly for her assistance in the preparation of this manuscript.

5. REFERENCES

1. ARMSON KA, M FUNG, WR BUNTING 1980 Operational rooting of black spruce cuttings. J For 78:341-343
2. BARNES RL 1961 Adenine metabolism in pine embryos and tissue cultures. Bot Gaz 123:141-143
3. BARNES RL, AW NAYLOR 1959 Studies on the ornithine cycle in roots and callus tissues of *Pinus serotina* and *Pinus clausa*. Bot Gaz 121:63-69
4. BARNES RL, AW NAYLOR 1962 Formation of β-alanine by pine tissues supplied with intermediates in uracil and orotic acid metabolism. Plant Physiol 37:171-175
5. BAUR PS, CH WALKINSHAW 1974 Fine structure of tannin accumulations in callus cultures of *Pinus elliottii* (slash pine). Can J Bot 52:615-619
6. BIRCHEM R, HE SOMMER, CL BROWN 1981 Comparison of plastids and pigments of *Pinus palustris* Mill and *Pinus elliottii* Engelm. in callus culture. Z Pflanzenphysiol 102:101-107
7. BONGA JM 1974 *In vitro* culture of microsporophylls and megagametophyte tissue of *Pinus*. In Vitro 9:270-277
8. BONGA, JM 1981 Haploid tissue culture and cytology of conifers. *In* IUFRO Colloque Intl sur la Culture "*In vitro*" de Essences Forestieres AFOCEL, Nangis, pp 283-293
9. BONGA JM 1977 Organogenesis in *in vitro* cultures of embryonic shoots of *Abies balsamea* (Balsam fir). In vitro 13:41-48
10. BONGA JM 1981 Organogenesis *in vitro* of tissues from mature conifers. In vitro 17:511-518
11. BONGA JM 1982 Shoot formation in callus from the stalks of young female strobili of *Larix decidua*. Can J Bot 60:1357-1359
12. BONGA JM 1983 In vitro propagation of conifers. *In* L Zsuffa, RM Rauter, CW Yeatman, eds, Proc 19th Canadian Tree Improv Assoc, Part 2:Symposium on Clonal Forestry:Its Impact on Forest Tree Improvement and our Future Forests, pp 75-83
13. BONGA JM 1984 Adventitious shoot formation in cultures of immature female strobili of *Larix decidua*. Physiol Plant 62:416 -421
14. BONGA JM, AH MCINNIS 1975 Stimulation of callus development from immature pollen of *Pinus resinosa* by centrifugation. Plant Sci Letters 4:199-203
15. BOULAY M 1979 Multiplication et clonage rapide du *Sequoia sempervirens* par la culture *in vitro*. *In* Micropropagation des Arbres Forestiers. Ann Assoc Foret - Cellulose No. 12, pp 49-55
16 BROWN CL, RH LAWRENCE 1968 Culture of pine callus on a defined medium. For Sci 14:62-69
17. BROWN CL, HE SOMMER 1977 Bud and root differentiation in conifer cultures. TAPPI 60:72-73

18. CAMPBELL RA, DJ DURZAN 1975 Induction of multiple buds and needles in tissue cultures of Picea glauca. Can J Bot 53:1652 -1657
19. CHAFE SC, DJ DURZAN 1973 Tannin inclusions in cell suspension cultures of white spruce. Planta 113:251-262
20. CHALUPA V, DJ DURZAN 1973 Growth and development of resting buds of conifers in vitro. Can J For Res 3:196-208
21. CHALUPA V, DJ DURZAN, C VITHAYASI 1976 Growth and metabolism of cells and tissue of jack pine (Pinus banksiana. 2. The quantitative analysis of the growth of callus from hypocotyls and radicles. Can J Bot 54:446-455
22. DAVID A 1982 In vitro propagation of gymnosperms. In JM Bonga and DJ Durzan, eds, Tissue Culture in Forestry. Martinus Nijhoff, The Hague pp 72-108
23. DAVID H, K ISEMUKALI, A DAVID 1978 Obtention de plantes de pin maritime (Pinus pinaster Sol.) a partir de brachyblastes ou d'apex caulinaires de tres jeunes sujects cultives in vitro. C R Acad Sci Ser D 287:245-248
24. DURZAN DJ 1970 Recognition of developmental patterns in cell cultures of Pinus banksiana Lamb. and Picea glauca (Moench) Voss. In Proc 12th Comm on For Tree Breeding in Canada, pp 79-83
25. DURZAN DJ, DR BENNETT 1968 Observations on single, freely-suspended single cells of jack pine growing in cell culture. In Proc 11th Comm on For Tree Breeding in Canada, pp 19-21
26. DURZAN, DJ, G BOURGON 1976 Growth and metabolism of cells and tissue of jack pine (Pinus banksiana). 7. Observations on cytoplasmic streaming and effects of ℓ-glutamine and its analogues on subcellular activites. Can J Bot 54:507-517
27. DURZAN DJ, SC CHAFE, SM LOPUSHANSKI 1973 Effects of environmental changes on sugars, tannins, and organized growth in cell suspension cultures of white spruce. Planta 113:241-249
28. DURZAN DJ, V CHALUPA, AJ MIA 1976 Growth and metabolism of cells and tissues of jack pine (Pinus banksiana). 1. The establishment and some characteristics of a proliferated callus from jack pine seedlings. Can J Bot 54:437-445
29. DURZAN DJ, V. CHALUPA 1976 Growth and metabolism of cells and tissue of jack pine (Pinus banksiana). 3. Growth of cells in liquid suspension cultures in light and darkness. Can J Bot 54:456-467
30. DURZAN DJ, V. CHALUPA 1976 Growth and metabolism of cells and tissue of jack pine (Pinus banksiana). 4. Changes in amino acids in callus and seedlings of similar genetic origin. Can J Bot 54:468-482
31. DURZAN DJ, V. CHALUPA 1976 Growth and metabolism of cells and tissue of jack pine (Pinus banksiana). 5. Changes in free arginine and sakaguchi-reactive compounds during callus growth and in germinating seedlings of similar genetic origin. Can J Bot 54:483-495
32. DURZAN DJ, V. CHALUPA 1976 Growth and metabolism of cells and tissue of jack pine (Pinus banksiana). 6. Free nitrogenous compounds in cell suspension cultures of jack pine as affected by light and darkness. Can J Bot 54:496-506
33. DURZAN DJ, FC STEWARD 1968 Cell and tissue culture of white spruce and jack pine. Bi Mon Res Notes, Can For Serv 24:30

34. DURZAN DJ, FC STEWARD 1970 Morphogenesis in cell cultures of gymnosperms: some growth patterns. Invited paper at IUFRO Workshop Helsinki,Finland 5/28-6/5 1970, 28p
35. FLINN LB, DT WEBB 1985 Adventitious shoot production by embryonic explants of Pinus strobus. In vitro 21(3):54A (Abstract)
36. FRAMPTON LJ, HV AMERSON, RJ WEIR 1983 Potential of in vitro screening of loblolly pine for fusiform rust resistance. In Proc 17th So For Tree Improv Conf, pp 325-331
37. FRIDBORG G, M PEDERSON, L LANDSTROM, T ERIKSSON 1978 The effect of activated charcoal on tissue cultures: adsorption of metabolites inhibiting morphogenesis. Physiol Plant 43:104-106
38. GRESSHOFF PM, CH DOY 1972 Development and differentiation of haploid Lycopersicon esculentum (tomato). Planta 107:161 -170
39. HAKMAN I, LC FOWKE, S VON ARNOLD, T ERIKSSON 1985 The development of somatic embryos in tissue cultures initiated from immature embryos of Picea abies (Norway spruce). Plant Sci Lett 38:53-59
40. HALL RH, PS BAUR, CH WALKINSHAW 1972 Variability in oxygen consumption and cell morphology in slash pine tissue cultures. For Sci 18:298-307
41. HASEGAWA M, T HIGUCHI, H ISHIKAWA 1960 Formation of lignin in tissue culture of Pinus strobus. Plant Cell Physiol 1:173-182
42. HELLERGREN, J 1983 Cold acclimation of suspension cultures of Pinus sylvestris in response to light and temperature treatments. Plant Physiol 72:992-995
43. HIGUCHI T, H ISIKAWA, M HASEGAWA 1960 Chemical nature of lignin in cultured tissues of eastern white pine and "kiri" (Paulownia tormentosa) tree. J Jpn Wood Res Soc 6:229-232
44. HUANG FH, S TSAI 1982 Attempts on organogenesis of Scots pine (Pinus sylvestris L.). In Proc 7th No Amer For Biol Workshop, pp 220-223
45. JORGENSEN E, D BALSILLIE 1969 Formation of heartwood phenols in callus tissue cultures of red pine (Pinus resinosa). Can J Bot 47:1015-1016
46. KARNOSKY DF 1981 Potential for forest tree improvement via tissue culture. BioScience 31:114-120
47. KARNOSKY DF, A DINER 1984 A cotyledon system for cloning Larix decidua and Pinus banksiana. In TAPPI 1984 Res and Devel Conf, TAPPI, Atlanta, GA, pp 13-15
48. KAUL K 1982 Establishment of tissue cultures of three species of Pinus. In Proc 7th No Amer For Biol Workshop, pp 231-235
49. KAUL K, TS KOCHHAR 1985 Growth and differentiation of callus cultures of Pinus. Plant Cell Rep 4:180-183
50. KONAR RN, YP OBEROI 1965 In vitro development of embryoids on the cotyledons of Biota orientalis. Phytomorphology 15:137-40
51. LASETER JL, GC LAWLER, CH WALKINSHAW, TD WEETE 1973a Fatty acids of Pinus elliottii tissues. Phytochemistry 12:817-821
52. LASETER JL, R EVANS, CH WALKINSHAW, JD WEETE 1973b Gas chroatography-mass spectrometry study of sterols from Pinus elliottii tissues. Phytochemistry 12:2255-2258

53. LAU YL, HW SCHELD, JR COWLES 1980 Phenylalanine ammonia-lyase activity in callus cultures of Pinus elliottii. Physiol Plant 49:299-303
54. LITVAY JD, MA JOHNSON, DC VERMA, D EINSPAHR, H KAUSTINEN 1981 Conifer suspension culture medium development using analytical data from developing seeds. Inst Paper Chem Tech Paper Ser 115, 17p
55. LOWENBERG JR, F SKOOG 1952 Pine tissue cultures. Physiol Plant 5:33-36
56. MCKEAND SE, RJ WEIR 1984 Tissue culture and forest producivity. J For 82:212-218
57. MEHRA-PALTA A, RH SMELTZER, RL MOTT 1978 Hormonal control of induced organogenesis. Experiments with excised plant parts of loblolly pine. TAPPI 61:37-40
58. MINOCHA SC 1980 Callus and adventitious shoot formation in excised embryos of white pine (Pinus strobus). Can J Bot 58:366-370
59. MOHAMMED, GH, DI DUNSTAN, TA THORPE 1985 Age and clonal effects on morphogenesis in bud cultures of white spruce. In Vitro 21(3):24A (Abstract)
60. MURASHIGE T, F SKOOG 1962 A revised medium for rapid growth and bioassays with tobacco cultures. Physiol Plant 15:473-499
61. NAGMANI R, JM BONGA 1985 Embryogenesis in subcultured callus of Larix decidua. Can J For Res 15: 1088-1091
62. NEWTON RJ, JP VAN BUIJTENEN 1984 Evaluation of stress resisance of loblolly pine with seedlings in controlled environment chambers and tissue culture. In TAPPI 1984 Res Devel Conf, TAPPI, Atlanta, GA, pp 39-42
63. PARHAM RA, HM KAUSTINEN 1977 On the site of tannin synthesis in plant cells. Bot Gaz 138:465-467
64. REINERT J, PR WHITE 1956 The cultivation in vitro of tumor tissues and normal tissues of Picea glauca. Physiol Plant 9:177-189
65. RIDING RT, J AITKEN 1982 Needle structure and development of the stomatal complex in cotyledons, primary needles, and secondary needles of Pinus radiata D. Don. Bot Gaz 143:52-62
66. RISSER PG, PR WHITE 1964 Nutritional requirements of spruce tumor cells in vitro. Physiol Plant 17:620-635
67. ROMBERGER JA, RJ VARNELL, CA TABOR 1970 Culture of apical meristems and embryonic shoots of Picea abies - approach and techniques. USDA Tech Bull 1409, 30p
68. RUMARY C, TA THORPE 1984 Plantlet formation in black and white spruce. I. In vitro techniques. Can J For Res 14:10-16
69. SCHENK RV and AC HILDEBRANDT 1972 Medium and techniques for induction and growth of monocotyledonous and dicotyledonous plant cell cultures. Can J Bot 50:199-204
70. SCHWAB DW, RW RITZERT 1970 Fine structural studies on Pinus elliottii cultured in vitro. Am J Bot 57:738 (Abstract).
71. SMITH DM 1980 The Forests of the United States. In JW Barrett, ed, Regional Silviculture of the United States, Second Edition, Wiley-Interscience, New York, pp 1-23
72. SMITH DR, KJ HORGAN, J AITKEN-CHRISTIE 1982 Micropropagation of Pinus radiata for afforestation. In A Fujiwara, ed, Plant Tissue Culture 1982, Proc 5th Int Cong Plant Tissue Cell Cult, Maruzen, Tokyo, pp 723-724

73. SOMMER HE, CL BROWN 1974 Plantlet formation in pine tissue cultures. Am J Bot 61 (Suppl. 5):11 (Abstract)

74. SOMMER HE, CL BROWN 1980 Embryogenesis in tissue cultures of sweetgum. For Sci 26:257-260

75. SOMMER HE, CL BROWN, PP KORMANIK 1975 Tissue cultures from the female gametophyte of *Pinus palustris* Mill. *In* KJ Kasha, ed, Haploids in Higher Plants, Advances and Potential, Univ Guelph, Guelph, p 143 (Abstract)

76. SOMMER HE, CL BROWN, PP KORMANIK 1975 Differentiation of plantlets in longleaf pine (*Pinus palustris*)tissue cultured *in vitro*. Bot Gaz 136:196-200

77. STINE M, HE SOMMER 1983 Adventitious shoot formation on embryos of *Pinus palustris* Mill., *P. elliottii* Engelm., and *P. palustris* x *elliottii*. *In* Proc 17th So For Tree Improv Conf, pp 349-350

78. TEASDALE RD, E RUGINI 1983 Preparation of viable protoplasts from suspension cultured loblolly pine (*Pinus taeda*) cells and subsequent regeneration to callus. Plant Cell Tissue Organ Cult 2:253-261

79. THOMPSON DG 1984 Clonal Reforestation: Forests of the Future? *In* ML Duryea, GN Brown, eds, Seedling Physiology and Reforesation Success, Martinus Nijhoff, Boston, MA, pp 3-28

80. WINTON LL, RA PARHAM, HM KAUSTINEN 1975 Isolation of conifer protoplasts. Inst Paper Chem,Genet Physiol Notes No 20,9p

81. WHITE PR 1967a Some aspects of differentiation in cells of *Picea glauca* cultured *in vitro*. Am J Bot 54:334-353

82. WHITE PR 1967b Sites of callus production in primary explants of spruce tissue. Am J Bot 54:1055-1059

83. WHITE PR, SN GILBEY 1966 Sources of nitrogen for spruce tissue cultures. Physiol Plant 19:177-186

84. WHITE PR, PG RISSER 1964 Some basic parameters in the cultivation of spruce tissues. Physiol Plant 17:600-619

6. TISSUE CULTURE OF DOUGLAS-FIR AND WESTERN NORTH AMERICAN CONIFERS

M. M. ABOEL-NIL

1. INTRODUCTION

Forests of the West Coast of North America contain a large share of the high value saw timber of the continent. In 1972, the total value of timber products harvested from the Pacific Coast region alone was about $3.0 billion; it followed at $2.1 billion in the Southern region, while the total for the U.S.A. was $6.3 billion (7). The projected soft wood supply from the Pacific Coast has dropped from 25.2 billion feet in 1976 to 19.6 billion board feet by 2030 (7). The main cause for this decline was suggested to be the exhaustion of the old growth on forest industry lands, which cannot be offset by harvests from second-growth stands for several decades. A reversal of this trend should be attainable by improved cultural practices, tree breeding and biotechnology applications (33).

The Northwestern region of North America is characterized by heavy rainfall and moderate temperatures which support the growth of dense forest stands in this vast region. A number of conifer tree species are growing in these forests, such as the Douglas-fir *Pseudotsuga menziesii*, Western hemlock *Tsuga heterophylla*, Western redcedar *Thuja plicata*, Coast redwood *Sequoia sempervirens*, true firs *Abies* spp., spruces *Picea* spp. and the pines *Pinus* spp.

In this chapter, I shall review tissue culture and micropropagation of the major timber producing species of the West Coast; namely, Douglas-fir, Western hemlock, and Western redcedar. Other species, important for the Western North

America such as: <u>Sequoia</u> spp. <u>Pices</u> spp., <u>Pinus</u> spp., are being discussed in other chapters.

2. DOUGLAS-FIR

2.1. <u>Distribution</u>

<u>Pseudotsuga</u> <u>menziesii</u> (Mirb.) Franco, the most important timber species in the world (53), is a member of the family Pinaceae. It is one of the biggest trees, occasionally attaining a height of 75 meters on the West Coast and up to 40 meters on the Rocky Mountains. Douglas-fir is one of the most common forest trees in North America, having a large north/south range of about 5,000 km. It has been introduced successfully into Europe, where it produced timber similar to the second-growth in North America (56). Douglas-fir is a fast growing tree, that is intermediately tolerant to adverse conditions within its habitat and produces more timber than any other species (53).

2.2. <u>Seed propagation</u>

Douglas-fir is naturally reproduced by seed. Male and female strobili open up during late winter and spring, about a year after their initiation as axillary bud primordia. Fertilization takes place about 10 weeks after pollination. Seeds develop throughout late spring and summer, and mature by August. For a complete review of the phenology and details of the life cycle of Douglas-fir, Allen and Owen's book (4) is highly recommended.

Seed production is erratic because of environmental factors. Abundant seed crops have occurred in 2- to 11-year cycles, which is common in many conifers. Variable seed production is complicating operational reforestation and tree improvement efforts. Douglas-fir seedlings are not adapted to being planted far in distance or elevation from their original sources. In 1966, tree seed zones were defined by the Western Forest Tree Seed Council (6) to assist foresters in collecting and planting seeds in their respective locations.

2.3. Tree improvement

Programs for genetic tree improvement are well in progress, by the USDA and some forest product companies in the Pacific Northwest. The intention is to identify superior trees for collecting first generation improved seed and scions for hybridization (33, 59). Details of one such program was discussed by Farnum, et al. (33), and also in Volume I in a chapter entitled, "Potential genetic gain through tissue culture." Farnum, et al. (33) found that the Douglas-fir productivity has increased about 70% over that in the natural forest on the same site class due to better silvicultural practices and genetic improvements. They predicted another large increase in productivity by optimizing the tree growing environment and by using biotechnological means to manipulate tree genetics.

2.4. Vegetative propagation

Douglas-fir can be propagated by stem cuttings. However, growth of Douglas-fir trees, that were propagated vegetatively by rooting of lateral branches, was plagiotropic (9, 16, 27, 51, 57, 61). Careful choice of genotypes and improved rooting techniques reduced plagiotropism and improves rooting capacity (9). Rooting potential of cuttings taken from up to 7-year old ortets is variable, and drops sharply afterward with aging of clones (9). This characteristic is a serious draw back to application of rooted cuttings in tree improvement because selection of superior clones is done at an adult age (52). A solution to this problem was envisaged through the use of sheared hedges (9, 57), successive micrografting (28, 34), severe pruning to stimulate epicormic bud growth (15) and foliar application of cytokinins (2). Hormonal treatments increased rooting percentages of some hard-to-root genotypes in certain seasons of the year (9, 12, 35, 46). Nevertheless , these advances in Douglas-fir cutting propagation technology are not sufficient to be of much use in tree improvement programs.

Operational Douglas-fir reforestation is done by seed, which has certain limitations in applying accelerated tree improvement concepts. A higher genetic gain can be achieved by vegetative propagation of superior individuals, in which additive and dominant gene effects will be better maintained than with the seed propagation, which will capture only the additive gene effects (33). Another advantage of vegetative propagation is the considerable saving in implementation time and reducing or eliminating production seed orchards (54). For these reasons, major efforts have been directed into vegetative propagation of Douglas-fir (33, 52, 57).

Micropropagation may overcome some of the problems associated with rooting of cuttings because morphogenesis is easier to control *in vitro* than *in vivo*. Micropropagation allows the use of precisely determined chemical and physical environments for growth and multiplication in which genes can express themselves without any environmental suppression. Tissue and cell culture allows for plant cells to fully express their totipotency, and thus offers the maximum potential for rapid multiplication rates. In addition, tissue culture offers the possibility of cryopreservation. This technique could provide long-term storage of Douglas-fir germplasm, and should be integrated into tree improvement and domestication activities to preserve unique and beneficial genes. Such genes may be needed in the future to breed for unique growth characteristics, and to maintain genetic pool diversity. Cryopreservation can also be used to prolong the physiological juvenility of a clone, while progeny tests are in progress. Once superior genotypes were identified, large-scale micropropagation of stored juvenile tissue could be resumed.

2.5. Tissue culture

Five decades ago, La Rue (47, 48) started culturing Douglas-fir embryos. Twenty-two years later, Blakely experimented with Douglas-fir callus initiated from the cambium, phloem and cortex of mature tree stem (10). The following year, Al-Talib and Torrey reported on the culture of Douglas-fir dormant buds and induction of callus (5). Substantial progress has been achieved since these early experiments, specially in the last 10 years since Tsai Cheng developed a system for propagating Douglas-fir from cotyledon cultures in 1975 (20). Refinement of medium formulations and culture manipulations produced a number of technologies aiming at micropropagation and tissue culture of Douglas-fir. Various regeneration routes of organogenesis and embryogenesis were reported using different types of explants of mature and juvenile specimens (1, 14, 18, 25, 36, 50). Advances in protoplast culture methods were also made to support possible implementation of genetic engineering techniques in tree improvements (44).

In spite of these advances, commercial large-scale application of Douglas-fir tissue culture is not yet at hand. Conifers are a unique group of plants in terms of their growth habit, life cycle and natural ability to regenerate vegetatively. The empirical research approach, using techniques that were developed for angiosperms, proved to have limited potential in developing long-sought technologies for gymnosperms. In spite of many efforts, regeneration from callus and protoplast, a fast rate of multiplication, and low cost of *extra vitrum* establishment are still not yet possible.

2.5.1. *Embryo culture*. Embryo culture was the first approach used to study the morphogenetic potential of Douglas-fir using various media formulations (8, 47, 48). *In vitro* bud induction of Douglas-fir was reported first by Cheng 1975, on extracted embryos that were cultured on modified MS medium supplemented with 5.0 μM of indole-3-acetic acid (IAA) and indole-3-butyric acid (IBA) combined with 0.5 to 1.0 mM N6-Benzyladenine (BA) or dimethyl allylamino purine (2iP) (20). Sommer, in 1975, induced adventitious bud formation on a medium low in nitrogen and phosphorous and containing 0.4 to 70 μM BA (60). Embryos formed callus when cultured on Linsmaier and Skoog's medium containing 5.3 to 10.6 μM naphthalene-1-acetic acid (NAA) and 4.4 μM BA, and then formed adventitious buds (17). Exposure to 660 nm light at 0.42 mW/cm^2 stimulated adventitious bud induction on cultured embryos, while blue light (420 to 467 nm) and near U.V. (371 nm) had no effect (40).

Rooted plantlets were obtained by Cheng using a 3-step embryo culture method (20). 1) Embryos were cultured on modified Murashige and Skoog's medium containing 5 µM of IAA, IBA, 2iP and BA. After cell division was initiated in about four weeks, embryos were irrigated with a sterile medium solution containing 0.5 - 1.0 mM BA. 2) Buds elongated into shoots on half-strength medium devoid of growth regulators. 3) Elongated shoots were rooted on the same medium containing 0.5% sucrose and IBA.

2.5.2. Callus and cell suspension culture. Callus was easily initiated from various types of explants such as whole embryos (17), excised cotyledons (43) cambium, phloem and cortex (10, 37), buds (5, 68), stem segments (66), shoot tips (14), and needles (68). Winton induced callus from 4- to 5-year-old stem segments on Brown and Lawrence's medium supplemented with 142.0 µM IAA or from 0.25 to 26.5 µM NAA combined with 0.4 to 4.4 µM BA. Douglas-fir cell suspension grew and doubled in volume weekly, while callus had died on agar solidified medium (66). Kirby studying nitrogen metabolism, initiated callus from cotyledon cultures on MS medium containing 15 µM NAA and 0.5 µM BA and found that organic nitrogen increased cell suspension growth rate. Glutamine, at 50 mM, increased cell dry weight, eliminated growth lag phase and caused cells to have a highly dense cytoplasm (43).

High concentrations of NAA or 2,4-dichlorophenoxy acetic acid (2,4-D) were used to maintain cell cultures growing in suspension (39, 43). Accumulation of phenolic compounds was observed in Douglas-fir cell suspension cultures even in presence of 2,4-D. Colorless proanthacyanidins accumulated in the absence of 2,4-D, and cell growth was reduced (39).

Reports of bud regeneration from subcultured, undifferentiated conifer calli or cells in suspension were limited. Douglas-fir embryos cultured on MS medium with 5.3 to 10.6 µM NAA and 4.4 µM BA produced green callus, and then buds (17). Subcultured cotyledon, needle and stem explant callus regenerated buds at low frequency when grown on a medium containing an auxin and a cytokinin (68).

2.5.3. Organogenesis of cultured tissues and organs. Embryo culturing was found to be a tedious process which yielded only a small number of buds because the embryo has a limited amount of organogenetic tissue. Organ culture systems were developed to overcome these problems. Direct formation of adventitious and axillary buds was observed in cultures of various explants from juvenile and mature trees.

2.5.3.1. Culture of juvenile tissue. Regeneration of conifer trees from juvenile specimens has limited applications in commercial forestry. Unfortunately, most of the Douglas-fir micropropagation technologies were developed for juvenile tissues. This is due to difficulties in culturing tissues from adult specimens, which were experienced in early trials to culture Douglas-fir (13). Cotyledon explants of germinated embryos were used and shown to be highly regenerative (22, 23). The morphogenetic potential of cotyledon segments was very variable from one genotype to the other (3, 60). A positive correlation was found between seed weight and ability of cotyledon cultures to form adventitious buds (3).

Detailed studies of Douglas-fir cotyledon cultures by Cheng (22, 25) indicated that adventitious bud induction is controlled by auxins and cytokinins. Optimum bud differentiation was induced by 5.0 μM BA combined with 0.5 to 5.0 nM NAA or 2.5 to 5 μM IAA and IBA. Adventitious buds originated from cells of epidermal and subepidermal cell layers in four distinct developmental stages: 1) meristemoid, 2) bud primordium, 3) early, and 4) late bud development (19). Synthesis of new low molecular weight proteins (16,000-20,000 daltons) was found to occur as early as day two *in vitro*, and reach a maximum at the 4th day. The rate of this synthesis was consistently correlated with morphogenetic process leading to adventitious bud development (38). A tissue culture method was designed for mass propagation of Douglas-fir. In this method the growing tissue was supported on a porous polyester fleece to facilitate and reduce the number of handling steps during culture media changing (24, 25).

Juvenile leaves of in vitro grown Douglas-fir shoots formed adventitious buds on modified Gresshoff and Doy medium containing 0.26 μM NAA and 44.0 μM BA (55). Boulay (14) proposed a system for micropropagation of juvenile germinants based upon the following sequence. Adventitious bud induction; a succession of axillary bud growth; elongation of these buds into shoots on hormone-free medium; and rooting of these shoots on a medium containing 49.0 μM IBA.

Seedling apices regenerated adventitious and axillary buds when cultured on Schenk and Hildebrandt's medium containing 4.4 μM BA or 9 to 23 μM zeatin (Z) combined with 1.1 μM 2,4-D or 1.3 μM NAA. Auxin concentration was found to be critical since too high concentrations induced callus and lack of auxin led to death of the apex. Buds elongated into shoots when apices with bud clusters were transferred into liquid medium without growth regulators (50). Unexpectedly, many plantlings produced from cotyledon or seedling apices were plagiotropic (50, 64).

Chromosome numbers of cells of seedlings and plantlets were determined from root tip squashes using Giemsa staining. Plantlets and seedlings had the diploid chromosome number of 26. This study indicated that Douglas-fir plantlets produced from cultured cotyledon segments may have a stable chromosomal complement typical of the species (70). Nuclear DNA contents of cells of cotyledon cultures and regenerated buds and shoots were measured and found to be stable around the 2 C value, which was the normal range for seedlings (G. Berlyn, pers. comm.).

2.5.3.2. Culture of mature tissue. Regeneration from mature genetically superior Douglas-fir trees is an essential requirement for large-scale application of tissue culture in forestry. Unfortunately, the potential for vegetative propagation utilizing current techniques, declines with aging of the ortet. Technologies, that were developed for embryos and juvenile seedling tissue culture, are not applicable to mature specimens. Empirical modification of the in vitro chemical and physical environment proved to be of limited value in morphogenetic induction of adventitious buds on mature Douglas-fir explants (69, and AboEl-Nil unpublished data). If we assume that all meristematic plant cells are totipotent (62), then the non-response of mature explants suggests a strong suppression by chemical or physical factor(s). A systematic approach to elucidate the complicated physiological interactions in maturing conifers requires integrated studies of phase change (13), rejuvenation (34), tree architecture (31), whole-plant morphogenesis (71) and in vitro cell and media interactions (43).

Wochok and AboEl-Nil induced adventitious and axillary bud formation on denuded shoot tips of 64-year-old ortets, cultured on the same medium as used for cotyledon bud induction (69). Adventitious bud formation was observed at low frequency on the needles of mature Douglas-fir, excised immediately before flushing of the buds (Caruso and Cheng, pers. Comm.).

Boulay (15 and his chapter in this volume) described in detail a micropropagation process for mature Douglas-fir based upon culturing of flushing buds. Two principal factors affect the morphogenesis of cultured buds: 1) age of ortet, and 2) time of explanting. He found that explants from young trees grew faster than explants from old trees and that buds explanted just before bud burst were the most responsive. An in vitro pruning method was employed to stimulate the growth of axillary bud meristems by removing the apical meristem and needle primordia. This method produced from 3 to 10 shoots per dormant bud that was cultured on MS medium devoid of ammonium nitrate and containing 5 μM IBA. Cytokinins were not required for this process. Buds were elongated on the same medium with 0.5% activated charcoal and were rooted in vitro on a medium containing from 57 to 114 μM IAA and from 5.3 to 26.5 μM NAA (15).

A rejuvenation process was applied to 11-year-old ortets of Douglas-fir consisting of repetitive foliar spraying with 0.88 μM BA aqueous solution, six weeks before bud break. The treated trees developed axillary buds on the dormant bud axis before flushing. These axillary buds elongated on a shoot growth medium containing 0.1% activated charcoal and 5 μM IBA (2).

Thompson and Zaerr (63) found that the high levels of ammonium and nitrate ions in MS medium were inhibitory to adventitious bud formation on shoot tips of Douglas-fir. Kinetin, Z and BA, but not 2iP were found to induce adventitious bud formation. Subsequent elongation of these buds was achieved on a modified MS medium devoid of any growth regulators.

Recently, a method for shoot multiplication of mature Douglas-fir trees was reported by Gupta and Durzan (36). In this method, stem segments from lateral branches containing preformed buds were cultured on DCR basal medium without growth regulators. Buds, which flushed within 4-5 weeks, were excised and cultured on half strength DCR basal medium with 0.3% activated charcoal (DCR-1). On DCR-1 medium, buds elongated into shoots. Shoot multiplication was achieved by subculturing elongated shoots on DCR basal medium containing 0.88 μM BA. This is the first report on high frequency bud growth, multiplication and elongation of shoots from 20-60 year-old trees.

Evers (31) found that environmental parameters, media composition, and incubation conditions optimal for growth and morphogenesis of shoot initials of Douglas-fir, vary according to the stage of bud development and its topophysical position. Buds from trees older than two years show poor *in vitro* growth and have narrow optima for nutritional salts and growth regulators.

2.5.3.3. *Rooting and extra vitrum establishment.* Rooting of Douglas-fir shoots has been a bottleneck in micropropagation technology. Although rooting of shoots has been achieved on an experimental scale (17, 25, 50, 64), large-scale rooting of shoots from many genotypes was sporadic, occurred at low percentages and required a long incubation time. Chalupa (17) rooted Douglas-fir shoots on a medium low in nutrient salts and sucrose and containing from 0.1 to 0.5 μM NAA.

Boulay (15) found that 10-20% of the juvenile shoots rooted while shoots from 4-year-old trees did not root when cultured on rooting medium containing from 57 to 114 μM IAA and from 5.3 to 26.5 μM NAA. Cheng and Voqui (25) induced rooting at high frequency on elongated shoots cultured at 19°C on a medium containing 1/3 strength MS salts, 0.5% sucrose and 0.25 μM NAA. When shoots were incubated at 24°C, NAA induced basal callus formation instead of roots.

Optimum conditions for rooting found by different researchers were not similar. Mapes, *et al.*, (50) found that the best rooting medium for juvenile Douglas-fir shoots was a high salt medium with 3% sucrose and a low growth regulator content of 0.5 μM NAA and 0.5 μM 2iP. Rooting occurred in all clones with no obvious genotypical differences in rooting ability.

Extra vitrum rooting was achieved by Chalupa in low frequencies when Douglas-fir shoots were dipped in a rooting powder containing 2.6 mM NAA, 4 mM nicotinic acid, and 30 μM thiamine, then planted in perlite under mist and fed twice a day with mineral nutrient solution (17). Depending upon shoot size, genotype, rooting treatment, terminal bud dormancy and overall vigor, higher rooting frequencies (about 50%) were achieved by utilizing a modified horticultural cutting-rooting procedure (32, 64). In this procedure, shoots were dipped in a commercial rooting powder and then planted in peat perlite mixture. Optimal rooting occurred at 21°C air and 24°C soil temperature and 100% R.H. Rooting started at 6-8 weeks and was complete at 20 weeks.

Transferring micropropagated plants from *in vitro* into *extra vitrum* is a critical step because plantlets usually have under-developed cuticles and stomata, which results in wilting in low humidity and bleaching in high light intensities. Roots formed *in vitro* are usally fragile and require careful handling during transplanting. Plantlets that were rooted *extra vitrum* were easier to transfer and to acclimatize to normal greenhouse growing conditions. Hardening off is achieved by gradual decreasing of humidity around plantlets when terminal growth resumes. This resumption of growth indicates successful establishment. Plantlets were grown in a horticulture soil mix in forestry containers to achieve the height and caliber required for field or nursery outplanting.

2.5.4. *Field Performance*. Field survival of micropropagated orthotropic plantlets was higher than the survival of plagiotropic ones or rooted cuttings. Orthotropic plantlets were similar to seedlings in morphology and growth rate after three growing seasons in the field. Plagiotropic plantlets recovered and grew orthotropically sooner than cuttings (Ritchie, pers. com.). These observations indicated the survial of *extra vitrum* established plantlets is not a major concern in the implementation of this technology. Uniformity among ramets and stability of characteristic growth parameters of the ortets are currently being studied.

2.5.5. *Embryogenesis*. Slow multiplication rates and high cost of production are two serious problems facing Douglas-fir micropropagation via organogenesis. Somatic embryogenesis offers a significant advantage to large-scale implementation of micropropagation in forestry. In embryogenesis a bipolar, ready to germinate embryo is formed, a process that will achieve a substantial reduction in production cost and complete rejuvenation of produced propagules (1). Techniques for growing cell suspension in shake cultures are well developed and were used to grow large volumes of Douglas-fir cell cultures at the Institute of Paper Chemistry in Wisconsin (49). Durzan (29) calculated that if 80% of Douglas-fir cells in suspension would regenerate plants by embryogenesis, only 100 liters of culture would reproduce enough seedlings in a period of three months to reforest 100,000 acres.

Embryo-like structures have been observed in Douglas-fir cell suspension cultures (30) and cotyledon cultures (1). The broad spectrum approach to systematic medium development, combined with an analysis of the nutrient solutions surrounding developing embryos, were used with encouraging early results by Verma, _et al_. (65). Durzan (30) found that transferring Douglas-fir cell suspension cultures into a liquid medium with a low calcium and NAA concentration, and devoid of nitrates induced globular and torpedo-shaped embryoids. Direct embryoid formation on cultured cotyledon segments was induced on MS medium supplemented with 5.3 μM NAA or 4.5 μM 2,4-D and no cytokinin (1). Histological examinations showed a well-developed vascular system, a radicle at one pole, and a chlorophyllous enlargement at the other pole, indicating well-structured bipolar embryoids. Complete plant development from embryoids has not been achieved yet.

2.5.6. _Protoplast culture_. Substantial yield improvement in Douglas-fir plantations is expected through biotechnological manipulations of the genetic make-up (33). Efforts in Douglas-fir domestication can be greatly facilitated by advances in cell selection, regeneration from protoplasts, somatic hybridization and genetic transformations (44).

Successful isolation of Douglas-fir protoplast from callus cells was by Winton, _et al_., (67). Protoplast isolation and callus regeneration was achieved by Kirby and Cheng (45), using cotyledon segments that had been pre-treated on a bud induction medium containing auxin and cytokinin. This pre-culturing of cotyledons was essential to change the physiological state of the tissue to one that favors production of a large number of uniformly sized protoplasts. The best enzyme mixtures contained cellulase and pectinase at concentrations up to 2.5% w/v in bud induction medium supplemented with sorbitol or mannitol to adjust the osmolarity. The protoplast were then washed with the same medium without enzymes but containing 120 μM glutamine. Subsequently, they were cultured on the surface of a non-woven polyester fleece to generate cell walls, dividing cells and callus colonies (42).

3. WESTERN HEMLOCK

Tsuga heterophylla (Ratn.) Sarg, is the tree following the Douglas-fir in importance in the west. It grows along the coast from southern Alaska to northern California, and is particularly abundant in the damp rainy areas of the northwest (53). It is a tall tree, frequently reaching a height of 45-60 meters and a diameter of one meter or more (56). The tree is highly tolerant to adverse conditions, and grows rapidly. Western hemlock produces an abundant seed supply, which is being used for commercial reforestation. Rooting of hemlock cuttings is easy and is not affected by age as in Douglas-fir and some other conifers (16). A number of tree improvement programs were initiated in the Pacific Northwest and British Columbia. Higher levels of genetic gain are expected for planted selected clonal propagules than for selected seed families, which emphasizes the importance of vegetative propagation (54).

3.1. Tissue culture of western hemlock

Cheng, in 1976, (21) developed a micropropagation system for western hemlock that produced adventitious buds on embryos or cotyledons on the Douglas-fir bud induction medium supplemented with 5.0 μM BA and 2.5 μM IAA and IBA. These buds elongated into shoots on the same medium devoid of growth regulators and containing activated charcoal. Rooting was achieved on low salt medium containing 0.5% sucrose and 0.25 μM NAA or *extra vitrum* in a non-sterile peat and vermiculite mixture.

4. WESTERN RED CEDAR

Thuja plicata Donn., also known as "arbor vitae," belongs to a group of evergreen conifers with tiny scale-like leaves. They are commonly called cedar, but since they belong to the Cupressaceae they are not true cedars. Western redcedar is an important timber tree in British Columbia and the Western United States. Its wood is exceptionally light in weight and highly resistant to decay (53). It produces large numbers of rapidly germinating seeds and fast-growing seedlings. Natural vegetative reproduction, probably by rooting of drooping basal branches, has been observed (58). Nursery production, for commercial reforestation and for landscape ornamental use, is by seed

4.1. Tissue culture of western red cedar

Harvey and Grasham, in 1969, (37) initiated callus cultures from the cortex of mature tree stems on Harvey's medium with 138 mM glucose, a mixture of amino acids, and 5.3 μM NAA.

Methods for successful micropropagation of *Thuja plicata* juvenile and mature trees were developed by Coleman and Thorpe in 1977 (26). Their technique followed four distinct steps: 1) meristemoid induction, 2) bud differentiation, 3) bud elongation into shoots, and 4) rooting of shoots. Cotyledon explants produced multiple dome-like swellings after 3-6 weeks on MS medium containing 1.0 μM BA with 0.1 μM NAA. Buds were formed on the surface of the dome-shaped meristemoids. These buds elongated into shoots in about 4-8 weeks. Rooting of these shoots was induced by first culturing them on half strength MS medium containing 50 μM IBA, and then transferring them *extra vitrum* into peat and vermiculite mixture. These rooted shoots have grown in the greenhouse and formed mature scale-like leaves similar to those of sexually produced seedlings.

Multiple adventitious and axillary buds were induced on lateral shoot tips of mature trees after 6-8 weeks on MS medium containing 50 μM BA and 0.1 μM NAA. These buds were elongated and rooted under the same conditions as the juvenile shoots from the cotyledon cultures (26).

The effect of different growth regulators on Western redcedar *in vitro* was studied. Application of GA_3 to cotyledons inhibited bud differentiation, while application to cultured shoot tips induced male strobili instead of adventitious buds (26).

5. CONCLUSIONS AND DISCUSSION

In the west of the North American continent, one cannot over-emphasize the value of conifers to the forest product industry and the economy of the region. Efforts to improve yield are progressing on many fronts, including the development of better silvicultural practices and tree improvement by breeding and biotechnology (33). Vegetative propagation combined with tree breeding, is expected to play a major role in improving the yield of well-managed forest plantations.

During the last ten years, tissue culture of Douglas-fir, Western hemlock, and Western red cedar has attracted the interest of many researchers, and a substantial progress in micropropagation techniques has been achieved. Methods for reproducible regeneration via organogenesis of embryonic tissues explanted from juvenile specimens have been developed. However, application of micropropagation techniques in forest tree breeding and genetic improvement, still suffers three serious deficiencies: 1) the low reproduction efficiency and high cost, 2) the difficulty to apply the techniques to mature trees, 3) regeneration from callus and cell suspension, a prerequisite for research on genetic engineering and biotechnology is still not possible (44). It is obvious, when assessing the potential and perspective of *in vitro* techniques to mass clonal micropropagation of forest trees, that the optimism of early workers (25, 29, 41) must cautiously be re-evaluated.

Major research problems have not yet been adequately addressed and progress is still lacking in areas such as: haplogenesis, regeneration from callus, selection at cell level, embryogenesis, plagiotropism and phase change. Some of these problems, e.g., plagiotropism have a serious impact on all applications of conifer micropropagation, others such as our present inability to induce haplogenesis for the production of homozygous lines may be less important. Whole-plant morphogenetic studies (71), instead of *in vitro* morphogenesis of excised organs, tissues and cells, may lead to a better understanding of factors affecting morphogenesis.

This could subsequently lead to the development of a successful means of in vitro differentiation and regeneration. New and innovative approaches have to be found in order to solve these complicated and inter-related problems, before research on conifer tissue culture will lead to technologies that are applicable in commercial forestry.

6. REFERENCES

1. ABOEL-NIL MM 1980 Embryogenesis of gymnosperm forest trees. US patent #4,217,730
2. ABOEL-NIL MM 1982 Method for asexual reproduction of coniferous trees. US patent #4,353,412
3. ABOEL-NIL MM, ZS WOCHOK 1978 Morphogenetic heterogeneity of cultured Douglas-fir cotyledons and its correlation with seed weight. Proc 4th Int Cong Plant Tissue Cell Cult, Univ Calgary, Alberta, Canada
4. ALLEN GS, JN OWENS 1972 The life history of Douglas-fir. Pac For Res Cent, Can For Serv, Victoria, BC, Canada
5. AL-TALIB KH, JG TORREY 1959 The aseptic culture of isolated buds of Pseudotsuga taxifolia. Plant Physiol 34:630-637
6. ANON 1966 Tree seed zone map for Oregon-Washington. W For Tree Seed Counc, adopted April 5, 1966, Portland, Oregon
7. ANON 1982 An analysis of the timber situation in the United States 1952-2030. USDA, For Serv, For Resour Rep No 23, 501p
8. BARTELS H 1957 Kulture isolierter Koniferenembryonen. Naturwissenschaften 44:290-291
9. BLACK DK 1973 Influence of shoot origin in certain pre- and post-severance treatments on the rooting and growth characteristics of Douglas-fir (Pseudotsuga menziesii (Mirb.) Franco) stem cuttings. PhD thesis, Oregon State Univ, Corvallis, 143p
10. BLAKELY LM 1958 Studies on the in vitro culture of Pseudotsuga menziesii and Arceuthobium douglasii. Proc Montana Acad Sci 18:21-25.
11. BLANKENSOP CM, RZ CALLAHAM 1960 Rooting cuttings from Douglas-fir, White fir, and California red fir Christmas trees. USDA For Serv, Pac Southwest For Range Exp Stn Res Note 160, 3p
12. BONGA JM 1977 Applications of tissue culture in forestry. In J Reinert, YPS Bajaj, eds, Applied and Fundamental Aspects of Plant Cell, Tissue and Organ Culture. Springer-Verlag, Berlin, Heidelberg, pp 93-108
13. BONGA JM 1982 Vegetative propagation in relation to juvenility, maturity and rejuvenation. In JM Bonga, DJ Durzan, eds, Tissue Culture in Forestry. Martinus Nijhoff/Junk Publ, The Hague, pp 387-412

14. BOULAY M 1976 Recherches sur la propagation du Douglas par culture in vitro. Ann Rech Sylvicol, AFOCEL, France, pp 84-145
15. BOULAY M 1979 Propagation in vitro du Douglas par micropropagation de germination aseptique et culture de bourgeons dormants. Etud Rech, AFOCEL, France, 12:67-75
16. BRIX H, R VAN DEN DRIESSCHE 1977 Use of rooted cuttings in reforestation, a review of opportunities, problems and activities, BC For Serv, Can For Serv Joint Rep No 6, Victoria, 16p
17 CHALUPA V 1977 Organogenesis in Norway spruce and Douglas-fir tissue cultures. Commun Inst For Cech 10:79-87
18. CHALUPA V, DJ DURZAN 1972 Growth and development of resting buds of conifers in vitro. Can J For Res 3:196-208
19. CHEAH K-T, T-Y CHENG 1978 Histological analysis of adventitious bud formation in cultured Douglas-fir cotyledon. Am J Bot 65:845-849
20. CHENG T-Y 1975 Adventitious bud formation in culture of Douglas-fir (Pseudotsuga menziesii (Mirb.) Franco. Plant Sci Lett 5:97-103
21. CHENG T-Y 1976 Vegetative propagation of Western hemlock (Tsuga heterophylla) through tissue culture. Plant Cell Physiol 17:1347-1350
22. CHENG T-Y 1975 Factors effecting adventitious bud formation of cotyledon culture of Douglas-fir. Plant Sci Lett 9:179-187
23. CHENG T-Y 1977 Recent advances in development of in vitro techniques for Douglas-fir. In WR Sharp, PO Larsen, EF Paddock, V Raghavan, eds, Plant Cell and Tissue Culture, Principles and applications. Ohio State Univ Press, Columbus, pp 493-508
24. CHENG T-Y, MM ABOEL-NIL 1977 Tissue culture methods designed for mass propagation of Douglas-fir. In vitro, 21:146
25. CHENG T-Y, TH Voqui 1977 Regeneration of Douglas-fir plantlets through tissue culture. Science 198:306-307
26. COLEMAN WK, TA THORPE 1977 In vitro culture of western redcedar (Thuja plicata Donn) I Plantlet formation. Bot Gaz 138:298-304
27. COPES DL 1976 Comparative leader growth of Douglas-fir grafts, cuttings and seedlings. Tree Planters' Note 27:13
28. DE LA GOUBLAYE DE NANTOIS T 1980 Rajeunissement chez le Douglas (Pseudotsuga menziesii) en vue de la propagation vegetative. Etudes sur la plagiotropie des parties aeriennes et racinaires. DEA, Phys Veg , Universite Paris 6, 44p
29. DURZAN DJ 1979 Progress and promise in forest genetics. Proc 50th Anniv conf "Paper Science and Technology--The Cutting Edge" Inst Pap Chem, Appleton, Wisc pp 31-60

30. DURZAN DJ 1982 Somatic embryogenesis and sphaeroblasts in conifer cell suspensions. In A Fujiwara, ed, Plant Tissue Culture 1982. Proc 5th Int Cong Plant Tissue Cell Cult, Tokyo, pp 113-114
31. EVERS PW 1981-1983 Growth and morphogenesis of shoot initials of Douglas-fir Pseudotsuga menziesii (Mirb.) Franco, in vitro. parts I-VI. Dorschkamp Res Inst For Landscape Planning, Dep Silvic and Hortic Agric Univ, Wageningen, The Netherlands
32. EVERS PW 1983 International conifer tissue culture work group, second workshop. Dorschkamp Res Inst For Landscape Planning, Wageningen, The Netherlands, Rep No 351
33. FARNUM P, R TIMMIS, JL KULP 1983 Biotechnology of forest yield. Science 219:694-702
34. FRANCLET A 1980 Rajeunissement et propagation vegetative des ligneux. Extrait Ann AFOCEL, France, pp 12-41
35. GRIFFITH BG 1940 Effect of indolebutyric acid, indoleacetic acid and alpha napthaleneacetic acid on rooting of cuttings of Douglas-fir and Sitka spruce. J For 38:496-501
36. GUPTA PK, DJ DURZAN 1985 Shoot multiplication from mature trees of Douglas-fir (Pseudotsuga menziesii) and sugar pine (Pinus lambertiana). Plant Cell Rep 4: 177-179
37. HARVEY AE, JL GRASHAM 1969 Procedures and media for obtaining tissue cultures of 12 conifer species. Can J Bot 47:547-549
38. HASEGAWA PM, T YASUDA, T-Y CHENG 1979 Effect of auxin and cytokinin on newly-synthesized proteins of cultured Douglas-fir cotyledons. Physiol Plant 46:211-217
39. JOHNSON MA, JA CARLSON 1982 Some redox considerations in conifer tissue culture. In A Fujiwara, ed, Plant Tissue Culture 1982. Proc 5th Int Cong Plant Tissue Cell Culture, Tokyo, pp 221-222.
40. KADKADE PG, H JOPSON 1978 Influence of light quality on organogenesis from the embryo-derived callus of Douglas-fir (Pseudotsuga menziesii). Plant Sci Lett 13:67-73
41. KARNOSKY DF 1981 Potential for forest tree improvement via tissue culture. BioScience 31:114-120
42. KIRBY EG 1981 The isolation and culture of conifer protoplasts. In Proc IUFRO Sect S2015 Int Workshop "Invitro" Cultivation For Tree Species, Fontainebleau, France, pp 349-354
43. KIRBY EG 1982 The effect of organic nitrogen sources on growth of cell culture of Douglas-fir. Physiol Plant 56:114-117
44. KIRBY EG 1982 The use of in vitro techniques for genetic modification of forest trees. In JM Bonga, DJ Durzan, eds, Tissue Culture in Forestry. Martinus Nijhoff/Junk Publ, The Hague, pp 369-386
45. KIRBY EG, T-Y CHENG 1979 Colony formation from protoplasts derived from Douglas-fir cotyledons. Plant Sci Lett 14:145-154

46. LANNER RM 1962 Some factors affecting rooting of Douglas-fir and true fir cuttings. USDA For Serv Pac Southwest For Range Exp Stn Res Note 202, 5p
47. LA RUE CD 1935 Culture of spermatophyte tissues. Am J Bot 22:914
48. LA RUE CD 1936 The growth of plant embryos in culture. Bull Torrey Bot Club 63:365-382
49. LITVAY JD, MA JOHNSON, D VERMA, D EINSPAHR, K WEYRANCH 1981 Conifer suspension culture medium development using analytical data from developing seeds. IPC Tech Paper No 115, 17p
50. MAPES MO, PM YOUNG, JB ZAERR 1981 Multiplication in vitro du Douglas (Pseudotsuga menziesii) par induction precoce d'un bourgeonement adventif et axillaire. IUFRO In Proc IUFRO Sect S2015 Int Workshop "Invitro" Cultivation For Tree Species, Fontainebleau, France pp 109-114
51. MCCULLOCH WF 1943 Field survival of vegetatively propagated Douglas-fir. J For 41:211-212
52. MOTT RL 1981 Trees. In BV Conger, ed, Cloning Agricultural Plants via in vitro Techniques, CRC Press Inc., Boca Raton, Fl, pp 217-254
53. PRESTON RJ 1976 North American Trees, 3rd ed, The MIT Press, Cambridge, Mass
54. REDISKE JH 1979 Vegetative propagation in forestry In KW Hughes, R Henke, M Constantin, eds, Propagation of Higher Plants through Tissue Culture. A Bridge between Research and Application, Conf 7804111, Nat Tech Inf Serv, Springfield, VA pp 35-43
55. REILLY K, CL BROWN 1976 In vitro studies of bud and shoot formation in Pinus radiata and Pseudotsuga menziesii. Georgia For Res Pap 86, 9p
56. RENDLE BJ 1969 World Timber, Vol 2, Ernest Benn Limited, London Univ, Toronto Press, Toronto
57. ROSS SD 1975 Production, propagation and shoot elongation of cuttings from sheared one year old Douglas-fir seedlings. For Sci 21:298-300
58. SCHMIDT RL 1955 Some aspects of western redcedar regeneration in the coastal forest of British Columbia. Can Dept Lands For, BC For Serv Res Note 29, 10p
59. SILEN RR 1966 A simple, progressive tree improvement program for Douglas-fir, US For Serv Res Note PNW-45, For Range Exp Stn, USDA, Portland, Oregon, 13p
60. SOMMER HE 1975 Differentiation of adventitious buds on Douglas-fir embryos in vitro. Proc Intl Plant Propagators Soc 25:125-127
61. STARBUCK CJ, AN ROBERTS 1982 Movement and distribution of C-indol-3-acetic acid in branches and rooted cuttings of Douglas-fir. Physiol Plant 55:389-394
62. STREET HE 1974 Embryogenesis and chemically induced organogenesis. In WR Sharp, PO Larsen, EF Paddock, V Raghavan, eds, Plant Cell and Tissue Culture: Principles and Applications. Ohio State Univ Press, Columbus, pp 123-153
63. THOMPSON DG, JB ZAERR 1981 Induction of adventitious buds on cultured shoot tips of Douglas-fir (Pseudotsuga

menziesii (Mirb.) Franco). *In* Proc IUFRO Sect S2015 Int Workshop "Invitro" Cultivation For Tree Species, Fontainebleau, France, pp 167-174

64. TIMMIS R, GA RITCHIE 1984 Progress in Douglas-fir tissue culture. Proc Int Symp Rec Adv For Biotech, Mich Biotech Inst, Traverse City, Michigan, pp 37-46
65. VERMA DC, JD LITVAY, MA JOHNSON, DW EINSPAHR 1982 Media development for cell suspensions of conifers. *In* A Fujiwara, ed, Plant Tissue Culture 1982, Proc 5th Intl Cong Plant Tissue Cell Cult, Tokyo, pp 59-60
66. WINTON LL 1972 Callus and cell culture of Douglas-fir. For Sci 18:151-154
67. WINTON LL, RA PARHAM, MA JOHNSON, DW EINSPAHR 1974 Tree improvement by callus, cell and protoplast culture. TAPPI 57:151-152
68. WINTON LL, SA VERHAGEN 1977 Shoots from Douglas-fir cultures. Can J Bot 55:1246-1250
69. WOCHOK ZS, MM ABOEL-NIL 1977 Conifer tissue culture. Proc Intl Plant Propagators Soc 27:131-136
70. WOCHOK ZS, J ANDREASSON, LM KLUNGNESS 1980 Giemsa banding in chromosomes of Douglas-fir (*Pseudotsuga menziesii*) seedlings and plantlets. Ann Bot (Lond) 46:249-254
71. ZAJACZKOWSKI S, TJ WODZICKI, J BRUINSMA 1983 A possible mechanism for whole-plant morphogenesis. Physiol Plant 57:306-310

7. LOBLOLLY PINE: MICROPROPAGATION, SOMATIC EMBRYOGENESIS AND ENCAPSULATION

P.K. GUPTA, D. SHAW and D.J. DURZAN

1. INTRODUCTION

Tissue culture in loblolly pine (Pinus taeda) has historically been viewed as a promising method for effective and economical vegetative propagation or micropropagation of this species. The potential value of reforestation using vegetative propagules of loblolly pine is well-documented (1,5). Briefly, the principal advantage conferred by micropropagation is that plants can be selected for superior performance that results from additive and non-additive genetic effects. Conventional seedling-oriented methods can effectively utilize only additive genetic variation. The presence of substantial non-additive genetic variation for important characters, such as volume production, are recognized by the numerous, ongoing tree improvement programs throughout the southeast U.S. (6,15). Use of such variation might improve yield by 5-10% above that provided from conventional methods (5). Experimental results obtained so far also suggest opportunities for improvement of quality and disease resistance characters. In this chapter we extend micropropagation to include somatic embryogenesis and encapsulation.

2. MICROPROPAGATION VERSUS ROOTING OF CUTTINGS

Tissue culture as a production method of vegetative propagation has attracted the attention of industry largely due to perceived difficulties encountered with alternative methods -- notably rooting of cuttings. However, recent advancements in loblolly pine rooted cuttings research (16), and some very effective economics-based arguments, have reinitiated a broad interest in rooted cuttings as an operational production method. Therefore, improvements in technology, together with economic considerations and a number of problems encountered in field trials of cotyledon-derived plantlets (1) have led to a reevaluation of currently available methods of tissue culture as an operational propagation tool. Nevertheless, there is a powerful incentive for micropropagation of loblolly pine, because of its potential for genetic and epigenetic manipulation. Proponents of genetic engineering have recently been encouraged by demonstrations of genetic transformation of pine using *Agrobacterium* (see, Sederoff, Volume 1). Methods for the clonal micropropagation of loblolly pine are available from the article by Amerson *et al*., (1).

Epigenetic manipulation of juvenile and mature characteristics using micropropagation has been demonstrated in loblolly pine. The combination of tissue culture methods with cultural and plant growth regulator treatments has resulted in apparently-juvenile plantlets from older materials (2 and Franclet, Volume 1). However, the opposite response (production of apparently mature plantlets) has been reported for cotyledon-derived plantlets (1). Field testing of apparently-juvenile plantlets is needed to confirm that they will remain juvenile. Despite these confusing discrepancies, results suggest that manipulation of maturity is feasible with appropriate treatments at least for some species (10).

Accumulation of experience with rooted cuttings and micropropagated plantlets has resulted in some interesting similarities in problems encountered. For example, the commonly observed clonal variations in response to rejuvenation and micropropagation may be paralleled in rooted cuttings by genetic variation in rooting potential (8,16). Such genetic variability reduces our ability to make direct use of highly field-tested genotypes of seedling origin. However, a common solution for plantlet and cutting programs may be to use only populations that have been selected to perform *in vitro* or to root respectively. A second parallel with micropropagation is evident from the growing concensus that many problems in field performance of loblolly pine plantlets and cuttings result from, and may be overcome by the post-rooting conditioning process (7). The message from the above discussion is that biotechnologists would do well to consult with more traditional horticulturists and foresters, and vice-versa.

FIGURE 1 White mucilaginous embryogenic mass develops around the female gametophyte (FG) on a MS-2 medium after 25 days in culture, X8.

FIGURE 2 Growth of polyembryogenic mass after subculture on MS-3 medium, X8.

FIGURE 3 Early stages of growth of somatic polyembryos on MS-3 medium (S = suspensor, e = embryonal end) after 12 days in culture, X20.

FIGURE 4 Development of globular embryos after three subcultures on MS-4 medium X8.

FIGURE 5 Somatic embryos with multiple cotyledons on MS-1 medium with 100 mg/l inositol after 60 days in culture.

FIGURE 6 Plantlet on MS-1 medium with 100 mg/l <u>myo</u>-inositol and 2.5% activated charcoal and without organic nitrogen (CH and gln).

FIGURE 7 Loblolly plantlets transferred to soil with the production of new shoot growth (<u>ca</u>. 5 cm height) after 3 months.

FIGURE 8 Encapsulated somatic embryos after 2 months storage at 4 $\pm$ 1 °C in the dark. Embryos remain viable and produce chlorophyll when returned to light.

3. SOMATIC EMBRYOGENESIS

Somatic embryogenesis (14) and more specifically somatic polyembryogenesis has been reported in *Pinus* sp. (11). This alternative clonal method is now feasible using mature stored seeds. It will be useful especially where controlled crosses are available and where seed material is scarce. The clonal populations represent the improved, hybrid generation because individuals are derived from the suspensor-embryonal mass of developing and mature embryos.

Somatic polyembryogenesis in loblolly pine has been achieved using protocols developed for sugar pine. The principal limiting factor is the conversion of well-developed somatic embryos to plantlets of suitable size and condition for outplanting. Thus far, the efficiency of conversion has been low, although the entire process is very new. More data are needed to verify the initial observations.

Nevertheless, somatic polyembryogenesis offers the potential for a cellular cloning cycle with numerous advantages, including: rapid mass propagation (e.g. we have observed 1057 somatic embryos in 100 ml of liquid culture medium), ease of handling by hydrodynamic methods, possibilities for artificial seed production (by encapsulation) and ease of genetic modification. We hope to apply many of the methods of agricultural microbiology to capture genetic gains, introduce new genetic variation and reduce risks in clonal systems (4).

The following provides an experimental protocol for somatic polyembryogenesis in loblolly pine.

3.1. Explant

1) Remove seeds from female cones collected 4-5 weeks after fertilization.
2) Sterilize the seeds as follows: 1) treat the seeds with 0.1% (w/v) Linbro (detergent) for 5 minutes and then wash them with distilled water 3-4 times; 2) treat the seeds with 30% (v/v) H_2O_2 for 10 min. and then wash them with distilled H_2O, 3 to 4 times; 3) sterilize the seeds with 0.1% (w/v) $HgCl_2$ for 10 minutes and wash them in sterile water 8 to 10 times.
3) Remove the seed coat and excise the female gametophyte with proembryo, embryo and suspensor. Inoculate all on MS-2 medium (see below) and culture in the dark at 23°C.

3.2. Culture Medium[1]

1) Basal medium (MS) (13) but modified to contain NH_4NO_3 at 550 mg/l, KNO_3 at 4676 mg/l, thiamine·HCl at 1.0 mg/l. Other formulations are:
MS-1: half-strength MS with casein hydrolysate (CH) (500 mg/l Difco casamino acids), L-glutamine (gln) (450 mg/l), myo-inositol (1000 mg/l), sucrose 3%.
MS-2: MS-1 + 2,4-D(5 x 10^{-5} M) + KN, BAP (2 x 10^{-5} M).
MS-3: MS-1 + 2,4-D(5 x 10^{-6} M) + KN, BAP (2 x 10^{-6} M).
MS-4: MS-1 + NAA (1 x 10^{-6} M).
Adjust all media to pH 5.7 with KCl and KOH, add Bacto agar 0.6% (w/v) and autoclave (1·1 Kg cm^{-2} at 121°C for 20 min.). Incubate the culture in dark at 23°C.

3.3 Culture procedures

1) After 5-6 weeks of culture a white embryonal-suspensor mass, develops around the female gametophyte.
2) Transfer this mass (Fig. 1) to MS-3 medium.
3) After 3-4 weeks the early stages of somatic embryos will develop. Each embryo (Fig. 3) has elongated cells at one end ("suspensor") and dense cytoplasmic cells with large nuclei at the embryonal end.
4) Subculture this proliferating mass (Fig. 2) every 10-12 days on MS-3 medium.
5) Transfer the mass to MS-4 medium.
6) After 3-4 subcultures on MS-4 medium, globular embryos (Fig. 4) will develop.
7) Transfer these globular embryos to a filter paper support in MS-1 liquid medium with a lower concentration of inositol (100 mg/l). Transfer to continous light (5·0, 2·0, 0·5 μW cm^{-2} nm^{-1} in the blue, red and far-red spectrum) and 24 to 25°C .
8) After 7-8 weeks the somatic embryos will elongate and develop multiple cotyledons (Fig. 5).

---

[1] 2,4-D 2,4-dichlorophenoxyacetic acid, KN kinetin, BAP N^6-benzyladenine.

9) Transfer the developed embryos to MS-1 basal agar medium with 0.25 (w/v) activated charcoal (E. Merck), 7% sucrose, and 100 mg/l inositol, but without CH and glu.
10) Within 5-6 weeks complete plantlets (Fig. 6) develop from the somatic embryos.

3.4. Cytological staining

Nonembryogenic - embryogenic cells can be distinguished from each other and from callus by a double-staining method. First, excise the embryonal-suspensor mass and submerge it in a few drops of 1% (wv) acetocarmine on a glass slide. Heat for few seconds over a flame and wash the cells with medium. Second, stain the cells with 0.5% Evan's blue for a few seconds and wash with medium. Embryogenic cells stain bright red, suspensors and nonembryogenic cells stain blue.

3.5. Planting

Transfer the plantlets to pots containing sterile peat moss, vermiculite and perlite (1:2:1 w/w/w) and place them in a greenhouse or nursery (Fig. 7).

3.6. Encapsulation for production of artificial seeds

Separate somatic embryos growing on MS-4 medium. Dip each embryo in a drop of sterile 1% sodium alginate and stir this drop in 100mM $Ca(NO_3)_2$ for 5 minutes. Wash the alginate drops containing encapsulated somatic embryos 4-5 times with sterile water to remove excess $Ca(NO_3)_2$. Store them at $4 \pm 1^{\circ}C$ in complete darkness. After storage for 50-60 days, transfer the embryos to light where they will produce chlorophyll after 6 to 10 days (Fig. 8). All embryos remain viable under these conditions. Studies are in progress to obtain plantlets from encapsulated somatic embryos.

4. DISCUSSION

Somatic polyembryogenesis has been achieved with *Picea abies* (12) sugar pine (11) and Douglas-fir (unpublished). As with other methods for tree improvement (e.g., use of polyploids and mutations, isozymes) it is likely that cell and tissue culture will aid classical breeding, testing and selection methods. While the successful application of new technologies may be limited only by the imagination and rigor with which we apply them, we must not be overcome by the temptation of simple solutions to difficult problems.

Most genetic characters that possess economic value are quantitatively inherited and are not subject to improvement by presently available gene transfer technologies. The effective application of gene transfer technolgies will require a knowledge of plant biochemistry and physiology that is not available yet for forest species. In this context, our laboratory, in cooperation with Dr. A. Dandekar now has evidence for transformation and foreign gene expression in tumor callus from micropropagated shoots from a 60-year-old elite Douglas-fir. Transformed cells displayed autotrophic growth in culture, synthesis of octopine and expression of a chimeral, bacterial, kanamycin resistance gene as demonstrated by presence and expression of aminoglycoside transferase activity (A. Dandekar, D.J. Durzan, P.K. Gupta, submitted for publication). This work enhances the potential of clonal systems employing somatic embryogenesis as described for loblolly pine.

Frampton *et al*. (9) have developed a fusiform rust-screening method based on the inoculation and scoring of excised embryos, which may result in an effective index of genetic resistance. It should be possible to produce clonal plantlets from the selected embryos by somatic polyembryogenesis. In some cases, plantlets derived from embryos could be sheared and rapidly propagated as rooted cuttings. The selected genotypes would be further tested using presently available greenhouse inoculation methods (3) or outplanted for field evaluation. After final screening, selected clones could be propagated for production planting or be infused into more traditional breeding programs.

In summary, programs that integrate the best of new and conventional genetic improvement methods are the most likely to be successful. An integrated future program might involve: conventional breeding, testing and selection; rejuvenation and rapid early propagation with plantlets of tissue culture origin; transfer of plantlets to a hedge orchard and even propagation by rooting of cuttings.

While the most appropriate technologies for Loblolly pine have yet to be defined, those available already offer promising avenues. Due to the advanced development of both conventional and laboratory methods, combined programs may be feasible earlier for loblolly pine than for any other North American conifer.

5. REFERENCES

1. AMERSON HV, LJ FRAMPTON, SE MCKEAND, RL MOTT, RJ WEIR 1985 Loblolly pine tissue culture: laboratory, greenhouse, and field studies. In Tissue Culture in Forestry and Agriculture, Plenum Press N Y , pp 271-287
2. ABO EL-NIL MM 1982 Method for asexual reproduction of coniferous trees. US patent No 4,353,184
3. CARSON MJ 1983 Breeding for fusiform rust in loblolly pine. PhD thesis, NC State Univ, NC 122 p
4. DURZAN DJ 1980 Progress and promise in forest genetics. In Paper Science and Technology: The Cutting Edge, 50th Anniversary, May 8-10, 1979, The Institute of Paper Chemistry, Appleton, WI, pp 31-60
5. FOSTER GS, FE BRIDGWATER, SE MCKEAND 1983 Mass vegetative propagation of loblolly pine: A reevaluation of direction. In Proc 16th Southern For Tree Improv Conf, pp 311-319
6. FOSTER GS, FE BRIDGWATER 1986 Genetic analysis of fifth-year data from a seventeen parent partial diallel of loblolly pine. Silvae Genet (In press)
7. FOSTER GS, CC LAMBETH, MS GREENWOOD 1986 Growth of loblolly pine rooted cuttings compared to seedlings. For Sci (submitted)
8. FOSTER GS, FE BRIDGWATER 1978 Cutting orchard design: A model for mass production of stem cuttings of loblolly pine. Weyerhaeuser Company Tech Rep No 042-3104178/81, 5 p
9. FRAMPTON LJ, HV AMERSON, RJ WEIR 1984 Potential of *in vitro* screening of loblolly pine for fusiform rust resistance. Proc 17th So For Tree Imp Conf, pp 325-333
10. GUPTA PK, DJ DURZAN 1985 Shoot multiplication from mature trees of Douglasfir (*Pseudotsuga menziesii*) and sugar pine (*Pinus lambertiana*). Plant Cell Rep 4:177-179
11. GUPTA PK, DJ DURZAN 1986 Somatic polyembryogenesis from callus of mature sugar pine embryos. Bio/Technology 4:643-645
12. GUPTA PK, DJ DURZAN 1986 Plantlet regeneration via somatic embryogenesis from subcultured callus of mature embryos of *Picea abies*. In Vitro, Rapid Communications (in press)
13. MURASHIGE T, F SKOOG 1962 A revised medium for tobacco tissue culture. Physiol Plant 15:473-479
14. NORSTOG K 1982 Experimental embryology of gymnosperms. In BM Johri, ed, Experimental Embryology of Vascular Plants, Springer-Verlag, Berlin, New York, pp 25-48
15. STONECYPHER RW, BJ ZOBEL, R BLAIR 1973 Inheritance patterns of loblolly pines from a non-selected natural population. NC Agric Exp Sta Tech Bul No 220
16. VAN BUIJTENEN JP, DV SHAW 1985 Vegetative propagation of loblolly pine. Int IUFRO Symp on Nursery Management of Southern Pines, Aug 4-9 1985. Alabama Expt Stn , Auburn, pp 10-11

8. IN VITRO CULTURE OF CRYPTOMERIA CALLUS AND ORGANS

H. ISIKAWA

1. INTRODUCTION

It is generally accepted that the genus *Cryptomeria* has only one species, *Cryptomeria japonica*, which occurs naturally from latitude 40°42' N to latitude 30°15' N and from 0 to 2,050 metres above sea-level in Japan. *Cryptomeria* in Japanese forestry is a most useful coniferous tree, supplying high-grade quality timbers for wooden houses.

In some regions of Japan, *Cryptomeria* has long been propagated by cuttings. Trees thus produced are used in the production of squared columns or polished small logs of high quality, mainly in several particular regions. Superior cultivars are selected on the basis of stem form or special wavy surface patterns on the stem.

Foresters also found rapidly growing trees in other regions, and succeeded in vegetative propagation of these. Recently about 49,000,000 cuttings of *Cryptomeria* have been produced per year. The area reforested by cuttings is nearly 12,000 ha in total, while the area reforested by seedlings covers about 48,000 ha.

In the near future selected trees resistant to insect pests, cold and snow damage will be propagated by cuttings and by sexual crossing. *In vitro* micropropagation techniques may provide a useful tool for propagation of some superior clones having poor rootability.

2. IN VITRO CULTURE OF CALLUS

Callus of *Cryptomeria* consistently showed better growth on Wolter and Skoog's medium (WS) (7) than on Murashige and Skoog's medium (MS) (5). The callus is generated originally on the cut surface of the basal part of cuttings of the youngest shoots from the crown of mature trees (Fig. 1), or on the basal cut surface of hypocotyls cultured on the WS medium supplemented with naphthaleneacetic acid (potassium salt) (NAA) or indolebutyric acid (sodium salt) (IBA) (Fig. 2).

With 10^{-5} M NAA and 10^{-6} M or 8×10^{-6} M N6-benzylaminopurine (BAP) supplements, the cultured callus formed roots, but not shoots or plantlets (Fig. 3) (4).

3. HYPOCOTYL CULTURE

Adventitious buds were generated in cultured Cryptomeria hypocotyl cuttings on WS medium (2, 3, 4). Hypotyl cuttings were obtained from 2-week-old seedlings grown from seeds under aseptic conditions. The seeds were sterilized with a 0.2% aqueous solution of Uspulun (methoxyethylene mercury chloride 4.2%, sodium bicarbonate 93.8%, coloring matter 2%) in a beaker on a magnetic stirrer at slow speed for 17 hrs. The culture medium contained the WS mineral constituents with the following additions: 0.1 mg/l thiamine, 0.5 mg/l of nicotinic acid, 100 mg/l of meso-inositol, 2 mg/l of glycine, 10 g/l of agar and 20 g/l of sucrose. The pH of the medium was 5.4 without adjusting after autoclaving.

Bud formation was promoted by the addition of BAP to this medium (Fig. 4) (3, 4).

4. EMBRYO CULTURE

Embryo culture of C. japonica in various media was studied (4, 6). The WS medium at 1/3 concentration, or Harvey and Grasham's medium (1) at 1/3 concentration was suitable for culture of Cryptomeria embryos. On the MS medium, even without NAA, the embryos developed into calluses.

5. IN VITRO CULTURE OF SHOOT APICES AND AXILLARY BUDS

Shoot apices and axillary buds of about 10-year-old trees of a famous cultivar named BOKA-SUGI were cultured (4).

5.1. Culture of shoot apices

Cuttings, 3 cm long, of shoot tips were surface sterilized in a 0.1% benzethonium chloride solution and brushed for 15 min while submerged in this solution. Then they were transferred to a 0.1%

corrosive sublimate solution. Finally they were rinsed by sterilized water. Shoot apices were exposed by removing the juvenile needles surrounding them with tweezers. Small tips of 1 mm in length, including the shoot apex, were excised from the cuttings. After excision the apices were put on the WS ager medium and exposed to 1,900 lux of white fluorescent light for 14 hr/day at 25°C.

As shown in Fig. 5, culture of the shoot apices on a medium supplemented with 0.1-1.0 mg/l IBA or 0.1-10.0 mg/l 2,4-dichlorophenoxyacetic acid (2,4-D) resulted in vigorous callus formation at the base of the shoot apex.

5.2. Culture of axillary buds

Sterilized cuttings, 3 cm long, were cultured on the same medium that was used to culture the shoot apices. A few axillary buds elongated at axillae of the top part of the cuttings, and reached 2-3 cm in length after 8 to 10 months exposure to conditions that were the same as those used for the apices (Fig. 6). Cuttings taken from shoots derived from axillary buds formed adventitious roots at the base of the cuttings after 1 to 3 months on medium supplemented with 0.3 mg/l IBA (Fig. 7). Two years after transfer of the rooted shoots to a potting medium they were planted in the field.

6. CONCLUSION

C. japonica is a responsive species *in vitro*. In particular the fact that regeneration has been obtained from explants from mature trees is encouraging. It still remains to be determined if this regeneration is true-to-type. If this would be the case for all cultivars, methods for large scale clonal propagation of selected superior trees could probably easily be developed.

7. REFERENCES

1. HARVEY AE, JL GRASHAM 1969 Procedures and media for obtaining tissue cultures of 12 conifer species. Can J Bot 47: 547-549
2. ISIKAWA H 1974 *In vitro* formation formation of adventitious buds and roots on the hypocotyls of *Cryptomeria japonica*. Bot Mag Tokyo 87: 73-77
3. ISIKAWA H 1978 Effects of cytokinin and morphactin on bud generation from *Cryptomeria* and *Chamaecyparis* hypocotyl segments cultured *in vitro*. *In* Akihama T, Nakajima K, eds, Long

Term Preservation of Favorable Germ Plasm in Arboreal Crops. Fruit Tree Res Sta, Tsukuba, pp 142-147
4. ISIKAWA H 1984 *In vitro* culture of forest tree calluses and organs. JARQ 18: 131-141
5. MURASHIGE T, F SKOOG 1962 A revised medium for rapid growth and bioassays with tobacco tissue cultures. Physiol Plant 15: 473-497
6. SATO T 1978 Search for the synthetic media suitable for embyo culture of cryptomeria and japanese black pine. J Jpn For Soc 60: 81-86
7. WOLTER K, F SKOOG 1966 Nutritional requirements of *Fraxinus* callus cultures. Am J Bot 53: 263-269

8. FIGURES

FIGURE 1. Callus formation at the basal cut surface and the basal region of a cutting of a branchlet from the crown of a mature *Cryptomeria* tree. Ca = callus.
FIGURE 2. Callus formation in the test-tube at the basal cut surface and the basal region of a *Cryptomeria* hypocotyl segment. Hy = hypocotyl, Ca = callus.
FIGURE 3. Adventitious root formation in the cultured callus on WS medium supplemented with NAA and BAP. Ro = root.
FIGURE 4. A *Cryptomeria* hypocotyl segment with buds formed on WS medium supplemented with BAP. Ad = adventitious bud, hy = hypocotyl.
FIGURE 5. A shoot tip with vigorous callus formation from the basal cut surface. Sh = shoot tip, Ca = callus.
FIGURE 6. Two shoots derived from the axillary buds of cuttings of a branchlet from the crown of a mature *Cryptomeria* tree. Sh = shoot.
FIGURE 7. Root formation on a shoot derived from the axillary bud in the test-tube. Ro = root.
FIGURE 8. Two year old plants grown from the rooted shoots shown in Figure 7, ready to be planted in the field.

1
Ca
3
cm 0
2
Hy
Ca
3
cm 0
3
Ro
Ro
3
cm 0
4
Hy
Ad
Ad
10
5
mm 0
5
Sh
Ca
5
mm 0
6
Sh
3
cm 0
7
Ro
3
cm 0
8
30
cm 0

9. VEGETATIVE PROPAGATION *IN VITRO* OF *CUNNINGHAMIA LANCEOLATA* (LAMB.) HOOK.

C. BIGOT and F. ENGELMANN

1. INTRODUCTION

In vitro mass propagation of trees of highest quality and rapid growth probably will become one of the main tools to improve forest productivity in the next 20 years (9). Today's forests are largely plant associations which are only partially accessible to genetic control and agricultural practices. Among species with a great potential for productivity, the gymnosperms are well represented. So far about 20 of them have been studied for their *in vitro* performance and their capacity to give viable and true-to-type plantlets (14).

In the *Taxodiaceae* family, only *Sequoia sempervirens* has been the object of extensive research aimed at inducing mass propagation *in vitro* (2, 13). However, *in vitro* propagation of another economically important member of the same family, *Cunninghamia lanceolata*, has not yet received the attention it deserves. This species is of common occurrence in large natural forests of eastern Asia, especially China, and has been successfully introduced into Africa, Brazil, and Malaysia. In Europe, only a few specimens are available in arboreta. In China it is used extensively for reforestation; it grows fast (up to 25 m high with a trunk diameter of 2 m in 20 years) and is of excellent quality for timber and pulp and paper manufacture.

Nevertheless, some ecological factors such as heliophily, a requirement for high rainfall (1500 mm/year), and siliceous clay soil with good drainage, limit its reforestation potential.

The ability to produce shoots from the base of the trunk is a remarkable characteristic of *Cunninghamia*, which is only rarely reported for other conifers (i.e., *Sequoia sempervirens*, *Pinus rigida*). These stump sprouts can be used in a short rotation system of clonal propagation similar to that used for some hardwood species (*Eucalyptus*, poplars).

Because of marked heterozygosity, conventional propagation through seeds gives rise to a heterogeneous population, and *in vitro* propagation has, therefore, been considered as an alternative to multiply desirable genotypes. In the present study the organogenetic potential, the rooting ability of shoots, and the growth habit after acclimatization of mature and juvenile material are compared. There are two phenomena associated with maturity that have, so far, rendered clonal propagation ineffective, i.e., the loss of rooting ability and the maintenance of plagiotropic growth. *In vitro* culture could be an experimental means to revert the material to juvenility, i.e., to an orthotropic growth habit, a juvenile morphology, and easy rooting (4, 6).

2. MATERIAL AND METHODS

In vitro culture has been started from 1) mature trees, grown in France for more than 50 years, selected for their resistance to cold (especially clone 78374); 2) the main stem of one 6-month-old tree (clone 78288); 3) seedlings raised from seeds from Brazil and Formosa.

In the case of mature trees, the explants consisted of apical sections or stem pieces without needles taken from stump sprouts. Similarly the explants that were taken from the 6-month-old tree and the seedlings consisted of needle-free apical sections or stem pieces. The explants were disinfected for 1 min in 90% alcohol and then during 20-30 min (15 min for seeds) in calcium hypochlorite (100 g l^{-1}) and Tween 80 (<1‰ w/v), and then were rinsed twice in distilled water.

Murashige and Skoog's (12) mineral medium was used, with vitamins (thiamine-HCl 10 mg l^{-1}, and myo-inositol 100 mg l^{-1}) and chelated iron (5 ml l^{-1} of a solution obtained by dissolving 7.45 g Na_2EDTA and 5.57 g $FeSO_4 \cdot 7H_2O$ in 1 litre of distilled water). This is called basal medium (BM), to which were added sucrose (30 g l^{-1}), activated charcoal (Merck 2186) in some experiments (2-3 g l^{-1}), a solution of growth regulators, and Bacto-Agar (7 g l^{-1}). The medium was poured into tubes and jars and autoclaved for 20 min at 110°C.

After adding the explants the tubes and jars were placed in a growth chamber where the following conditions prevailed: 16 h light (20 Wm^{-2} provided by a mixture of 40 W Phillips TL, Sylvania Grolux and True-lite Durotest fluorescent tubes), and 26°C ± 1°C day temperature and 22°C ± 1°C night temperature. In different experiments, subcultures were prepared at 4, 6, or 8 week intervals. Acclimatization was carried out at 20°C ± 2°C, 90% relative humidity in small plastic "greenhouses" (miniserres).

3. RESULTS

3.1. In vitro behavior of stem explants with emphasis on shoot production

In all of our experiments, new explants were excised from a stock of mother plants kept under *in vitro* conditions on BM + activated charcoal and the auxins indoleacetic acid (IAA) (1-2 mg l^{-1}), or indolebutyric acid (IBA) (3 mg l^{-1}). Each mother plant consisted of a small, rooted basal shoot cutting which had developed basal shoots after two months in culture. The explants (1 cm long) were taken from the principal axis of these mother plants; the stock of mother plants was restored every 8 weeks by rooting cuttings taken from the apical zone of the main stem.

Preliminary assays (not reported here) demonstrated that benzylaminopurine (BAP), with or without an auxin, stimulated the growth of axillary meristems of the needles (Fig. 1). However, shoots (Fig. 2) that developed from these meristems did not elongate enough to root. This lack of rooting was subsequently overcome; firstly, by induction of axillary budding by supplying BAP for two months (Table 1); secondly, by elongation of shoots and production of more shoots by 2 months of subculture of BM supplemented with charcoal and IAA 2 mg l^{-1}. Orientation of the explant on the medium was important because when placed horizontally three times more axillary shoots were obtained then when placed vertically (5.9 instead of 2 per primary explant of the mature clone). Thus a sustained production of shoots (Fig. 3) was established and studied over a 1-year period.

Table 1 shows that BAP 1 mg l^{-1}, and α-naphthaleneacetic acid (NAA) 0.03 mg l^{-1}, when used together, stimulate organogenesis and

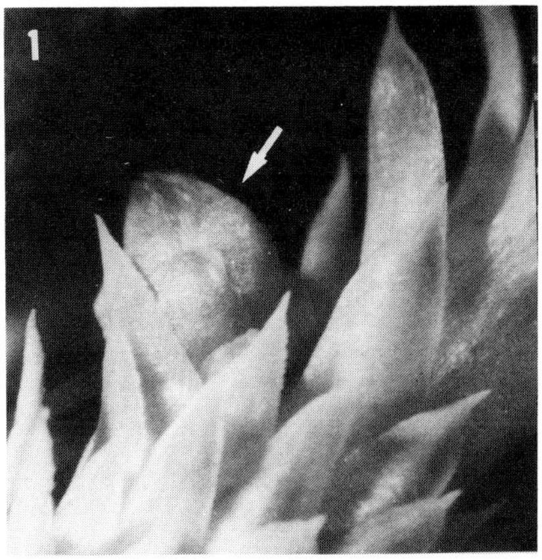

FIGURE 1. Induction of axillary buds (mature clone 78374).
FIGURE 2. Shoots induced by BAP 1, and NAA 0.03 mg l^{-1} at the base of a primary explant placed vertically on the BM.

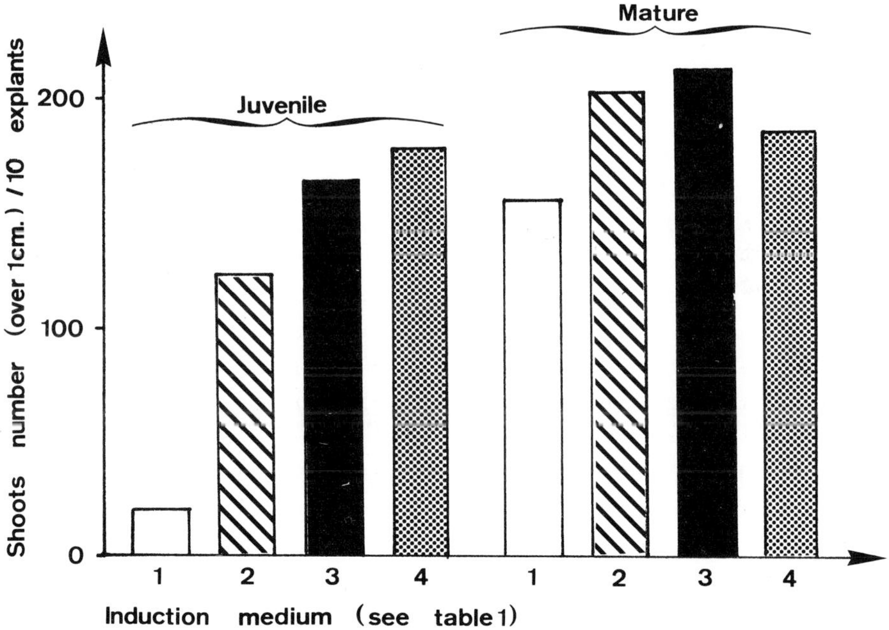

FIGURE 3. Number of shoots (over 1 cm long) produced per 10 explants after 12 subcultures (for composition of induction medium (number 1, 2, 3, and 4) see Table 1).

Table 1. Induction of axillary shoots on primary explants cultured for 2 months on BM

Clones and induction media[a]	Number of explants	Explants with shoots(%)	Shoots per 10 explants	Survival of explants (%)
78288 (juvenile)				
1. Control	12	25	20	25
2. BAP 1 IAA 2	28	78	19	64
3. BAP 1 NAA 0.03	16	81	22	87
4. BAP 2 IAA 2	12	92	21	83
78734 (mature)				
1. Control	12	41	18	41
2. BAP 1 IAA 2	36	83	37	87
3. BAP 1 NAA 0.03	18	88	42	92
4. BAP 2 IAA 2	12	83	25	91

[a]All concentrations in mg l^{-1}.

survival during the shoot induction and production phase. Initially all the shoots originated from pre-established axillary meristems, but after several successive subcultures many of them started to arise from the base of the shoots. In comparison with the control (Fig. 3), addition of growth regulators had a stimulatory effect on the number of shoots produced per year and on the percentage of explants surviving.

3.2. Rhizogenesis

All shoots selected to provide mother plants had rooted spontaneously after 22 weeks of serial subculture on an auxin-free medium, provided that no lateral shoots were excised prior to rooting of the main shoot. Shoots and roots also arose from the more or less callused base of the mother plants, but no apparent vascular connections were established between these shoots and roots.

When shoots (2 cm long) were taken from the primary explant to induce rhizogenesis (Fig. 4) it was found that IBA between 1 to 5 mg l^{-1} or IAA 5 mg l^{-1} in the medium stimulated the formation of short, thick roots without root hairs (Fig. 5 and 6). IAA was more favourable to subsequent root elongation.

In later experiments, IBA (3 mg l^{-1}) was used to test the rooting potential of shoots of the juvenile and mature clones obtained from the 7th subculture of the primary explant. After 6 weeks all

FIGURE 4. Shoots developing from a primary explant (juvenile clone 78288).
FIGURE 5. Shoots rooting *in vitro* after 4 weeks (IBA concentration: a = 1; b = 2; c = 3 mg l^{-1}).
FIGURE 6. Rooted shoots before acclimatization (juvenile clone 78288).

non-rooted shoots were subjected again to fresh IBA containing rooting medium, and after another 6 weeks shoots that had still not rooted were treated again with IBA. The following results were obtained (Table 2). The rooting percentage varied greatly in the two clones (4 to 76%). Shoot and root formation was not correlated with the number of subcultures. There was extreme variability in the number of roots per explant (1 to 31) and of the mean number of roots per shoot (1 to 9.5). After three subcultures on rooting medium no differences were observed between juvenile and mature clones (Fig. 7). The total percentage of rooted shoots has reached 70% on average, but 100% was recorded in some experiments.

Table 2. Rhizogenesis of shoots excised from primary explants after 6 weeks on BM + IBA 3 mg l^{-1} (the primary explants have been subcultured 7 times; $p = 0.05$).

Clones	Subculture	Number of shoots	Rooted shoots (%)	Mean number of roots per shoot
Juvenile (78288)	I	19	10	1.00
	II	24	30	3.57 ± 2.72
	III	33	57	3.95 ± 1.81
	IV	12	25	3.33 ± 1.05
	V	24	4	2.00
	VI	31	22	2.71 ± 2.19
	VII	28	75	5.57 ± 2.78
Mature (78374)	I	30	50	3.27 ± 1.84
	II	65	65	5.50 ± 0.92
	III	119	47	3.91 ± 0.75
	IV	63	65	3.30 ± 0.64
	V	116	21	4.75 ± 1.33
	VI	87	35	4.29 ± 1.33
	VII	135	76	9.51 ± 1.27

In subsequent attempts to obtain more uniform rooting, shoots were pretreated with an auxin solution and then cultured on auxin-free medium or subjected to a dark period of 15 days. Both methods failed to induce uniform rooting of the shoots.

On the other hand, direct rooting by dipping the base of the shoots for 5 hours in a sterile solution of IBA (3 mg l^{-1}) followed by storage in a moist chamber for one day before planting has been effective. All shoots of juvenile clone 78288 rooted after three months in rooting medium in the greenhouse under mist (5 1.4 roots per shoot, $p = 0.05$), but the method failed with the mature clone 78374 because of callusing.

3.3. Acclimatization

After several tests, the following acclimatization method was established. The plantlets were maintained in a closed environment

(20°C ± 2°C, 90% R.H.) on a freely draining medium made up of pine bark (5 mm chips), perlite, and enriched TKS_2-type peat (3:1:1 v/v) for 6 weeks, under low intensity natural light complemented by artificial light (8 hours per day, Phytoclaude 400 W Lamps, 10 Wm^{-2}). Liquid fertilizer (3) was applied at weekly intervals.

In all experiments combined, about 23% of the plantlets of the juvenile clone 78288 and 19% of the plantlets of the mature clone 78374 were lost during the acclimatization phase. No correlation was found between the initial root number and the successful establishment of the plantlet. At the end of the acclimatization period the plantlets of the mature clone showed a high variability in growth rate (Table 3).

After acclimatization, plantlets were planted in pots with the earlier described pine bark, perlite, peat (3:1:1) mix and transferred to a greenhouse. Complementary artificial lighting (10 Wm^{-2}) was supplied during 6 hours every day. Liquid fertilizer was applied as before. After 4-5 months the plants formed basal shoots (1 to 3 per plant) and subapical branches (1 to 2 per plant) (Fig. 8). The plants were then planted in containers (3 liters or more) with the pine bark, perlite, peat mix to which

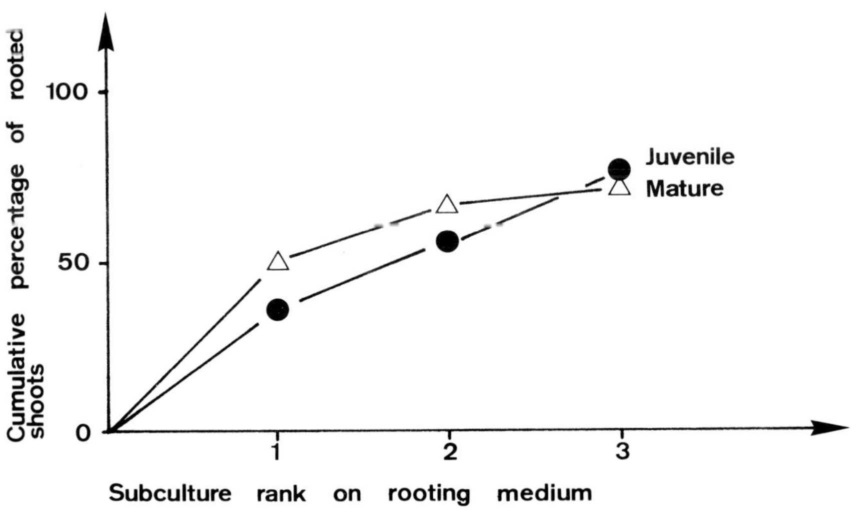

FIGURE 7. Rooting percentage after three subcultures on BM with IBA 3 mg l^{-1}.

a slow release fertilizer ('Osmocote' 15-12-15) had been added. These plants as well as those obtained from seeds (from Sao Paulo, Brazil) were transferred from the greenhouse to outdoors. Table 4 shows the growth characteristics of plants propagated _in vitro_ from the juvenile clone (Fig. 9) and plants derived from seeds after two years of growth. The micropropagated plants had grown more in height and had produced more basal shoots than the plants obtained from seed. Variability within the two groups was similar.

Table 3. Effects of 6 weeks of acclimatization on plantlets of juvenile and mature clones (samples from 5 subcultures of primary explants, $p = 0.05$).

Condition / Clone	Number of plantlets	Roots number per plantlet	Mean length of stem (cm)	Variance (stem)
Not acclimatized				
Juvenile	200	6.51 ± 0.88	5.03 ± 0.24	2.8
Mature	490	4.75 ± 0.33	5.14 ± 0.15	2.6
Acclimatized				
Juvenile	154	8.61 ± 1.09	6.83 ± 0.43	7.07
Mature	394	6.63 ± 0.41	7.44 ± 1.89	353.44

3.4. Plagiotropic behavior

All 265 plants derived from mature clone 78374 were plagiotropic (Fig. 8b), whereas of the 156 plants obtained from juvenile clone 78288, 93% showed orthotropic growth (Figs. 8a and 9) (similar percentages of orthotropic plants were obtained _in vitro_ from other juvenile clones).

During the culture of shoots successively sampled (subculture I-VII in Table 2) from the mature clone, no reversion in needle morphology from the mature pattern (short, spiny) to the juvenile pattern (long, flexible), nor in the pattern of needle implant from asymmetrical to symmetrical occurred.

Attempts were made to modify the plagiotropic growth by decapitation of the shoot, by grafting on seedlings, by cold treatment (8°C, 8 h/day at low light intensity), by frequent subculture

FIGURE 8. Young plants after 3 months acclimatization in the greenhouse (a = juvenile clone; b = mature clone).
FIGURE 9. Plant obtained from juvenile clone 78288 after two years in the greenhouse.

FIGURE 10. Orthotropic shoots obtained from the mature clone after 6 weeks in darkness on BM + IBA 3 mg l^{-1}.
FIGURE 11. Reversion to plagiotropy after 8 weeks exposure to light following 8 weeks in the dark (mature clone).

FIGURE 12. Control shoots of the mature clone maintained in continuous light (8 weeks of culture).

Table 4. Comparison between plants obtained by *in vitro* propagation (juvenile clone 78288) and from seeds after two years growth (variance is shown in brackets, p = 0.05).

Origin	Number of plants	Mean of stem height(cm)	Mean number of layers	Mean number of basal shoots
In vitro 78288	13	61.4 ± 9.4 (240.56)	5.54 ± 0.73 (1.44)	4.08 ± 0.97 (2.59)
From seeds	9	42.4 ± 7.18 (87.04)	5.33 ± 0.94 (1.48)	2.33 ± 1.22 (5.42)

(every 15 days) and by culture in darkness. Only culture in darkness led to a temporary modification of growth of plants derived from the mature clone. As is shown in Fig. 10, plants obtained *in vitro* from isolated shoots or basal shoots of mother plants were orthotropic at the end of the dark period. However, after exposure to light, they quickly (after one month) reverted to plagiotropic growth (Fig. 11), similar to that of the control (Fig. 12). It appears that in plants derived from the mature clone plagiotropic behaviour is induced by light.

4. CONCLUSIONS

The *in vitro* propagation of *Cunninghamia lanceolata* is feasible as evidenced by this study and this species can now be added to the list of conifers cultured *in vitro* compiled by Thorpe and Biondi (14). Regeneration through axillary budding was performed several times during a one year period. However, viable plantlets maintained the behaviour of the original clone, i.e., there was no obvious rejuvenation when mature clones were used. Plants derived from juvenile clones *in vitro* were compared with those obtained from seeds in a 2-year field test. The micropropagated plants were somewhat more vigorous than the seed plants. This could be a reflection of the clone being a selected superior genotype and the seed plants having varying genotypes due to the heterozygosity of the species.

Unlike *Sequoia sempervirens*, which rejuvenated *in vitro* (7), the mature clone of *Cunninghamia* that we used failed to do so. Therefore, other methods such as micrografting on seedlings (5, 8), pruning (10, 11), etc., will have to be applied. However, no difference in rooting ability was observed between shoots obtained from explants from mature and juvenile clones (mother plants) maintained *in vitro*.

We have observed orthotropy in some of our mature *Cunninghamia* clones. In 1985, several 4-year-old container grown plants each produced one vigorous dominant orthotropic shoot with axial symmetry, but bearing mature needles. Black (1) observed such orthotropy in rooted cuttings of 3-8 year old Douglas fir. In addition, we found that shoots in darkness become orthotropic, but revert to plagiotropic growth when light is applied. This environmental factor could be a determining one for the maintenance of plagiotropy, at least in *Cunninghamia*. Studies should be started to determine if phytochrome or specific metabolic modifications are linked to these changes in tropism.

5. ACKNOWLEDGEMENTS

Dr. A. Franclet and Dr. M. Boulay (AFOCEL) are thanked for the gift of *Cunninghamia* seeds and culture stocks, Miss B. Godin, Mrs Casenave and Mr J. Leboeuf for their technical assistance, Mr. H. Gopaul for his help and manuscript translation, and Mrs E. Nawoj for typing the text.

6. REFERENCES

1. BLACK D 1973 Influence of shoot origin and certain pre and post severance treatments on rooting and growth characteristics of Douglas fir (*Pseudotsuga menziesii* Franco) stem cuttings. Ph D Thesis, Oregon State Univ
2. BOULAY M 1979 Multiplication et clonage rapide du *Sequoia sempervirens* par la culture *in vitro*. *In* Micropropagation d'Arbres Forestiers, Ann AFOCEL 12: 49-56
3. COIC Y, C LESAINT 1979 La nutrition minérale et en eau des plantes en horticulture avancée. Doc Tech FCPA n° 23, 22 pp
4. ENGELMANN F 1982 Multiplication végétative *in vitro* de clones juvénile et mature de *Cunninghamia lanceolata* (Lamb.) Hook. DEA Physiol Vég, Univ Pierre et Marie Curie, Paris, 50 pp
5. FRANCLET A 1977 Manipulation des pieds-méres et amélioration de la qualité des boutures. Etud Rech AFOCEL, France, 8: 1-20
6. FRANCLET A 1981 Rajeunissement et micropropagation des ligneux. *In* Proc IUFRO Sect S2 01 5. Int Workshop "*In Vitro*" Cultivation For Tree Species, Fontainebleau, France, pp 55-64
7. de la GOUBLAYE de NANTOIS H 1980 Vieillissement et rajeunissement chez le *Sequoia sempervirens* Endlicher en relation

avec la propagation végétative. Thèse Doct, Univ Pierre et Marie Curie, Paris 6, 170 pp

8. de la GOUBLAYE de NANTOIS T 1980 Rajeunissement chez le Douglas (Pseudotsuga menziesii) en vue de la propagation végétative. Etude sur la plagiotropie des parties aériennes et racinaires. DEA Physiol Vég, Univ Pierre et Marie Curie, Paris, 41 pp
9. KARNOSKY DF 1981 Potential for forest tree improvement via tissue culture. BioScience 31: 114-120
10. KLEINSCHMIT J, J SCHMIDT 1977 Experiences with Picea abies cuttings propagation in Germany and problems connected with large scale application. Silvae Genet 26: 5-6
11. LIBBY WJ, JV HOOD 1976 Juvenility in hedged radiata pine. Acta Hortic 56: 91-98
12. MURASHIGE T, F SKOOG 1962 A revised medium for rapid growth and bioassays with tobacco culture. Physiol Plant 15: 473-497
13. POISSONNIER M, A FRANCLET, MJ DUMANT, JY GAUTRY 1980 Enracinement de tigelles in vitro de Sequoia sempervirens. Ann Rech Sylvicol, AFOCEL, France, pp 231-253
14. THORPE TA, S BIONDI 1984 Conifers. In WR Sharp, DA Evans, PV Ammirato, Y Yamada, eds, Vol 2, Handbook of Plant Cell Culture, MacMillan Publ Co, New York, pp 435-470

10. PINUS RADIATA

KATHRYN HORGAN

1. INTRODUCTION

During the last two years two reviews have been published on the micropropagation of *Pinus radiata* D. Don. The first, by Aitken-Christie and Thorpe (3), detailed the techniques used in the micropropagation of juvenile radiata pine. The second, by Smith (25), described the genetic improvement of *Pinus radiata*, procedures used to micropropagate juvenile radiata pine in a small production laboratory, and a cost-benefit analysis of the use of micropropagated planting stock. The following chapter describes new developments in the propagation of radiata pine and their effect on the direction of tissue culture research at the Forest Research Institute (FRI) New Zealand. Only brief details of the methods used to micropropagate juvenile radiata pine are given, since these have been well covered in the two previous reviews. Over the past 4 years research on the micropropagation of mature radiata has also been undertaken. Progress made and problems encountered during this research are described.

Micropropagation is being researched as a method which could increase the availability of genetically superior stock. In radiata pine it has been found that large genetic improvements can be made by the use of specific controlled-pollinated seed (7,30). This superior seed is in short supply. There are two options for ensuring that large quantities of these improved genotypes are readily available for forest plantings. The first is to produce more of the superior seed. This involves setting up large orchards of the specific parental clones and carrying out controlled-pollinations. To establish these orchards, a fast, reliable method of obtaining large quantities of ramets of the specific mature progeny-tested clones is required.

Micropropagation of mature radiata pine may meet this requirement (see Section 3). The advantages, problems, and bottlenecks in the use of this large-scale controlled-pollination strategy are being evaluated at present. The second option, also under investigation, is to produce smaller amounts of controlled-pollinated seed and amplify the embryos or seedlings using vegetative propagation. In 1977 when it had been shown that Pinus radiata could be micropropagated from embryos (20) it was proposed that perhaps micropropagation could be used for this amplification of juvenile material.

2. MICROPROPAGATION OF JUVENILE RADIATA PINE

2.1. Our experience and changing rationale

In 1979 a small production laboratory was established at the FRI, Rotorua, New Zealand. The main functions of the laboratory were to produce planting stock for field trials and to collect data on the costs of production (24). In 1980 the first small field trial of tissue-culture plantlets was established. Two years later field measurements indicated that plantlets were growing as fast as seedlings (24). It was noted that most of the plantlets displayed an advanced maturation state, forming secondary needles earlier in the season than seedlings and displaying a reduced tendency to form open buds characteristic of seedlings. In 1981 a second more extensive field trial designed to compare plantlets from cultured embryos with seedlings from the same seed source was established. Four years on from trial establishment, the micropropagated trees are now significantly shorter and stem volumes of plantlets are estimated to be 78% of those of seedling controls (Smith pers. comm.). Because it is general practice in New Zealand forests to have a waste thinning, it is too early to confidently predict if this observed growth difference will translate to a difference in final harvest volume. The plantlets' present growth disadvantage may be related to their previously observed advanced maturation state. This needs to be determined and the factors affecting plantlet maturation state identified and controlled.

Costs of producing plantlets ready for outplanting in the forest were found in 1983 to range from NZ$300 to NZ$1000 per thousand plantlets. For a large-scale operation, a realistic cost per thousand plants was considered to be $450, of which 80% would be due to labour costs. (Plant costs of $450 per thousand are approximately ten times the cost of normal seedlings.) This cost was considered to be economically attractive if trees were planted at low density and if volume growth increments of 10 to 20% more than normal seedling planting stock could be assured (24). With these results (but before the seedling/plantlet growth differences were observed), FRI, in 1983, decided that further research into ways of reducing the labour content of micropropagation and improving the rooting potential and handling of plantlets was warranted. Tasman Forestry Limited, a division of a large New Zealand public company, decided that available results were sufficiently promising to justify establishing a commercial operation micropropagating superior seed.

In the period since 1983 research into other methods of vegetatively propagating seedling radiata pine has also been undertaken. Tremendous progress has been made and methods to vegetatively propagate seedling ortets in the nursery have been developed (10,15). It is now possible to obtain an average of 80 fascicle cuttings from a 20-month-old seedling and set these as cuttings in the open nursery bed. Twelve months later over 90% of these propagules have developed good root systems and are ready for planting in the forest. Other cutting systems (with multiplication factors of 8 to 80 depending on size of cutting and time in nursery bed) may also be used. The cost per propagule has been estimated to be approximately 1.5 times that of a normal seedling (15).

It now seems unlikely that micropropagation, with its much higher costs, could compete with this cutting system when used as a general method of multiplying up progeny from controlled pollinations for forest establishment. However, it remains to be seen if micropropagation costs can be considerably reduced by the use of new methods under development such as the multiplication of meristematic tissue before shoot formation and the automatic feeding of cultures (Aitken-Christie pers. comm.).

The main research effort in the micropropagation of juvenile radiata pine (except for work on automation and somatic embryogenesis) has now shifted to various aspects of cold storage. A unique advantage of micropropagation over propagation by cuttings is that plant material in culture can be kept in cold storage for a number of years (see chapter by Aitken-Christie and Singh, Volume 2). The cheap storage of large numbers of clones will allow the introduction of clonal testing (to select the best genotypes from already selected superior families) while maintaining all clones in culture in the cold store, rather than in hedged archives as previously described (14,28). Hopefully, cold-stored shoots would remain juvenile over the storage period. The maintenance of some degree of juvenility is thought to be important to ensure good growth rates of the propagules, particularly in the first 3 to 5 years after planting in the forest (27).

An experiment to test the effects of cold storage on clonal material was initiated by Aitken-Christie in 1983. Two hundred clones are being kept in the cold store for 6 to 8 years, during which time the same clones are under field evaluation. The growth, proliferation, and rooting characteristics of the various clones in culture will be recorded. The experiment will help answer a number of questions:

(a) Can radiata pine cultures be cold-stored for 8 years?

(b) Do the cultures in the cold-store age, or is juvenility maintained?

(c) How do propagules from cold-stored cultures compare with propagules from the ortet and/or the same clone which have been hedged and maintained outside as a clonal archive?

(d) Are there any correlations between characteristics of the clone in culture and its performance in the field?

(e) What are the hazards and economics of maintaining cultures in the cold store compared to maintaining juvenile clonal archives by repeated top pruning either of the parent seedling ortet or of their ramets?

If cold storage is a feasible and successful method of maintaining clones in the juvenile state while the same clones are being field tested, it is envisaged that a two-step multiplication system could be used for the propagation of the 'elite' tested and proven clones. After clones have been reactivated from the cold store, further proliferation of the shoots could be induced and micropropagules rooted and grown in the nursery. Cuttings could then be taken from these micropropagules for stool beds. Such a system would spread the expensive micropropagation costs over a large number of plants, resulting in a plant cost little more than that of a rooted cutting (15).

2.2. Methods used to micropropagate juvenile radiata pine

Plantlets can be formed from a variety of explants including embryos, cotyledons from 5- to 7-day-old germinated seed, seedling shoot tips and induced fascicle shoots from 1-year-old seedlings. A number of publications (2, 3, 4, 12, 19, 20, 22, 24) describe in detail the steps involved in the micropropagation process. A brief summary is given below.

2.2.1. Shoot initiation. Explants are surface-sterilized in a 3% sodium hypochlorite solution for 5 to 15 minutes and, after rinsing in sterile water, they are cultured for 3 weeks on a Le Poivre (LP) (17) nutrient medium containing 5 mg/litre of the cytokinin benzylaminopurine (BAP) and solidified with 0.8% Difco Bacto agar.

2.2.2. Shoot elongation. After the cytokinin treatment, explants are cultured on LP medium without cytokinin. Several transfers (two to six) are necessary for shoots to elongate to 15 to 20 mm. Usually clones with large numbers of shoots (200 or more) require at least six transfers. Each time clumps of shoots are transferred they are cut into smaller pieces and the newly-cut surface placed in contact with the medium. Adventitious shoots develop on embryos and cotyledons, while axillary shoots form from already existing meristems on the fascicle explants. Some explants consistently produce many adventitious shoots which have little surface needle wax and the needles tend to stick together (wet). Other explants produce only a few adventitious shoots. These are frequently waxy. Axillary shoots are waxy and look like normal field-grown shoots. There is better survival of the waxy shoots than wet shoots during rooting (4).

2.2.3. Shoot multiplication. Shoots with stem lengths approximately 2 cm or more can be multiplied in culture by excising the terminal portion and placing both shoot tip and base back into the growth medium. The shoot tip grows to form a normal shoot, while the basal portion forms axillary shoots. On average a four-fold multiplication is achieved for each multiplication cycle of 12 weeks.

2.2.4. Rooting. For root initiation, shoots are given a 5-day auxin treatment in water agar containing 0.1 mg/litre indole butyric acid (IBA) and 0.5 mg/litre naphthaleneacetic acid (NAA). After this they are washed and planted in a non-sterile peat/pumice/perlite (2:1:1) mix. Trays of shoots are kept in a high humidity chamber, which is misted and aerated daily until roots are formed, generally in 2 to 8 weeks. The plantlets are then transferred to a plastic tent in the glasshouse, hardened off, and lined out in the nursery bed under shade cloth. After the shade cloth is removed the plantlets receive the standard forest nursery treatment given to seedlings.

The multiplication rate of juvenile material is dependent on the time *in vitro*. A multiplication rate of 1000-fold would be feasible over 22 months from seed to plantlets at the nursery gate (Aitken-Christie pers. comm.). Higher multiplication rates would be possible with more *in vitro* remultiplication cycles (see 2.3.3.). At present (October 1985) there are 8400 juvenile plantlets in forest plantings (Aitken-Christie pers. comm.). The number will increase considerably when plantlets from Tasman Forestry's commercial micropropagation operation become available.

3. THE MICROPROPAGATION OF MATURE RADIATA PINE

3.1. Background

One of the main reasons for investigating the micropropagation of mature radiata pine is to provide a fast reliable method of obtaining large numbers of plants from selected mature trees (usually aged about 10 years) for seed orchard establishment. Traditionally, grafting of scions from parent trees onto seedling root stock was the main means of asexual propagation. However, in radiata pine there is a high incidence of delayed incompatibility between root stock and scion, resulting in the death of the graft (26). More recently, rooted cuttings have also been used. Cuttings taken directly from old parent ortets rarely root satisfactorily, necessitating grafting as the first step in the propagation procedures for newly selected mature trees. Three years after grafting the plants can be used as a source for a small number of cuttings for seed orchards, or for establishing permanent archives. Archives can be used for controlled pollination, or to provide more cutting material 2 to 3 years later. There is thus a delay of at least 5 years, and usually more like 8 years, between the selection of a superior tree and its establishment in quantity in a seed orchard. Micropropagation has the potential of reducing this time lag to about 2 years. This time saving means that economic gains from tree breeding can be obtained by the forest industry earlier.

Vegetative propagules from selected superior trees could conceivably be used for direct plantation establishment of clonal tests or even for plantations (28). However, field trials have indicated that cuttings from mature trees do not grow as fast in diameter as do cuttings from young trees or seedlings, especially in the first few years after establishment (15, 27). Because of the growth loss this option for improving the genetic quality of new forest plantings has not been pursued using conventional propagation techniques. As yet, it is not known if micropropagules from mature radiata pine will behave in the same way as cuttings. If rejuvenation of the clone or the restoration of seedling growth rates could be obtained through the use of micropropagation techniques (see Section 4) direct clonal propagation would offer considerable benefits and flexibility to the tree improvement programme of radiata pine. For further information on the role of vegetative propagation in tree breeding strategies see Burdon (9).

3.2. Progress in micropropagation from mature radiata pine

The micropropagation of mature radiata pine is more difficult than that of juvenile radiata pine. To be successful with mature material, all steps of the procedure, the exact developmental state of the cultures and the environmental conditions within the culture jar need to be defined and optimized to a much greater degree than is necessary for juvenile micropropagation (5). Growth, proliferation, and rooting are all more severely affected by suboptimal conditions using material from mature radiata pine than using juvenile material.

3.2.1. Explant selection and culture establishment. For reliable and consistent results, the enclosed buds from mature trees are preferred over other types of explants. These buds, enclosed in scales, can be obtained during most periods of the year. It is best to choose buds which have visible needle fascicle primordia present when the scales are removed [Fig. 1a]. Buds with only sterile cataphylls or barely visible primordia enclosed by scales do not respond well in culture. Some authors studying other conifer species (1) recommend using buds from grafted shoots and pretreating the buds with weekly cytokinin sprays before placing in culture. Such a treatment, fortunately, seems to be unnecessary in radiata pine. We aim to propagate directly from selected forest trees in progeny trials rather than having to graft the buds and then wait 1 to 2 years for the plant to grow to a suitable size before spraying and culturing.

The buds are surface-sterilized in a 3% sodium hypochlorite solution for 15 minutes and then rinsed twice in sterile water. The bud scales are then peeled off and the bud cross-sectioned into discs 5 to 10 mm in thickness. These are placed on LP (17) nutrient media containing 5 ppm BAP and solidified with 0.8% Difco Bacto agar (Fig. 1b). Various media formulations have been tested,

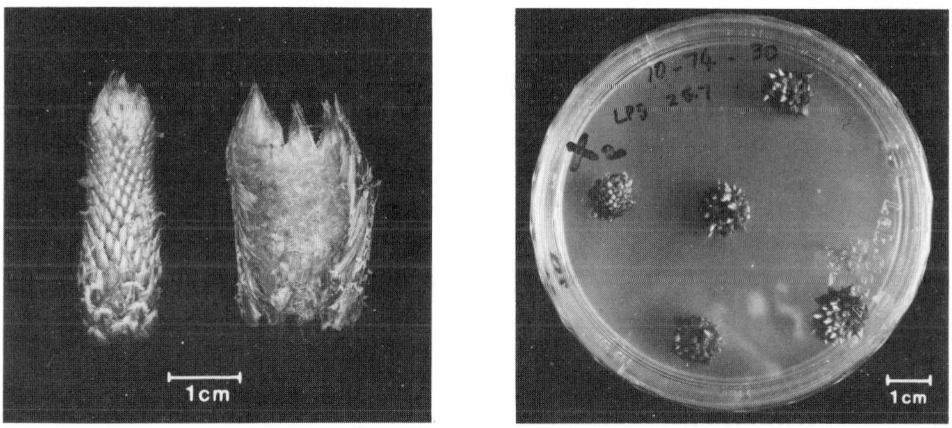

a b

FIGURE 1. a) Bud (left) has had its protective scales (right) removed and is now ready for cross-sectioning and placing in culture. b) Needle fascicle primordia on the cross sectioned bud swell after approximately 2 weeks in culture.

but that of Le Poivre has been found to be most effective in terms of number of buds and the number of needle fascicle primordia per bud stimulated to grow (Horgan & Aitken-Christie, unpublished results). After approximately 4 weeks on this medium, the needle fascicle primordia (which on the tree would have developed into three needles) burst open to reveal a number of very small primary needles.

3.2.2. <u>Shoot elongation</u>. Shoot elongation occurs when the cultures are transferred to a fresh LP medium containing 0.5% activated charcoal without the cytokinin (Fig. 2). When some elongation of the stem axis of the shoot has occurred, the shoots can be either rooted or remultiplied.

a b

FIGURE 2. Needle fascicle primordia on bud discs form shoots on a charcoal medium. Bud sections after (a) 8 weeks and (b) 12 weeks in culture.

3.2.3. <u>Shoot proliferation</u>. New cycles of axillary shoots are obtained by reculturing the shoots on the BAP medium and elongating them on the charcoal medium. The stage of development of the shoot when placed on the BAP medium is critical to the success or failure of the treatment. Shoots which have only primary needles visible die on the BAP medium. However, shoots with primary needles and very small needle fascicle primordia visible will survive and develop new shoots from the fascicle primordia. Cutting the needles of the shoots back to about one quarter their original length also seems to aid survival of the shoot on the BAP medium, perhaps because desiccation of the shoot is reduced. It also makes the shoots easier to handle.

Shoots are transferred to a fresh medium every 3 to 4 weeks. This is a very slow procedure because, even within the one jar of shoots originating from the same explant piece, shoots will be at various stages of development and decisions must be made whether the shoots can survive a BAP treatment or whether they require a further elongation treatment first.

3.2.4. Root formation. The condition of the shoot before rooting is attempted is of paramount importance. Large variations in rooting response occur between batches of shoots. This is probably caused by variation in the condition of the shoots, rather than variation in conditions for root formation. Photoperiod may influence the condition of the shoots. In one experiment shoots grown under an 8-hour photoperiod had higher survival and rooting percentages than shoots grown under a 16-hour photoperiod.

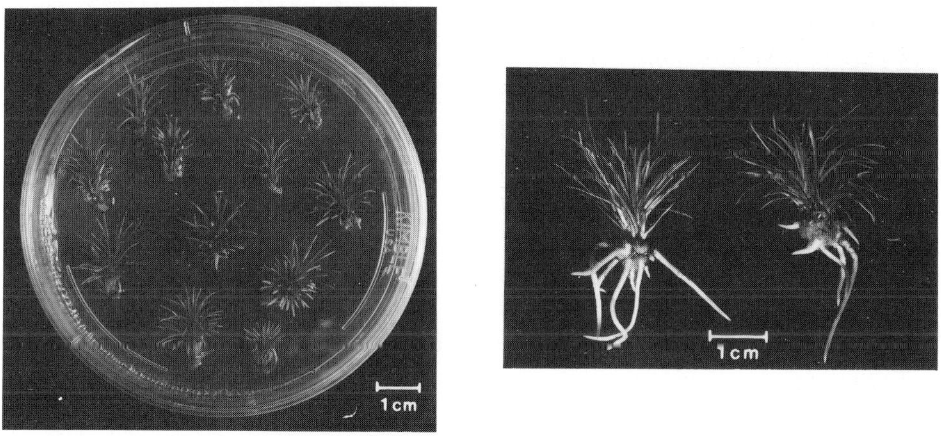

a b

FIGURE 3. a) Shoots are ready for an auxin root-inducing treatment b) Roots form approximately 6 weeks afer the root-inducing treatment.

The standard procedure used for inducing roots on shoots is to place shoots with a visible stem axis on an LP or Gresshoff-Doy (GD - as modified by Sommer *et al.*) (25) medium containing 0.5ppm

NAA and 1.0 ppm IBA for 2 weeks. During this period a callus forms around the base of the shoot. The shoots are then placed in a peat/pumice/perlite mix under similar light conditions to those under which the shoots were grown. A high humidity is maintained inside a propagation chamber. Shoots take from 1 to 4 months to form roots (Fig. 3). Once rooted, the plantlets are slowly acclimatized to glasshouse, then to outside conditions (Fig. 4).

FIGURE 4. Plantlets (foreground) originating from 8- to 17-year-old trees are growing in the nursery bed. Mean plantlet height 20 cm.

Following the procedures outlined, approximately 20 plants can be obtained in an 18-month period from one bud in culture. However, the procedure is not entirely reliable, optimum conditions need to be defined and many problems have yet to be solved.

3.3. Problems

3.3.1. Repeatability of results. The biggest problem in studying the culture of mature radiata pine has been obtaining repeatable results. One can expect varying results when repeating experiments using first one clonal mix, then another. Different clones respond differently to the same medium and environmental conditions during establishment, proliferation, and rooting *in vivo*. This widespread genotypic variation can be found even within full-sib families. It is thus very important to ensure that a large number of clones are tested, or, if interested in only particular clones, that the same clonal mix and proportions are used when carrying out experiments to optimize conditions.

If this was the only source of variation, experiments would be manageable. However, other variabilities, e.g., covert bacterial contaminants and little understood cultural dynamics, greatly affect the repeatability of results and thus frustrate efforts made toward optimizing conditions for the micropropagation of mature radiata pine.

3.3.2. Bacterial contaminants. A persistent problem in culturing radiata pine buds has been the often sudden appearance of a white bacterial infection around the base of the explant, or on the medium where a needle is cut or broken and it touches the agar surface. This bacterium (as yet unidentified) is usually associated with browning of the lower needles of a shoot. Cultures displaying this symptom, although showing no sign of bacterial growth, are now being screened on a peptone/yeast extract agar. However, this is not an entirely reliable test. James and Thurbon reported that 25% of tissue sections from *in vitro* cultured apples, screened monthly on peptone/yeast agar with negative results for five consecutive times, contained bacterial contaminants when screened a sixth time (13). The presence of contaminants leads to variable experimental results, e.g., we have had conflicting results in assessing the effect of growing cultures on supports in liquid as opposed to growing them on agar. It would appear that if cultures are free of contaminants they grow almost twice as fast in liquid as on agar. However, if the shoots are contaminated they grow better on agar than in liquid. We have also noticed that contamination and/or the browning symptoms tend to be more pronounced on the cytokinin medium than on the elongation medium containing activated charcoal. Bacterial growth not noted previously may also become visible at the base of shoots which have been cold stored for approximately 6 months or more.

3.3.3. Cultural dynamics. The vessels used to propagate radiata pine, the agar to air ratio, and the number of shoots per jar, all seem to play an important role in determining the growth of the cultures. It has often been noted that when shoots have been crowded in a jar and are subcultured and divided among two to three jars, the needles of the shoots brown and the whole shoot may die. On the other hand, if the crowded shoots of two jars are subcultured into three jars all shoots survive and grow well. This effect is particulary noticeable when shoots are grown under high light intensity. It has always been assumed that the humidity in a sealed agar jar is very high. We have reason to doubt this and are currently investigating the humidity in jars with varying numbers of shoots and under differing light conditions.

3.3.4. Hard buds. At times shoot cultures tend to form large buds enclosed by hard scales (Fig. 5) similar to buds from mature trees in the field. Once these buds are formed in culture it is difficult to induce them to flush into active growth and the buds eventually die. Occasionally the buds do flush but we have been unable to relate this flushing to a specific treatment. Placing cultures at 4°C for 6 to 8 weeks and/or transfer to a BAP medium have not resulted in renewed growth.

FIGURE 5. Shoots from the same 10-year-old clone displaying both juvenile type shoots (left) and the more mature type of hard enclosed buds (right).

It has been noted that hard buds are formed more frequently under high light intensity (100 $\mu E\ m^{-2}s^{-1}$) than under lower light intensity (60-80 $\mu E\ m^{-2}s^{-1}$) (Horgan and Holland unpublished results). Perhaps it is better to grow cultures at this lower light intensity and avoid the problem rather than to attempt to keep hard buds alive.

FIGURE 6. Radiata pine plantlet from 6-year-old tree. The plantlet has been in the field for 9 months (height 1 metre).

Despite all these problems, thousands of shoots of mature radiata pine have been multiplied up and approximately one hundred plantlets are now growing in the nursery bed.

In culture the shoots take on a more juvenile form. They have well developed primary needles and many have an open type of terminal bud. In the mature tree, primary needles are reduced to brown scales and the open type of bud is replaced by a hard closed bud. The juvenile form of the shoots is normally short-lived, and when the shoots are rooted and placed outside they quickly assume the more mature form. This phenomenon has also been found by

Zimmerman in apples and other crop plants (31). The future timing and occurrence of flowering in the micropropagated radiata pine clones will be recorded with interest.

4. SOMATIC EMBRYOGENESIS

Complete rejuvenation takes place through either somatic or gametic embryogenesis (cf Tulecke Volume 2). Rejuvenation probably would not be desirable for seed orchard establishment because flowering would be delayed, but plantlet formation through somatic embryogenesis of mature and juvenile trees would add an amazing amount of flexibility to the tree improvement program in radiata pine. The establishment of clonal forest directly from proven superior trees or clones of trees would become possible. The need for large-scale seed orchards would no longer exist and the cool storage of clones to maintain juvenility while the clone is under test would no longer be necessary.

Various authors (e.g., 7, 11, 23, 29) have advocated somatic embryogenesis as a method which promises to be less labour-intensive and much cheaper than the organogenic methods now in use. It has been suggested that somatic embryos could be artificially coated and dispensed directly into the nursery bed (21). Although somatic embryogenesis has been described for many plants, it is not yet in general use as a method of rapid propagation (31). Many problems remain. These include problems in inducing all species and even all genotypes of the one species to form somatic embryos (8) and in ensuring genetic stability and phenotypic uniformity of the plants produced (6, 31). New developments in automating and mechanizing organogenic micropropagation may for some species reduce plantlet costs to an extent close to that promised by the use of the somatic embryogenic route.

Somatic embryogenesis from radiata pine embryos is being studied at FRI Rotorua by Dr D. Smith and at Calgary by Prof. T.A. Thorpe and Dr K. Patel. Initially seed embryos are being studied as it is thought that the younger the tissue the easier it is to produce somatic embryos. Recently, Smith (pers. comm.) reported that proembryonic tissue can proliferate and form many somatic embryo initials. This is the first step towards realizing somatic embryogenesis *in vitro*. The biggest breakthrough will come when mature radiata pine can be induced to form somatic embryos. In a number of species, nucellus tissue from the mother plant has been used for the induction of somatic embryos (16, 18). Nucellus tissue and mature shoot meristems will also be used at FRI to study the induction of embryogenesis.

5. CONCLUSION

The micropropagation of *Pinus radiata* is being researched so that improved genotypes may be quickly made available for forest plantings.

Micropropagation can be used to multiply up superior control-pollinated seed. However, using present technology, stem cuttings which give large numbers of propagules have been shown to be a cheaper method. Micropropagation techniques may be combined with the cold storage of shoots to hopefully maintain juvenility while ramets of the same clone are being field tested. The proven superior clones should then be reactivated from the cold store and proliferated and grown-on for clonal plantings. This option will allow the selection of specific clones from superior families and then the evaluation of larger scale clonal forestry.

The micropropagation of mature radiata pine will considerably reduce the time lag between identification of superior genotypes in the field and their establishment in bulk in a seed orchard. Plantlets have been formed from mature trees and are now growing in the field. A procedure of explant establishment, shoot proliferation and rooting is given. Problems have been experienced with the repeatability of experiments, latent bacterial contaminants, cultural dynamics, and the formation of hard buds.

Many scientists have suggested that somatic embryogenesis is the method which will significantly reduce the cost of producing micropropagules. Research in this field is underway. It remains to be seen, with new developments in the automation of the organogenic systems, which propagation procedure will work out to be the cheapest and most reliable.

ACKNOWLEDGEMENTS

Jenny Aitken-Christie is thanked for her input into some of the results outlined in this paper. The technical assistance of Cathy Jones and Lyn Holland is gratefully acknowledged. Harold Hemming and John Barran carried out the photographic work.

REFERENCES

1. ABO EL-NIL MM 1982 Method for asexual reproduction of coniferous trees. US Patent 4,353,184, to Weyerhaeuser Company, Tacoma, Washington
2. AITKEN J, KJ HORGAN, TA THORPE 1981 Influence of explant selection on the shoot forming capacity of juvenile tissue of Pinus radiata. Can J For Res 11: 112-117
3. AITKEN-CHRISTIE J, TA THORPE 1984 Clonal propagation. Gymnosperms. In IK Vasil, ed, Cell Culture and Somatic Cell Genetics of Plants. Academic press Inc, Florida, Vol 1 pp 82-95
4. AITKEN-CHRISTIE J, C JONES, S BOND 1985 Wet and waxy shoots in radiata pine micropropagation. Acta Hortic 166: 93-100
5. ANON 1979 Report of Forest Research Institute, New Zealand pp 8-12
6. BLAKE J 1983 Tissue culture propagation of coconut, date and oil palm. In John H Dodds, ed, Tissue Culture of Trees. pp 29-50
7. BROWN CL, HE SOMMER 1982 Vegetative propagation of dicotyledonous trees. In JM Bonga DJ Durzan, eds, Tissue Culture in Forestry. Martinus Nijhoff / Dr W Junk Publishers, pp 109-149
8. BROWN DCW, A ATANASSON 1985 Role of genetic background in somatic embryogenesis in Medicago. Plant Cell Tissue Organ Culture 4: 111-122
9. BURDON, RD 1982 The roles and optimal place of vegetative propagation in tree breeding strategies. Proc IUFRO Joint Meet Working Parties Genet Breeding Strategies including Multiclonal Varieties. Sensenstein, Sept 6-10, 1982
10. CLARKE FB, MU SLEE 1985 Prospects for clonal forestry with radiata pine. Aust For 47: 266-271
11. FARNUM P, TIMMIS R, KULP JL 1983 Biotechnology of forest yield. Science 219: 694-702
12. HORGAN K, J AITKEN 1981 Reliable plantlet formation from embryos and seedling shoot tips of radiata pine. Physiol Plant 53: 170-175
13. JAMES DJ, IJ THURBON 1979 Culture in vitro of M9 Apple. Rep East Malling Res Sta For 1978, pp 179-80
14. LIBBY WJ, AG BROWN, JM FIELDING 1972 Effects of hedging radiata pine on production, rooting and early growth of cuttings. NZ J For Sci 2: 263-283
15. MENZIES MI, T FAULDS, M DIBLEY, J AITKEN-CHRISTIE 1985 Vegetative propagation of radiata pine in New Zealand. In press

16. MULLINS MG, C SRINIVASAN 1976 Somatic embryos and plantlets from an ancient clone of the grapevine (cv Cabernet-Sauvignon) by apomixis in vitro. J Exp Bot 27: 1022-1030
17. QUOIRIN M, P LEPOIVRE 1977 Etudes de milieux adaptés aux cultures in vitro de Prunus. Acta Hort 78: 437-442
18. RAWGAN TS, T MURASHIGE, WP BITTERS 1968 In vitro initiation of nucellar embryos in monoembryonic citrus. Hortscience 3: 226-227
19. REILLY K, CL BROWN 1976 In vitro studies of bud and shoot formation in Pinus radiata and Pseudotsuga menziesii. Georgia For Res Pap 86: 1-9
20. REILLY K, J WASHER 1977 Vegetative propagation of radiata pine by tissue culture: Plantlet formation from embryonic tissue. NZ J For Sci 7: 199-206
21. SHARP WP, DA EVANS, MR SONDAHL 1982 Application of somatic embryogenesis to crop improvement. In A Fujiwara, ed, Plant Tissue Culture 1982. Proc 5th Int Congr Plant Tissue and Cell Cult, Tokyo pp 759-764
22. SMITH DR, KJ HORGAN, J AITKEN-CHRISTIE 1982 Micropropagation of Pinus radiata for afforestation. A Fujiwara, ed, Proc 5th Intl Cong Plant Tissue and Cell Culture. pp 723-724
23. SMITH DR 1984 Tissue culture for clonal forestry: present and future technologies. Proc 16th NZ Biotech Conf, Massey Univ, Palmerston North, May 1984
24. SMITH DR Pinus radiata. In YPS Bajaj, ed, Biotechnology of Tree Improvement. Springer Verlag, In press
25. SOMMER HE, CL BROWN, PP KORMANIK 1975 Differentiation of plantlets in long leaf pine (Pinus palustris Mill) tissue cultured in vitro. Bot Gaz 136: 196-200
26. SWEET GB, IJ THULIN 1973 Graft incompatibility in radiata pine in New Zealand. NZ J For Sci 3: 82-90
27. SWEET GB, LG WELLS 1974 Comparison of the growth of vegetative propagules and seedlings of Pinus radiata. NZ J For Sci 4: 399-409
28. THULIN IJ, T FAULDS 1968 The use of cuttings in the breeding and afforestation of Pinus radiata. NZ J For Sci 13: 66-77
29. VERMA DC, EINSPAHR DW 1983 Conifer tissue culture and how it may impact the pulp and paper industry. Tappi 66: 25-27
30. WILCOX MD, CJA SHELBOURNE, A FIRTH 1975 General and specific combining ability in eight selected clones of radiata pine. NZ J For Sci 5: 219-225
31. ZIMMERMAN RH 1984 Rapid clonal propagation by tissue culture In Lowell D Owens, ed, Genetic Engineering: Applications to Agriculture, Beltsville Symposium 7, pp 305-316

11. TISSUE CULTURE MULTIPLICATION OF *SEQUOIA*

ERNEST A. BALL

1. INTRODUCTION

Because of its rapid growth and low microbiological contamination, the coast redwood provides outstanding material for *in vitro* cloning. The persistence of the juvenile state in some organs (e.g., the basal shoots, young leaves and stems from the upper bole) adds to its desirability as a source of explants for experimental work. Also, the strongly excurrent rapid growth of the donor tree is carried over in the development of orthotropic plantlets in tissue culture.

2. SURFACE-STERILIZATION OF THE EXPLANTS

The most rapid and efficient method for buds, leaves, and stem pieces is treatment for 1 to 2 minutes in 10% hydrogen peroxide, washing in sterile water, followed by 1 to 2 minutes in 20% commercial bleach (5.25% sodium hypochlorite). Washing in water prior to inoculation is unnecessary. The mutual breakdown of the active components releases gases (oxygen and chlorine) that are lethal to most microorganisms, but only slightly damaging to the plant tissues. The sequential application of these two oxidizers is much more efficient in sterilization than either one applied uniquely. Concentrations of hydrogen peroxide higher than 10% are, however, damaging to the tissues. Thirty percent of this oxidizer, as used by Gupta and Durzan for explants from mature trees of *Pseudotsuga* and *Pinus* (14) is lethal to *Sequoia*.

Some explants of redwood, e.g., the interior of vegetative buds (cf. 2) and female gametophytes, are without microbiological contamination. Such organs, therefore, require no surface sterilization; indeed, the slightest application of any toxic solution to them kills their tissues.

3. CULTURE MEDIA

Sterilization of all components of the culture media, except the agar, was by filtration. The minerals, sugar, inositol, vitamin, cytokinin, and auxin, were dissolved in glass-distilled water, brought to the volume of 255 ml, then passed through a Morton ultrafine (fritted glass) filter, or a millipore filter (cellulose derivative) of 0.45 micrometer pores. Thus the hydrolyzing of sugars and their acid-formation (3, 18, 19) and their changed biological activity after autoclaving in the presence of the inorganic constituents of culture media, are avoided (cf. 13, 21).

The regeneration and growth from explants of stem and of leaves have been most efficient in the mineral mixture of Wolter and Skoog (22). Optimum regeneration and growth were obtained when the standard concentrations of sodium phosphate and ammonium nitrate were increased 10-fold.

The universally used mineral mixture of Murashige and Skoog (20) was less efficient than the Wolter-Skoog mixture for regeneration and growth from explants of stems and leaves. It was, however, excellent for the growth of terminal and axillary buds which readily sprouted and developed shoots. Since this development originated from already-present shoot apices, no true regeneration was involved. It was also the preferred mineral mixture for the regeneration of callus and plantlets from female gametophytes that had been excised from mature seeds. However, the medium here was laced with fresh coconut water and a high concentration of auxin-herbicide (2,4-D). Trials without coconut water gave no development.

Most media were not adjusted for pH. Measurements indicated that their pH was between 4 and 5, a range of hydrogen ion concentration that is more efficient in making iron available to plant tissues than conditions nearer neutrality (cf. 15).

3.1. Cytokinin

SD 8339 [6-(benzylamino)-9-(2-tetrahydroxypyranyl)-9H-purine], of the six cytokinins tried in the culture media, SD 8339, commercially available as ACCEL, is the most efficient stimulator of regeneration and growth from explants of stem and leaf. This cytokinin did not stimulate growth of the female gametophyte in culture; the suitable one was kinetin (6-furfurylaminopurine).

3.2. Auxin

A wide variety of auxins and auxin-like herbicides were tested for stimulation of regeneration and growth of explants. In stem pieces the desideratum was avoidance of callus as an intermediary in the development of adventitious buds. A low concentration of 2,4,5-T (2,4,5-trichlorophenoxyacetic acid) brought about this end. In leaf explants, where intermediary callus was a prerequisite to regeneration of buds, a stronger concentration of 2,4-D (2,4-dichlorophenoxyacetic acid) was necessary.

3.3. Sugar

Although it has been suggested that glucose is the most suitable carbohydrate for nutrition of conifer explants, and has been applied as the standard in this work, tests with other sugars were made. In collaboration with K. Doan (unpublished) it was found that stem explants of redwood grew well on fructose, mannose, galactose, sucrose, maltose, cellobiose, lactose, and raffinose. Data on these experiments will be published elsewhere. Since some of these sugars are not present in the normal tissues of the tree, it is assumed that redwood cells produce appropriate enzymes to metabolize certain foreign carbohydrates.

3.4. Agar

The utilization of Difco's "new" agar, which may contain little or no *Gelidium*, but derivatives of other red algae, has given comparable growth to that which was obtained on the "old" agar. The new agar was used at 1.2% (w/v). The 25 x 150 mm culture tubes with the complete culture medium were slanted. Although the reason for the beneficial effect of slanted agar is unknown, it can be conjectured that the water exuded from agar by the process of syneresis collects on the surface of unslanted medium and lowers the concentration of the agar to the explant. Debergh *et al.* (11) showed that a concentration of 0.6% agar brought about 50% vitrification (glassiness) of leaves developed from shoot explants of globe artichoke. Progressively higher concentrations of agar reduced the percentage of glassified leaves until 1.1% agar allowed all leaves to develop normally.

Our early experiments on growing stem-piece explants of *Sequoia* on 1% transversely solidified agar yielded low numbers of adventitious buds. Experiments with slanted agar media yielded encouraging increases in numbers of buds. In order to achieve stable slants in the vertical orientation, it was necessary to increase the concentration of the agar to 1.2%. All slanted media at this concentration of agar show syneresis, and ca. 1.0 ml of liquid accumulates at the bottom of the slant. Because this liquid is lost to the solidified suspension of agar, the concentration of the latter is increased to ca. 1.28%. In each tube the stem-piece explant (1 cm long) is pushed 0.3 cm into the lower third of the agar slant. From such a portion of an internode the numbers of regenerated buds varies from 100 down to zero, with an average of ca. 30 for a group of 50 cultures that utilized one liter of medium distributed at 20 ml/tube. If the relatively high concentration of agar that is used here has an inhibiting effect on growth and regeneration for *Sequoia*, as recorded by Debergh *et al.* (11) for *Cynaria*, and by Von Arnold and Ericksson (1) for *Picea*, it is adequately reversed by the cytokinin SD 8339 for *Sequoia*.

The inoculated culture tubes were maintained in an air-conditioned room at 20 ± 5°C under light from equal wattage of fluorescent and incandescent sources at ca. 1000 foot candles, with the photoperiod of 18 h light, 6 h dark.

4. STEM PIECES AS EXPLANTS

Because of the multiple regeneration of buds, the 1 cm long stem pieces (Fig. 1) from the orthotropic basal shoots were the most productive explants. The buds were regenerated in longitudinal rows (Fig. 2) and the maximum number achieved was more than 100 per inoculum. The presence of the cytokinin (SD 8339) in the medium brought about this productivity. Each bud, even with a lengthened period of growth, i.e., up to 2 or 3 months in the original medium, produced a rosette shoot as long as it was attached to the stem piece. The shoots showed elongation only after they were excised and placed upon an agar medium lacking cytokinin and containing activated charcoal (Fig. 5), and auxin. The function of the charcoal is unknown; it has been suggested that it adsorbs inhibitory substances.

After appropriate elongation (to ca. 5 cm) both by stretching of already present internodes and the production of additional leaves and internodes, the shoots were transferred to rooting medium. The ROOTING I consisted of half-strength Wolter-Skoog medium containing NH_4NO_3 (cf. 17) and a high concentration of indolebutyric acid (30 mg/l) and without cytokinin and agar. Each shoot's basal one centimeter was immersed for 12 h in 5 ml of this solution and maintained in the dark. It was then transferred to ROOTING II consisting of half-strength Wolter-Skoog medium, without NH_4NO_3, with 2 g/l activated charcoal and 15 g/l agar and kept in the dark ten days, then grown in the light. Eighty to ninety percent of the shoots bore one to several roots after 30 days. All shoots, the rooted and the non-rooted, were transferred to soil mix, and most of the latter developed roots.

5. LEAVES AS EXPLANTS

The leaves borne on the lateral branches of basal shoots, or from the newest lateral branches of the upper bole (at heights of 15 meters or more) during the months of March and April, provide inocula that undergo a different sort of development. While the research achieved a method for regeneration of buds from stem pieces without the precursor of callus, it was found that leaf inocula required a callus for regeneration of buds. This result was related to the necessity of cutting the leaves transversely in half (Fig. 10); the cut surfaces produced abundant callus as the first growth response (Figs. 11, 12).

Upon this callus several buds regenerated (Figs. 13, 14). Such buds were later excised and placed upon elongation medium similar to that used for buds regenerated upon stem pieces. The buds from leaf callus elongated in a manner comparable to that of stem buds. The elongated leaf sprouts were rooted, planted in pots, out-planted in the field, where they grew into trees (Fig. 15) no different in general appearance from those produced from stem-derived buds (Fig. 9). Aside from constituting another method of multiplying redwood trees via tissue culture, these leaf-formed trees are of interest in that they are produced from callus. D'Amato (9) expressed the opinion that, in general, variations in ploidy, chromosome number, and karyotype are common in plant callus, either of somatic (non-meristematic) or germinal (pollen grain) origin. His Table I lists ten examples in which callus has other chromosome numbers than that of the original plant, and in which some of the regenerated plants had different chromosome numbers than those of the original plant. Also, four examples are listed where callus had chromosome numbers other than that in the original plant, but the regenerated plants had only the same chromosome number as in the original plant. D'Amato (10) deprecated regeneration from callus as a method of plant propagation because variation in chromosome number can occur. However, he listed four examples of plants where their calli had different ploidy levels and/or aneuploidy, yet only, or mostly, diploid plants regenerated from these calli. Therefore, he expressed the opinion that the diploid line has a selective advantage in organogenesis.

Chromosome counts in root-tip smears were made of our Sequoia trees regenerated from stem pieces and from leaf inocula. In both cases the chromosome number was the $2n = 66$. I conclude tentatively that regeneration of plants from cortical cells as well as from leaf callus derives from the diploid cells and not from the rarely present cells of different chromosome numbers.

6. REGENERATION OF TREES FROM THE FEMALE GAMETOPHYTE

In the mature unshed seed the female gametophyte is a layer of 10 to 15 cells thick surrounding the embryo (Fig. 16). At this stage the cells are still alive, and each contains cytoplasts with numerous starch-containing plastids, and a nucleus (Fig. 18). The cone and then each seed is surface-sterilized by soaking them in full-strength bleach. After washing in sterile water, the wing on one side of the seed is excised and then it is possible to spread the seed coat apart and extract the female gametophyte. The gametophyte is then slit along one side and the embryo is withdrawn, leaving a hollow cavity that, in section, shows an oval tube made up of cells of the gametophyte (Fig. 17).

Prior to inoculation, the edges of the slit along one side of the gametophyte are spread apart and the inner surface of the wall are examined carefully for extra embryos. In the hundreds of gametophytes that I have excised, only one had two embryos--a normal-sized one and a smaller one. Both were inoculated in culture medium and grew rapidly into seedlings.

The female gametophyte is naturally axenic and is so delicate that if the bleach contacts it, death of its cells results. It grows well on the following medium: the macro- and microelements of Murashige-Skoog (20), 20 g/l sucrose, 1 mg/l each of kinetin and of 2,4-D, plus 150 ml of coconut water. The last was previously filtered through Whatman No. 1 filter paper. Water was added to bring the volume to 255 ml. After solution of the sugar, the whole is filter-sterilized by passing through a Millipore filter of 0.4 micrometer pore-size. The filtrate is then added to 750 ml of just autoclaved, and cooled to 40°C, suspension of 12 grams of agar. After mixing thoroughly, the now complete medium is distributed at 20 ml to each of 50 culture tubes. Each tube is plugged with cotton and covered with a fingered stainless-steel cap. The tubes are appropriately tilted to allow the medium to solidify to form slants ca. 10 cm in length in each 2.5 x 20 cm tube.

Prior to inoculation, the cap and plug of a tube are removed, the plug is rested upon the open end of the cap, and a 30 cm-long Potts-Smith dressing forceps (V. Mueller, Chicago) is inserted to the lower third of the agar slant and a small trough is dug into the gel. The forceps are excerted and used to pick up gently the hollow female gametophyte and to insert it into the trough with the slit on the outside. Great care must be taken not to squeeze the gametophyte. The culture tube is maintained in the culture room under the lights (cf. above) and after a month callus begins to develop at the broad end of the gametophyte from areas on the granular inner surface of the hollow tube. The callus soon develops a faint green color (Figs. 19, 20). After about two months the callus develops extensively outside the gametophyte (Fig. 21) and has regenerated several buds. After ca. three months the entire gametophyte has callused, and several of the buds have developed leafy shoots (Fig. 22). These shoots are then taken through the system of elongation medium (containing no kinetin or coconut water) to rooting media (Fig. 23), to pots of soil mix, and out-planting in the field. Two hundred of them have developed into small trees now four years of age (Fig. 24). These trees differ in subtle ways from seedlings of the same age (Fig. 25). Perhaps their growth rate is the clearest difference from that of the stem piece regenerated plantlets (29.2 cm/yr vs. 62.0 cm/yr). It is noted that the growth rate of gametophyte trees is similar to that of seedlings. The karyological smears of root tips of the gametophyte trees have given exclusively diploid chromosome counts (2n = 66). Studies of the chromosome counts of the living female gametophytes of seeds have given various counts ranging from haploid to 16x and higher and various aneuploids. It is assumed that the occasional diploid cells divide to produce the callus from which the buds and shoots regenerate.

Because of the slow growth rate of the gametophyte trees, it is tempting to consider them dihaploids that are showing "inbred depression."

An alternate interpretation is that the gametophyte trees have been derived from buds on callus that developed from very underdeveloped embryos. Several serial sections of entire female gametophytes at the stage of inoculation have revealed not such embryos. Such histological studies were made in recognition of the numerous archegonia and possible multiple embryos developed in the ovule of Sequoia (cf. 6).

7. TRANSFER TO SOIL

The difficulties resulting after the transfer of tissue-culture-multiplied plantlets from the conditions of the culture tube to soil are reviewed by Conner and Thomas (7). Their conclusion is that poor control of water loss and the change from hetero- to autophytic nutrition are responsible for the losses. My method of control of water loss from rooted plantlets of Sequoia was accomplished by first transferring their root systems (after removal of the agar by washing) to a mixture of (2:1:1) sphagnum, perlite, vermiculite in white plastic growth tubes (35 mm x 200 mm) with drain holes. These growth tubes were irrigated with a solution of Peter's 20-20-20 minerals (25 g in 20 l). The tubes are stored in a special rack maintained in a specially constructed transparent plastic "greenhouse" (62 cm long x 36 cm wide x 37 cm high) that is topped by transparent plastic normally used for wrapping. The "greenhouse" is kept in a large aluminum tray containing water about 1 cm in depth; the growth tubes in the rack do not touch this water. The growth of the Sequoia plantlets in this system shows a survival rate of 90%. The losses are due mainly to the death of shoots that had shown no visible roots at the time of the transplanting. Minor losses occur as a result of accidental breakage of the extensive root system (Fig. 6) when the plants are pulled gently from the agar in the culture tubes.

It is my opinion that the common practice of transferring tissue-culture-multiplied shoots to soil prior to their rooting is the major cause of their death. While such shoots could obtain sufficient water from the culture medium while they were in the agar culture medium, their water absorption from soil-mix is inadequate to their survival. The advantage of my system is the establishment of a large root system on each plantlet before it is transferred to soil. There is no shock from transfer from the axenic to the septic environment; the established root system immediately absorbs water from the soil. Finally, the starting material is non-vitrified plantlets that do not contain excessive water and whose rate of water loss is not beyond control.

Most of the trees that I have out-planted in the field have been multiplied from stem piece explants of a single donor tree (e.g., Fig. 9). The average rate of growth in height during eight years of development has been 62.0 cm/yr. In comparison, trees developed from seedlings growing in the same field have shown an average rate of growth of 29.2 cm/yr. Harlow and Harrar (16, p. 190-191) wrote that sprout trees may attain heights of 50 feet (1500 cm) in 20 years, with a growth rate of 75 cm/yr, and that the leaders of seedling redwood trees 4 to 10-years old commonly elongate 2 to 6 feet (60 to 180 cm) per year.

8. DISCUSSION

Boulay (5) published the first comprehensive study of production of rooted plantlets by tissue culture of Sequoia. Explants were stems from basal shoots and from the upper bole, and were grown upon modifications of the medium of Murashige and Skoog (20). The low production of adventitious buds from the inocula shows the restriction of this medium. The high rate of contamination of inocula after surface sterilization, and the frequent death of plantlets after transfer to soil were characteristic. No distinction was made between buds arising from the axils of leaves and those from the interior of an internode, and all of them were termed adventitious (5, p. 49). No consideration was given to the probability of an accumulation of variations from successive periods of *in vitro* growth. Deshayes (12) showed increases in variations after each successive cycle of an *in vitro* growth of tobacco.

All of the present redwood trees were produced by tissue-culture multiplication of buds solely from the original inoculation (stem, leaf, or female gametophyte). No subcultures were made. This principle was maintained with the aim of limiting variations that could be caused by prolonged exposure of the inocula to the culture media.

In contrast, the increase in the final numbers of rooted plantlets was achieved by Boulay (5) by reculturing explants from the first *in vitro* shoots.

Boulay (5) used a low concentration of agar (0.8%) in all media, and no consideration was given to the possibility that the shoots grown in the resulting comparatively high availability of water could have been susceptible to vitrification and death upon transfer to soil. Indeed, some of the microphotographs show the leaves to be of excessive length; this is characteristic of vitrified leaves (e.g., frontispiece and photo 13, p. 67). Debergh _et al_. (11) established this observation on leaves of vitrified plants of _Cynaria_.

The important publication of Bekkaoui _et al_. (4) concerns mainly the rooting of tissue-culture-regenerated shoots of redwood. Rooting is considered the criterion on rejuvenescence. They introduce the work by noting that rooting by classical methods of redwood shoots is quite easy, but that the process is difficult in tissue-cultured shoots.

In describing the culture medium it is noted that the concentration of agar was 0.35% (w/v). At this level the shoots of _Picea_ were vitrified (1), as were those of _Cynaria_ (11). That level of agar must have been unsatisfactory, for on page 12, the medium used in studies of mineral nutrition was described as containing 0.7% agar. Indeed, photo 2, p. 8, shows shoots grown presumably on 0.35% agar with extremely long leaves, which is an indication of vitrification, according to Debergh _et al_. (11).

The auxin treatment of the shoots to stimulate rooting by Bekkaoui _et al_. (4) utilizes one week's exposure to $5 \cdot 10^{-5}$ of IAA (8 mg/l), IBA (10 mg/l), or NAA (9 mg/l). Their results can be considered excellent; more than 75% of the shoots from the young donor, and 30-50% of those from the old, rooted _in vitro_. Their concentrations are ca. one-fourth that used in my laboratory, and are miniscule in comparison to the 10,000 ppm that Dr. W.J. Libby uses (personal communication).

The high rooting percentage achieved by Knop's medium is unexpected. A probable reason for this efficiency is the absence of NH_4NO_3 in Knop's medium; the relationship resembles that for the medium of Murashige and Skoog (2) used without ammoniacal nitrogen for increased rooting of _Vicia faba_ (17). Bekkaoui _et al_. (4) consider Knop's medium to be the best one for rooting of redwood shoots. They fail to correlate its efficiency with the lack of ammoniacal nitrogen.

The finding of Bekkaoui _et al_. (4) that activated charcoal inhibits rooting is, at first view, surprising. When it is noted that the charcoal is used at a relatively high concentration (20 mg/l) in media with auxin, the result is understandable on the basis of adsorption of the auxin. An analogous situation was described by Constantin _et al_. (8) in which tobacco callus failed to grow on MS medium that contained sucrose, auxin, cytokinin, and 10 g/l activated charcoal. They interpreted the failure to be due to the adsorption of these hormones by the activated charcoal, and substantiated their view by removing the activated charcoal from the medium prior to culturing the callus. The pre-adsorbed medium supported no growth of the callus.

I have avoided the adsorption of hormones by activated charcoal in my rooting media by introducing auxin to the bases of the shoots in a liquid medium that lacks activated charcoal; this can be termed, in the language of Bekkaoui et al. (4, p. 16) the induction phase. After this several-hour exposure, the shoots are then transferred to agar medium that contains a much lower concentration of activated charcoal (2 g/l), the expression phase, where rooting occurs during the next few days.

9. REFERENCES

1. ARNOLD S, VON T ERIKSSON 1984 Effect of agar concentration on growth and anatomy of adventitious shoots of *Picea abies* (L.) Karst. Plant Cell Tissue Organ Cult 3: 257-264
2. BALL EA 1946 Development in sterile culture of stem tips and subjacent regions of *Tropaeolum majus* L. and of *Lupinus albus* L. Am J Bot 33: 301-318
3. BALL EA 1946 Hydrolysis of sucrose by autoclaving media, a neglected aspect in the technique of culture of plant tissue. Bull Torrey Bot Club 80: 419-411
4. BEKKAOUI F, Y ARNAUD, C LARRIEU, E MIGINIAC 1983 Étude comparative de la rhizogenèse in vitro du *Sequoia sempervirens* chez deux clones d'age différents. AFOCEL, pp 1-25
5. BOULAY M 1978 Multiplication rapide du *Sequoia sempervirens* en culture in vitro. AFOCEL, pp 37-65
6. BUCHHOLZ JT 1939 The embryogeny of *Sequoia sempervirens* with a comparison of the sequoias. Am J Bot 26: 248-257
7. CONNER AJ, MB THOMAS 1981 Reestablishing plantlets from tissue culture: a review. Proc Int Plant Propagators Soc 31: 342-357
8. CONSTANTIN MJ, RR HENKE, MA MANSUR 1977 Effect of activated charcoal on callus growth and shoot organogenesis in tobacco. *In Vitro* 13: 293-296
9. D'AMATO F 1977 Cytogenetics of differentiation in tissue and cell cultures. *In* J Reinert and YPS Bajaj, eds, Plant Cell, Tissue and Organ Culture. Springer-Verlag, New York, pp 343-375
10. D'AMATO F 1978 Chromosome number variation in cultured cells and regenerated plants. *In* T Thorpe, ed, Frontiers of Plant Tissue Culture. Int Assoc Plant Tissue Cult, Univ Calgary, Canada, pp 287-295
11. DEBERGH P, Y HARBAOUI, R LEMEUR 1981 Mass propagation of globe artichoke (*Cynara scolymus*): Evaluation of different hypotheses to overcome vitrification with special reference to water potential. Physiol Plant 53: 181-187
12. DESHAYES A 1976 Effets de cycles successifs de néoformation de bourgeons in vitro sur l'aptitude à la variation somatique chez un mutant chlorophyllien de *Nicotiana tabacum*. Mutat Res 35: 221-246
13. GRIFFITHS DJ, CL THRESHER, HE STREET 1960 The heterotrophic nutrition of *Chlorella vulgaris* (Brannon No. 1 strain). Ann Bot NS 24: 1-11
14. GUPTA PK, DJ DURZAN 1985 Shoot multiplication of mature trees of Douglas-fir (*Pseutotsuga menziesii*) and sugar pine (*Pinus lambertiana*). Plant Cell Repts 4: 177-179
15. HALVORSON HO, RL STARKEY 1927 Studies on the transformations of iron in nature. I. Theoretical considerations. J Phys Chem 31: 626-631

16. HARLOW WM, ES HARRAR 1941 Textbook of Dendrology. McGraw-Hill, New York
17. MARTIN C, M CARRÉ, G DUC 1979 Note sure les cultures de tissus de féverole (Vicia faba L.) Bouturage, culture de cals, culture de méristèmes. Ann Amelior Plant 29: 277-287
18. MCALPINE JG 1923 The influence of autoclave sterilization on carbohydrates in culture media. Abs Bact 7: 5
19. MUDGE CS 1917 The effect of sterilization upon sugars in culture media. J Bact 2: 403-415
20. MURASHIGE T, F SKOOG 1962 A revised medium for rapid growth and bioassays with tobacco tissue cultures. Physiol Plant 15: 473-497
21. WESTON GD 1970 Growth of excised tomato roots in glucose; an effect on ageing. Can J Bot 48: 2193-2197
22. WOLTER KE, F SKOOG 1966 Nutritional requirements of Fraxinus callus cultures. Am J Bot 53: 263-269

FIGURES

FIGURE 1. Stem-piece, 1 cm long, inoculated one month previously, showing the longitudinal rows of incipient buds, now indicated by swellings beneath the epidermis. x2.
FIGURE 2. Stem-piece, 1 cm long, inoculated 40 days previously, showing the longitudinal rows of short shoots developed from the regenerated buds in the cortex. Ca. 100 buds have developed. x2.
FIGURE 3. Five stem-pieces, each 1 cm long inoculated two months previously. Each inoculum had regenerated numerous buds. All of the latter were adventitious; the uppermost inoculum bore an axillary bud, the already-present shoot apex of which developed a long shoot that had not been inhibited by the cytokinin. x1/2.
FIGURE 4. Stem-piece, 1 cm long, inoculated two months previously, showing the vigorous development of a single axillary bud into a long leafy shoot. The already-present shoot apex was not suppressed by the cytokinin. This mode of multiplication is restricted to the production of one shoot per axillary bud; it is numerically inefficient, as compared to the regeneration of adventitious buds (average 1:30), and has not been discussed in this work. x1/2.
FIGURE 5. Eight leafy shoots developed from buds, as shown in Fig. 2, on elongation medium during 25 days. Each shoot is ca. 4 to 5 cm in length, and is ready for transfer to the rooting media. This rapid growth by the production of leaves and stem tissue demonstrates the overcoming of the previous inhibition to growth in length (cf. Fig. 2) of the adventitious shoots by the adsorption of substances by the activated charcoal. x1/2.
FIGURE 6. Rooted plantlet in ROOTING II medium two months. The leafy shoot was regenerated from a stem-piece inoculum. The abundant root system is characteristic. x1/3.
FIGURE 7. Plantlet from a stem-piece inoculum after one month in ROOTING II medium. The root system was removed from the agar medium in order to show the absence of hairs. x1/2.
FIGURE 8. Root-tip smear of a potted plant grown from a stem-piece inoculated October 14, 1979. Processed August 6, 1981. This figure of the 2n = 66 chromosomes is characteristic of all tissue-culture cloned plants. x1000.
FIGURE 9. A tree that was begun by an inoculum of a stem-piece on August 3, 1980, growing in UCSC Arboretum. Photo May 28, 1985, when the tree was 3.5 meters tall.

FIGURE 10. The two halves of a leaf after transverse division and inoculation on an agar slant. x2.
FIGURE 11. The two halves of a leaf after 12 days in dark, plus 15 days in light. Callus has developed on the cut surface of the upper half and on the base of the lower half. x2.
FIGURE 12. The two leaf-halves after two months on the agar slant. Callus is restricted to the cut surfaces. x2.
FIGURE 13. The upper leaf-half after 3 months + 20 days showing the origin of buds on the callus on the cut surface. Callus started at the leaf tip, i.e., originally the most mature part. x2.
FIGURE 14. Eight leaf-halves three months after inoculation. The four leaves were excised from the same shoot; variation in the growth of adventitious shoots is evident. x1/2.
FIGURE 15. Tree grown from bud on callus of the leaf-half that was inoculated April 13, 1979. Photo September 25, 1985, after 2356 days of growth. Height at time of photo, 3.8 meters.
FIGURE 16. Median longitudinal section of a female gametophyte enclosing an embryo. x3.
FIGURE 17. Longitudinal section of a female gametophyte from which the embryo was excised, prior to inoculation on the agar slant. x4.
FIGURE 18. Section of the female gemetophyte showing the cells' thick walls, the numerous starch-containing plastids, and the irregularly-shaped nuclei. x400.
FIGURE 19. A female gametophyte after one month on culture medium. The edges of the slit had spread apart, and the callus has started to develop at the lower end. x4.
FIGURE 20. Female gametophyte after five weeks on culture medium. Bright green callus has spread outside the cavity.
FIGURE 21. Female gametophyte with callus bearing buds. Six months on medium. x4.
FIGURE 22. A fast-growing gametophyte of which even the surface cells had divided to contribute to the callus. Some of the buds had produced short leafy shoots. Three months on medium. x4.
FIGURE 23. A single shoot from a female gemetophyte had regenerated several shoots, some of which had produced roots. Twenty-one days on ROOTING II medium. x1/2.
FIGURE 24. Tree grown from a bud on callus of a female gametophyte inoculated four years previously. Duration of growth from time of inoculation to the date of photographing was 1264 days, when the tree's height was 1.2 meters.
FIGURE 25. Tree grown from seed planted 1639 days prior to date of photographing when its height was 1.2 meters.

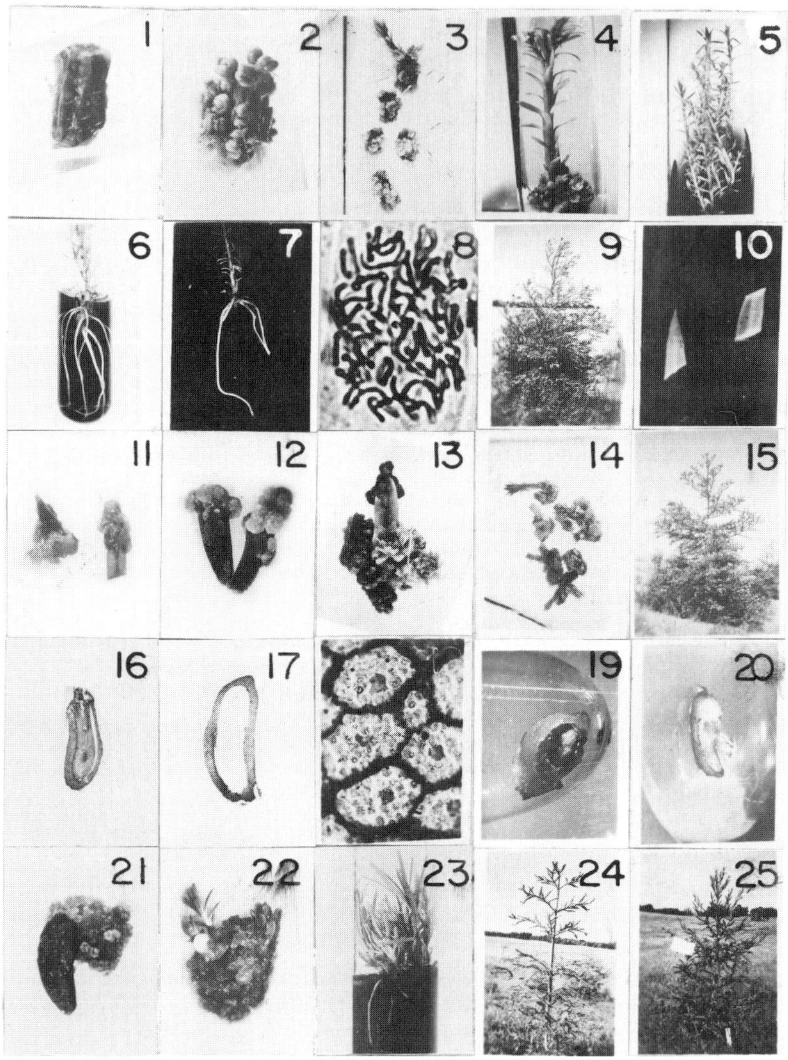
1
2
3
4
5
6
7
8
9
10
11
12
13
14
15
16
17
19
20
21
22
23
24
25

12. TISSUE CULTURE OF HIMALAYAN CONIFERS AND ALLIED GYMNOSPERMS

R. NAGMANI

1. INTRODUCTION

The "Himalayas" have some of the world's highest peaks, including Mount Everest. These peaks form the core of a mountain system which extends over 1,500 miles along the northern part of the Indian sub-continent. The Himalayas increase in height towards northern latitudes and, therefore, have tropical, temperate, and arctic type floral zones.

Conifers constitute a noteworthy part of the Himalayan forests, although the proportion of conifers decreases progressively eastward (3). They are most abundant in the Western, Central, and Eastern regions at elevations ranging from 5,000 to 11,000 ft.

In the western Himalayas, *Juniperus macropoda* and *Pinus gerardiana* are found throughout the Kashmir Valley, at elevations between 8,000 and 11,000 ft. Groups of *Cedrus deodara* and silver fir (*Abies pindrow*), occur on the outer Himalayas at elevations of about 7,000-8,000 ft.

In the central Himalayas, *Pinus roxburghii* occupies the hotter southern slopes and occurs more or less gregariously down to about 2,500 ft. In the next higher zone (7,000-9,000 ft), which is somewhat cooler and drier, *Abies pindrow*, *Picea morinda* are found, and *Pinus wallichiana* replaces *Pinus roxburghii*. Occasionally, associations of cypress (*Cupressus torulosa*) colonize dry sites.

In the eastern Himalayas (7,000-9,000 ft) *Tsuga dumosa* (=*brunoniana*) occurs on the warmer slopes and *Abies densa* at high elevations. *Taxus baccata* is found associated with broad leaved species. In the zone from 9,000 to 11,000 ft, *Picea spinulosa*, *Pinus wallichiana*, *Larix griffithii*, *Juniperus wallichiana* and hemlock, *Tsuga dumosa*, constitute the major conifer species.

The maintenance and rapid propagation of these Himalayan conifers are important for the economy of India. So far India has been dependent mainly on Sweden and Finland for imports of paper, timber, and wood pulp (9), but the Himalayan spruces and firs could be used for the establishment of an indigenous wood-pulp industry on a modest scale.

In vitro propagation of some of these naturally occurring Himalayan conifers and other gymosperms has been attempted over the past two decades. Considerable progress has been made towards establishment of callus cultures and plantlet regeneration with some conifer species, as is described below.

2. TISSUE CULTURE OF *PINUS* SPECIES

In vitro cloning of the Himalayan pines, *Pinus gerardiana*, *P. roxburghii* and *P. wallichiana* was attempted by Konar and co-workers (4, 10). In 1963, Konar initiated callus from hypocotyl segments of *P. gerardiana* (nut-yielding pine) on White's basal medium supplemented with coconut milk (CM), casein hydrolysate (CH) and 2,4-dichlorophenoxyacetic acid (2,4-D). He also established suspension cultures and defined some of the growth parameters of these cultures.

Cytological preparations of 10-week-old callus showed only diploid cells, i.e., the presence of various growth regulators, including 2,4-D, in the nutrient medium did not change the ploidy of cells (8).

In their continued efforts to induce shoot buds and roots in callus cultures of *P. gerardiana*, Konar *et al*. (5, 6) transferred hypocotyl callus to Wood and Braun's medium containing 3160 mg/l of ammonium sulfate. After 4 weeks, shoot-like structures appeared. Both roots and shoots differentiated in callus cultured on White's medium supplemented with indolebutyric acid (IBA) at 10, 5, 2 and 1 mg/l and kinetin at 0.5 mg/l.

Mehra and Anand (11) established fast growing callus cultures from different parts of 3- to 5-week-old seedlings of *P. roxburghii* cultured on Murashige and Skoog's (MS) medium supplemented with naphtaleneacetic acid (NAA) or 2,4-D at 4 mg/l, kinetin 1 mg/l and CM (15 %). The callus obtained was maintained by subculturing for 1 year. Chromosome numbers of cells of 4- to 12-week-old calli were determined in aceto-orcein squashes (8). Most of the cells were found to be diploid with 2n = 24 chromosomes, though a few polyploid and aneuploid cells were also observed. Chromosome bridges and laggards were common.

In an attempt to induce roots and shoot buds in callus cultures of *P. roxburghii*, Bhatnagar *et al.* (1) cultured mature embryos on MS medium supplemented with N^6-benzylaminopurine (BA) at 1 mg/l and NAA at 2 mg/l. The embryos turned green within 10 days. The cotyledons and hypocotyl formed callus where they were in contact with the medium. However, this callus could not be subcultured. After 3 weeks, shoot buds and primary needles appeared on the surface of the swollen cotyledons in 60% of the cultures.

Konar and Singh (10) induced shoot buds in callus cultures of *P. wallichiana*, mainly by using embryos as explants. These embryos were cultured on modified MS medium supplemented with NAA (0.1 mg/l). Calli formed and were transferred to medium containing BA (1 mg/l). Shoot buds developed on this medium. After replacing the cytokinin by CM the shoot buds elongated (Fig. 1).

3. TISSUE CULTURE OF *CEDRUS DEODARA*

Bhatnagar *et al.* (1) cultured embryos and female gametophytes of *Cedrus deodara* separately on MS medium supplemented with auxins, cytokinins, and CM.

On MS medium containing kinetin (2 mg/l), shoot buds developed in the hypocotyl region of the cultured embryos. After 3 weeks in culture the portions of hypocotyl and cotyledons in contact with the medium formed compact green callus. Shoot buds and needles differentiated from this callus after 5 weeks.

The mature female gametophytes of *C. deodara* formed green and friable callus on MS medium supplemented with CM (10%), 2,4-D (2 mg/l) and kinetin (2 mg/l). This callus was subcultured on the same medium through several passages.

4. TISSUE CULTURE OF *PICEA SMITHIANA*

Mehra and Verma (12) obtained callus and shoot buds from cotyledons of *Picea smithiana* (Western Himalayan spruce) by culture of embryos on MS medium supplemented with auxin and cytokinin. The callus cells were predominantly diploid even after repeated subculture.

5. TISSUE CULTURE OF BIOTA ORIENTALIS

In 1965, Konar and Oberoi (7) cultured embryos of *Biota orientalis* on Butenko's medium. After 8 weeks in culture, protrusions

FIGURE 1. *Pinus wallichiana*, embryo culture. A. Forty-day old culture on BM + 1 ppm BAP showing shoot bud formation. x3; B. Four-month-old culture on MS + 2% sucrose + 10% CM (transferred from A) showing shoot formation. x2; C. Thirty-day-old culture on BM + 1 ppm BAP showing shoot bud formation along the surface of the hypocotyl. x4; D. Four-month-old culture on MS + 2% sucrose + 10% CM transferred from BM + 1 ppm BAP + 2 ppm NAA. x1.5. After Konar and Sigh (10).

appeared on the adaxial surface of the cotyledons which they described as embryoids, but which subsequently differentiated into shoots. Later studies by Thomas et al. (15) confirmed that these structures were adventitious buds.

6. TISSUE CULTURE OF EPHEDRA SPECIES

Ephedra is a gymnosperm which is phylogenetically related to Gnetum and grouped under the Gnetales. Several species of Ephedra

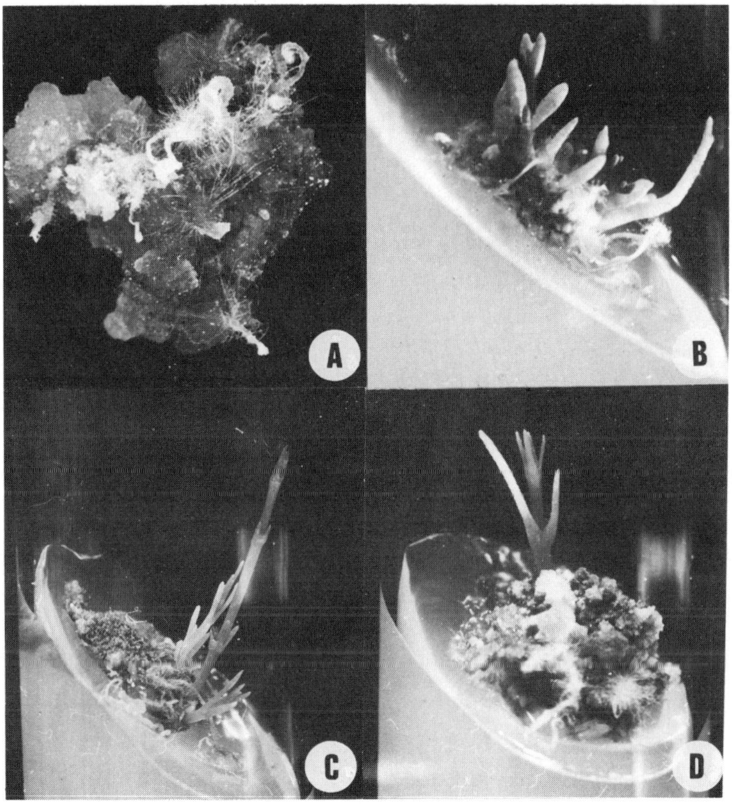

FIGURE 2. Ephedra foliata, female gametophyte culture. A. Four-week-old culture on BM + 0.05 ppm NAA + 0.05 ppm BAP, showing a mass of callus with roots. x2.5; B. Six-week-old culture on BM + 0.05 ppm NAA + 6 ppm BAP showing formation of roots and shoot buds. x2.5; C. Same on BM + 0.5 ppm NAA + 2 ppm BAP. Note the callus mass with roots and shoots. x2; D. Same on BM + 0.5 ppm NAA + 4 ppm BAP showing differentiation of roots and shoots. x2.5. After Bhatnagar and Singh (2).

occur as bushy shrubs on hotter slopes. Ephedra, like Equisetum, has leaves that are reduced to scales. The plant is the source of the alkaloid ephedrene, which is of medicinal importance.

Embryogenesis was induced in cultures of somatic tissues of Ephedra gerardiana (13) on a nutrient medium supplemented with BA, kinetin and NAA.

Haploid callus cultures were raised from female gametophytes of E. foliata by Singh et al. (14) and Bhatnagar and Singh (2). The callus originated from longitudinal and transverse halves of gametophytes with young archegonia. By adding NAA and BA to the medium

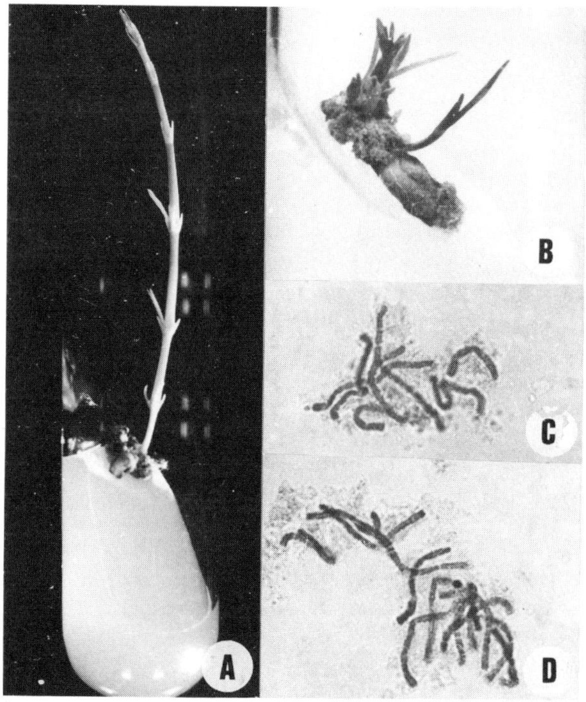

FIGURE 3. Ephedra foliata, female gametophyte culture. A. Six-week-old culture on BM + 0.05 ppm 2,4-D + 1 ppm kinetin showing a shoot and root that developed from the callused portion of the female gametophyte. x1; B. Same on BM + 0.05 ppm 2,4-D + 6 ppm kinetin. Note large number of shoot buds. x2.5; C. An acetocarmine squash of a p-dichlorobenzene treated root tip of a root, formed in vitro from the callus, showing a haploid number of chromosomes, n = 7. x3500; D. A cell from the same root tip showing a diploid number of chromosomes, 2n = 14. x3500. After Bhatnagar and Singh (2).

at varying concentrations, they induced roots and shoot buds in these callus cultures (Fig. 2). Cells of the callus and the root tips had the haploid number of chromosomes, n = 7. Only rarely did a few cells of the callus show a diploid chromosome number (Fig. 3). Meristemoids appeared at the surface of the callus or embedded in it. The deep seated meristemoids organized root primordia, while the peripheral ones gave rise to root as well as shoot bud primordia. Initially, there were no vascular connections between the shoot buds and the callus, but these were subsequently established.

7. REFERENCES

1. BHATNAGAR SP, MN SINGH, N KAPUR 1983 Preliminary investigations on organ differentiation in tissue cultures of Cedrus deodara and Pinus roxburghii. Indian J Exp Biol 21: 524-526
2. BHATNAGAR SP, MN SINGH 1984 Organogenesis in the cultured female gametophyte of Ephedra foliata. J Exp Bot 35: 268-278
3. CHARTURVEDI MD 1956 India. In S Haden-Guest, JK Wright, EM Teclaff, eds, A World Geography of Forest Resources. Am Geogr Soc Special Publ 33, pp 455-482
4. KONAR RN 1963 Studies on submerged callus structures of Pinus gerardiana Wall. Phytomorphology 13: 165-169
5. KONAR RN 1974 In vitro studies on Pinus. I. Establishment and growth of callus. Physiol Plant 32: 193-197
6. KONAR RN 1975 In vitro studies on Pinus. II. The growth and morphogenesis of cell cultures from Pinus gerardiana. Phytomorphology 25: 55-59
7. KONAR RN, YP OBEROI 1965 In vitro development of embryoids on cotyledons of Biota orientalis. Phytomorphology 15: 137-140
8. KONAR RN, R NAGMANI 1972 Chromosome number in callus cultures of Pinus gerardiana Wall. Curr Sci 41: 714-715
9. KONAR RN, R NAGAMANI, S BHUTIANI 1972 Final report of USDA Pl.480 project on "Tissue and cell culture of pines and allied conifers" A7-FS-36, in -250, Dep Bot Univ Delhi, India
10. KONAR RN, MN SINGH 1980 Induction of shoot buds from tissue cultures of Pinus wallichiana. Z Pflanzenphysiol 99: 173-177
11. MEHRA PN, M ANAND 1983 Callus of Pinus roxburghii (chir pine) and its cytology. Physiol Plant 58: 282-286
12. MEHRA PN, V VERMA 1981 Callus induction and shoot bud formation on cotyledons in West Himalayan spruce (Picea smithiana). Phytomorphology 31: 60-69
13. RAMAWAT KG, HC ARYA 1976 Growth and morphogenesis in callus cultures of Ephedra gerardiana. Phytomorphology 26: 395-403
14. SINGH MN, RN KONAR, SP BHATNAGAR 1981 Haploid plantlet formation from female gametophyte of E. foliata Boiss, in vitro. Ann Bot 48: 215-220
15. THOMAS MJ, E DUHOUX, J VAZART 1977 In vitro organ initiation in tissue cultures of Biota orientalis and other species of Cupressaceae. Plant Sci Lett 8: 395-400

13. IN VITRO REGENERATION OF CENTRAL AND SOUTH AMERICAN CONIFERS

DT WEBB and VM VILLALOBOS

1. INTRODUCTION

Tropical forests comprise approximately one third of the world's woodlands and contain 80% of the total land vegetation. The original extent of these forests is estimated at 1.6 x 10^7 km^2. Approximately one third of this is located in Latin America where 40% of the primary forest has been lost to human activities (20). For instance, in Guatemala the pine forests are being diminished by 2.5% each year (11).

In addition to yielding approximately 1.2 x 10^9 tons yr^{-1} of fuel wood, tropical forests provide 10% of the world's non-fuel wood supply (20). It is certain that demand in both of these areas will increase dramatically in the near future. Since hardwoods form the dominant component of most tropical forests, they are currently being most heavily exploited (20). Tropical hardwoods comprise one seventh of the global trade in forest products. Most of this involves high quality lumber and veneer, but there is an increasing demand for pulp and paper products. Currently, 7% of the world's paper and cardboard originates in the tropics, and it is estimated that the demand for pulp from the tropics will double to 4 x 10^8 tons yr^{-1} by the year 2000 (20). While tropical forests are highly productive and can yield as much as 70 tons ha^{-1} yr^{-1}, they are ecologically diverse, complex and delicate. Therefore, sustained yield forestry does not work well, and productivity usually declines markedly after initial logging.

To improve timber productivity on a long-term basis, efforts are underway to use plantations of native and exotic species (20). Clonal forestry practices, including tissue culture, will no doubt play a role in this process as they have with angiosperms

(5) like Populus and Eucalyptus (10) and with Pinus radiata as described elsewhere in this volume.

The forests of Central and South America contain an extremely wide array of coniferous species and varieties which occur over a vast range of ecological conditions (7, 8, 11, 16, 18). For instance, over seventy species and varieties of Pinus have been identified in Mexico and Central America alone, and new ones are being documented regularly (11). However, expanding human populations with their need for agricultural land and forest products have endangered many provenances and some species such as Abies guatemalensis (11), thus depleting potentially valuable genetic resources.

Presently, the low elevation species Pinus caribaea var. hondurensis and P. oocarpa are being planted over thousands of ha in Latin America and other tropical regions. At mid-elevations (1500-2000 m), P. patula and Cupressus lusitanica are proving useful for reforestation in Latin America and Africa (1, 11). Other species of current and future significance for tropical forestry include Pinus greggii, P. pseudostrobus and Araucaria angustifolia (25, F Wadsworth; DB Webb, pers. comm.). In addition, there are many other coniferous species of local significance such as P. hartwegii and P. maximartinezii which may have direct value or may be useful in tree improvement programs.

The Central American and Mexico Coniferous Resources Cooperative (CAMCORE) was formed in 1980 to save and evaluate endemic conifer populations. Centered at North Carolina State University, Raleigh, under the direction of WS Dvorak, CAMCORE has member organizations in Mexico, Guatemala, Honduras, Colombia, Venezuela, Brazil, Africa and the United States. CAMCORE has two current sets of objectives. Phase I involves seed collection from species and/or provenances in Mexico and Central America that are being degraded or endangered. Phase II includes the establishment of conservation banks and progeny tests in Mexico, Central America and Africa under a wide range of environmental conditions (11, 12). Eventually, seeds will be available for general distribution, as well as for experiments on asexual propagation, including tissue culture (WS Dvorak, pers. comm.). Species currently under study by CAMCORE are listed in Table 1, and further species will be added pending future investigations.

Table 1. Conifers under evaluation by CAMCORE (11, 12).

Species	Elevation (m)
Pinus ayacahuite	2000-3000
P. caribaea	0-750
P. chiapensis	850-1800
P. maximinoi	850-1800
P. oocarpa	745-1780
P. pseudostrobus	1000-1800
P. tecunumanii	1000-2400

While callus formation and vascular differentiation have been studied with cambial explants of *C. lusitanica* (2, 3), with the exception of *P. radiata* (this volume), and *Araucaria* (15, this volume), there have been few regeneration studies with Central and South American conifers (4, 13, 17, 24, 26). Since *P. radiata* and *Araucaria* are treated elsewhere in this volume, we will summarize past and present *in vitro* propagation studies with other tropical conifers.

2. METHODS AND RESULTS

2.1. Seed sources

For the studies described herein (Table 2), seeds were obtained from the following: International Forest Seed Company, Birmingham, Alabama; Seed Export, Guatemala City, Guatemala (13); Institute of Tropical Forestry, U.S.D.A., Rio Piedras, Puerto Rico; Ministry of Natural Resources, Forestry Department, Belmonpan, Belize (26); and Banco de Germoplasma Forestal, Instituto Nacional de Investigaciones Forestales, Mexico (24).

2.2. Decontamination

For *P. caribaea* (26), *P. oocarpa* and *C. lusitanica* (13), seeds were soaked in NaOCl (commercial bleach) at 0.5% or 1.05% (w/v) active chlorine, plus 0.1% or 0.2% Tween 20, respectively, for 30 min followed by 3 rinses in sterile distilled H_2O. With *P.*

caribaea (26) seeds were first rinsed in 95% ethanol for 1 min. Ethanol has also been used with P. maximartinezii where seeds were first scarified, submerged in 70% ethanol (1 min), soaked in 6% NaOCl (10 min), then rinsed in sterile H_2O (3 times). Sterility has been difficult to achieve with P. hartwegii and P. patula. With P. hartwegii, seeds were soaked in soapy H_2O (5 min), then placed in 2% NaOCl (5 min), and rinsed 3 times with sterile H_2O. For P. patula, seeds were placed in 0.1% $HgCl_2$ (5 min), rinsed in tap H_2O, soaked in sterile H_2O (24 h), rinsed in 70% ethanol (1 min), submersed in 0.1% $HgCl_2$ (5 min), and finally washed 3 times in sterile H_2O (24). Treatment with 1% H_2O_2 also gave good results with P. hartwegii.

2.3. Explant preparation and selection

To date, only mature embryos or young (10-28 day-old) seedlings have been used as explant sources. In general, seed pretreatment or germination has promoted adventitous shoot induction. With P. caribaea (26), best results were obtained after seeds were incubated on 50% basal medium in darkness for 7 days, prior to embryo excision and culture on shoot-induction medium. For P. patula, embryos were excised after 10 days of incubation, while cotyledons and shoot tips were taken after 15-22 days (24). P. oocarpa and C. lusitanica seeds were germinated in light at 2 μ Em-2 s^{-1} on a 1:1 mix of sand and vermiculite, watered with sterile distilled H_2O for 10 or 20 days respectively prior to explant culture (13). With P. maximartinezii and P. hartwegii, seeds were germinated on vermiculite moistened with H_2O prior to excision.

While whole embryos have been suitable explants for adventitious bud formation, isolated seedling cotyledons or hypocotyl sections with attached cotyledons have proven best overall. With isolated shoot tips, lateral bud induction has been observed with P. patula (Table 2). For rooting trials, adventitious shoots 5 mm or longer have been routinely employed.

2.4. Culture environment

Culture temperatures have ranged from 23 ± 2°C for P. patula (24) to 27 ± 2°C for P. caribaea (26). In all cases, fluorescent tubes were used with either 12 h (P. caribaea) or 16 h photoperiods. The same light conditions have been used for shoot and

Table 2

Summary of Tissue Culture Experiments with Latin American Conifers

	Pinus caribaea	Pinus hartwegii	Pinus maximartinezii	Pinus oocarpa	Pinus patula	Cupressus lusitanica
Explants						
	E[a]	E, SC[b], SH[c]	E, SC	SC	E, SC, ST[d]	HC[e]
Temperature ($\pm$ 2°C)						
	27	23	23	25	23	25
Photoperiod						
	12	16	16	16	16	16
Irradiance						
SI[f]	10^{-4}[g]	8.5[h]	8.5[h]	60[i]	8.5[h]	60[i]
SE[j]	10^{-4}	--	8.5	60	8.5	40
RI[k]	--	--	8.5	60	8.5	30
Culture Media						
SI	SHR[l] MS[m]	MS SHR	SHR	MSM[n]	GDS[o] MSC[p]	CBM[q]
SE	SHR	--	--	MSM	1/2 GDS	CBM
RI	--	--	--	CBM-1/2-GDS	--	PVS[r]
pH						
	5.7	5.5	5.5	5.5	5.5	5.5
Growth Regulators						
SI	BA	BA, K, 2iP	BA	BA, K, 2iP NAA	BA, IAA IBA, NAA	BA, K, 2iP IBA
RI	--	--	NAA	NAA	NAA, IBA	NAA

Table 2. Continued.

	Species					
	Pinus caribaea	Pinus hartwegii	Pinus maximartinezii	Pinus ocarpa	Pinus patula	Cupressus lusitanica
	Agar (% w/v)					
SI	0.8	0.7	0.7	1.0	0.7	0.7
SE	0.8	--	0.7	1.0	0.7	0.7
RI	--	--	0.7	0.7	0.7	PVS[r]
	Organ Formation					
	AS[s]	AS	AS,R[t]	AS,R,P[u]	AS,AX[v],R,P	AS,R,P
	Reference					
	26	w	x	13	24	13

[a]Excised embryo. [b]Seedling cotyledon. [c]Seedling hypocotyl. [d]Seedling shoot tip. [e]Seedling hypocotyl with attached cotyledons. [f]Shoot induction. [g]Wm^{-2}. [h]klx. [i]$\mu Em^{-2}s^{-1}$. [j]Shoot elongation. [k]Root induction. [l]Schenk and Hildebrandt (22), modified by Reilly and Washer (21). [m]Murashige and Skoog (19). [n]Modified Murashige and Skoog (13). [o]Greshoff and Doy (14), modified by Sommer et al (23). [p]Murashige and Skoog (19), modified by Cheng (6). [q]*Cupressus* basal medium (13). [r]Mixture of peat:vermiculite:sand (1:1:1). [s]Adventitious shoots. [t]Roots. [u]Plants. [v]Axillary shoots. [w]Orea and Villalobos, unpubl. data. [x]Villalobos and Robledo, unpubl. data.

root induction. In most cases, caulogenesis and rhizogenesis occurred *in vitro* but, with *C. lusitanica*, rhizogenesis occurred in pots under plastic (13).

2.5. Culture media

In general, a low salt medium like SHR (Schenk and Hildebrandt, 1972; modified by Reilly and Washer, 1977) or GDS (Greshoff and Doy, 1972; modified by Sommer *et al*., 1975) have been best for shoot induction. However, MS (Murashige and Skoog, 1962) medium was better than SHR with *P. hartwegii*, and a slightly modified MS medium (MSM) was suitable for *P. oocarpa* (13). Also, the *Cupressus* basal medium (CBM) used with *C. lusitanica* (13) was very similar to MSM. In both latter cases, the concentrations of NH_4NO_3 and KNO_3 were lower than that of MS medium, and CBM contained biotin plus folic acid in addition to the other MS vitamins. Both MSM and CBM lacked glycine, and the inositol content of MSM was 250 mg l^{-1} rather than the usual 100 mg l^{-1} of MS medium (13). Villalobos and Robledo (unpubl. data) have found that while inositol was not necessary for shoot induction with *P. maximartinezii* cotyledons, it improved color and growth when present.

For shoot elongation, 50% dilution of the salts and reduction of the sucrose concentration from 3% to 1% yielded good results with *P. patula* (24), and the sucrose was reduced to 2% with *P. oocarpa* (13). For rooting of *P. oocarpa*, elongated shoots were transferred to GDS medium with 1% or 0.5% sucrose (13). Non-axenic rooting of *C. lusitanica* shoots occurred in a peat:sand:vermiculite (4:2:1) mixture (13).

Agar was used at 0.7, 0.8 or 1%, and medium pH was set at 5.5 or 5.7 prior to autoclaving (Table 2).

2.6. Growth regulators and organogenesis

In all cases, a cytokinin alone was sufficient to induce shoot formation. Of the cytokinins tested, N^6-benzylaminopurine (BA) was the most potent. However, kinetin (K) and N^6-isopentyl adenine (2iP) were also caulogenic (Table 2). With *P. patula*, indoleacetic acid (IAA) caused a limited amount of shoot formation, and at 0.1 to 0.5 mg l^{-1} IAA promoted the elongation of excised, seedling shoot tips (24). For shoot induction, combinations of cytokinin and auxin were no better than the application of cytokinin alone (13, 24). Some spontaneous rooting of shoots was observed with *P. patula* (24) and *P. maximartinezii* during shoot elongation, but auxins, especially naphthaleneacetic acid (NAA), were effective at stimulating root formation. With *P. patula*, a combination of indolebutyric acid (IBA) (1.0 mg l^{-1}) and NAA (0.05 mg l^{-1}) was very effective, although only one root developed on each shoot after 3 weeks (24). With *P. oocarpa*, application of NAA alone for 2 weeks in agar medium resulted in a 46% rooting efficiency. Watering shoots of *C. lusitanica* with NAA twice weekly caused 36% rooting with an average of 2.6 roots per explant (13).

By combining caulogenesis and rhizogenesis, plantlet production has been achieved with *P. oocarpa* (13), *P. patula* (24) and *C. lusitanica* (13). Additionally, plantlets of *P. oocarpa* and *C. lusitanica* have been acclimated to greenhouse conditions and, with *C. lusitanica*, some field planting has been achieved. Furthermore, with *P. caribaea*, soil adapted plants have been obtained (O Schwarz, pers. comm.).

3. DISCUSSION

Based on the results above, Latin Amerian conifers respond like their temperate zone counterparts (9). This was the conclusion reached by Franco and Schwartz (13) in their work with *P. oocarpa* and *C. lusitanica*. These trees suffer from a lack of systematic, detailed research. While work is in progress at Chapingo, Mexico (VM Villalobos) and Knoxville, Tennessee (OJ Schwartz), more effort is needed to meet the challenges caused by the expanding management of tropical forests for fuel, lumber and fiber production. Given the progress that is being made with temperate conifers (see other chapters in this volume), it is hoped that research funding will become available for *in vitro* regeneration studies of Central and South American conifers.

ACKNOWLEDGEMENTS

This work could not have been done without grants from the Natural Sciences and Engineering Research Council of Canada to DTW. The authors gratefully acknowledge the help of OJ Schwartz during the preparation of this manuscript.

4. REFERENCES

1. BARRET WGH 1972 Variacion de caracteres morfologicos en poblaciones naturales de Pinus patula Schl. et Cham. en Mexico. IDIA, Suplemento Forestal 7: 9-35
2. BORCHERT R 1967 The cultivation of tissues of Cupressus lusitanica in vitro. Physiol Plant 20: 608-616
3. BORCHERT R 1968 Histogenese des in vitro kultivierten Phloems von Cupressus lusitanica. Beitr Biol Pflanzen 44: 435-446
4. BROWN CL, HE SOMMER 1975 An Atlas of Gymnosperms Cultured In Vitro: 1924-1974. Georgia For Res Counc, Macon
5. BROWN CL, HE SOMMER 1982 Vegetative propagation of dicotyledonous trees. In JM Bonga, DJ Durzan, eds, Tissue Culture in Forestry. Martinus Nijhoff, The Hague, pp 109-149
6. CHENG TY 1975 Adventitious bud formation in culture of Douglas Fir (Pseudotsuga menziesii [Mirb.] Franco). Plant Sci Lett 5: 97-102
7. CRITCHFIELD WB, EL LITTLE Jr 1966 Geographic Distribution of the Pines of the World. USDA, Washington
8. DALLIMORE W, AB JACKSON 1966 A Handbook of Coniferae and Ginkgoaceae, Ed 4. Edward Arnold, London
9. DAVID A 1982 In vitro propagation of gymnosperms. In JM Bonga, DJ Durzan, eds, Tissue Culture in Forestry. Martinus Nijhoff, The Hague, pp 72-108
10. DURAND-CRESSWELL R, M BOULAY, A FRANCLET 1982 Vegetative propagation of Eucalyptus. In JM Bonga, DJ Durzan, eds, Tissue Culture in Forestry. Martinus Nijhoff, The Hague, pp 150-181
11. DVORAK WS 1983 Strategy for the development of conservation banks and breeding programs for coniferous species from Central America and Mexico. Proc 17 S For Tree Improvement Conf, Athens, Georgia, pp22-29, Paper No 8943, N Carolina Agric Res Serv, Raleigh
12. DVORAK WS 1984 Seed collections of coniferous species in Central America and Mexico: A summary of activities of the CAMCORE cooperative since 1980. Paper No 9250, N Carolina Agric Res Serv, Raleigh
13. FRANCO EO, OJ SCHWARZ 1985 Micropropagation of two conifers: Pinus oocarpa Schiede and Cupressus lusitanica Miller. In RJR Henke, KW Hughes, A Hollander, eds, Tissue Culture if Forestry and Agriculture. Plenum Press, New York, pp 195-213
14. GRESSHOFF PM, CH DOY 1972 Development and differentiation of haploid Lycopersicon esculentum tomato. Planta 107: 161-170
15. MAENE LJ, PC DEBERGH 1981 Preliminary results on organ culture of Araucaria excelsa (Lamb) R. Br. In Proc IUFRO Sect S2 01 5. Int Workshop "In Vitro" Cultivation For Tree Species, Fontainebleau, France, pp 155-157
16. MARTINEZ M 1948 Los Pinos Mexicanos, Ed 2. Ediciones Botas, Mexico
17. MASON E, FW MAINE 1984 Tissue Culture and Micropropagation for Forest Biomass Production: Literature Review. Div Energy, Nat Res Counc Canada, Ottawa
18. MIROV NT 1967 The Genus Pinus. Ronald Press Cy, New York
19. MURASHIGE T, F SKOOG 1962 A revised medium for rapid growth and bioassays with tobacco tissue cultures. Physiol Plant 15: 473-497

20. MYERS N 1984 The Primary Source: Tropical Forests and our Future. WW Thornton & Cy, New York
21. REILLY K, J WASHER 1977 Vegetative propagation of radiata pine by tissue culture: Plantlet formation from embryonic tissue. N Z J For Sci 7: 199-206
22. SCHENK RU, AC HILDERBRANDT 1972 Medium and techniques for induction and growth of monocotyledonous and dicotyledenous plant cell structures. Can J Bot 50: 199-204
23. SOMMER HE, CL BROWN, PP KORMANIC 1975 Differentiation of plantlets in longleaf pine (*Pinus palustris* Mill.) tissue cultured *in vitro*. Bot Gaz 132: 196-200
24. VARGAS HJ 1982 Morfogenesis *in vitro* de *Pinus patula* Schl. et Cham. BS Thesis, Universidad Autonoma Chapingo, Mexico
25. WEBB DB, PJ WOOD, JP SMITH, GS HENMAN 1984 A Guide to Species Selection for Tropical and Sub-Tropical Plantations. Tropical Forestry Papers, No 15, Ed 2, Commonw For Inst, Oxford
26. WEBB DT, O DIAZ SANTIAGO 1983 Cytokinin induced bud formation on Caribbean pine (*Pinus caribeae* Morelet) embryos *in vitro*. Plant Sci Lett 32: 17-21

14. ARAUCARIA

L. MAENE and P.DEBERGH

1. INTRODUCTION

The natural distribution of the two genera, Araucaria and Agathis, belonging to the family of the Araucariaceae is limited to the Southern hemisphere.

In the genus Araucaria, only three species are currently important for timber, i.e., A. angustifolia, A. cunninghamii and A. hunsteinii. These are grown in plantations mainly in the countries of their natural occurrence. A. araucana and A. excelsa are ornamentals which are in high demand in Europe and in the countries around the Mediterranean sea.

The Araucarias are used for structural and general construction, treated poles and piles, plywood, furniture, joinery, moldings, planks, ladders, boats crates and pallets, particle board, hard board and pulp and paper (17, Handro, pers. comm.).

Natural forests of A. angustifolia have been over exploited. In 1964 the timber from these natural forests was still the fourth most important export product in Brazil (16), but today it is available only from commercial sources, i.e., from large plantations in Southern Brazil and Argentina (Handro, pers. comm.).

A. cunninghamii is extensively planted on former rainforest sites in Southern Queensland (Australia). The productivity of this species has been greatly improved through the use of breeding programs in which seed for plantation establishment was obtained from grafted clones of superior genotypes (10).

Plantations of A. hunsteinii are established in different regions of the world (Handro, pers. comm.).

A. araucaria is grown as an outdoor ornamental in Europe, and is mainly propagated by seeds.

In most European countries *A. excelsa* is in high demand as a house plant. For more than 100 years, plants with a peculiar decorative value have been selected for propagation by rooting of cuttings. Over the last decade, however, production rates have been reduced by poor rootability and rotting.

The high value of the forest species and of the ornamental species justifies the efforts to develop large scale clonal propagation of superior genotypes by the use of tissue culture techniques.

2. TISSUE CULTURE OF *ARAUCARIA* SPP.

2.1. Explant choice

The unique feature of species belonging to the *Araucariaceae* is that the orthotropic stems with their bud traces, retain these traits, regardless of any orientation or treatment imposed on them (10, 14).

The aim of clonal propagation is to obtain trees like the parental tree. Explants which are mainly used in the tissue culture experiments with *Araucaria* are isolated from the orthotropic stem of young and older trees. In the case of *A. angustifolia*, the explants were obtained from 30 day-old dark grown seedlings (11), in *A. cunninghamii* from 2 year-old (3) and 18 months-old (10) trees, in *A. araucana* from segments of megagametophytes, with and without embryonic tissue (5).

In our research with *A. excelsa* (13) we attempted to rejuvenate plagiotropic shoots as described in the next section.

2.2. Induction of juvenility

Different approaches were investigated to induce juvenility in adult *Araucaria excelsa*.

The mother-plants in our experiments were 70 to 100 years old. They were regularly pruned for the harvesting of cuttings. The system is comparable with the hedged Radiata pine pruning process proposed by Libby and Hood (12). On these pruned mother-plants we never observed a reversion from plagiotropic to orthotropic growth and the flat disposition of the branches (typical of plagiotropic shoots) never changed.

Another approach to induce juvenility was to spray these mother-plants with a solution of 250 or 500 mg/l N_6-benzyladenine (BA) in 0.1% DMSO, weekly for 8 successive weeks. Within the year following the last treatment no juvenile characteristics were observed. Weekly sprays with 250 mg/l BA were not effective. With 500 mg/l BA, all the treated plants became gradually necrotic without showing any visible reaction. The same approach however seemed to be effective to induce juvenility in other Gymnosperms such as Araucaria cunninghamii (Abo-El-Nil, pers. comm.), Pseudotsuga menziesii and Pinus maritima (Abo-El-Nil, Chaperon and Franclet, mentioned in 9).

In other experiments, the grafting procedure, successfully used at AFOCEL with Eucalyptus camaldulensis and Cupressus dupreziana (7) was investigated with A. excelsa. Because the diameter of orthotropic shoots (6-10 mm) harvested from the mother-plants was larger than that of the seedlings (3 mm) it was necessary to use plagiotropic shoots as grafts. The technique of cascade-grafting as proposed by Franclet (8) was used.

One to two-centimetre-long terminal grafts from plagiotropic branches coming from adult donors were grafted on one-year-old seedlings. They were re-grafted on seedlings as soon as the graft was more or less 5 cm long, with the terminal 1-2 cm being used as graft. After 4 successive graftings an improved growth rate was manifest. The graft became more and more orthotropic, but the disposition of its branches remained in one plane, which is typical for the plagiotropic state (Fig. 1).

These experiments confirm the classical work of Massart (14) who claimed that it was extremely difficult to induce orthotropy in plagiotropic shoots of A. excelsa.

Recently we have grafted small orthotropic epicormic branches, found on the naked stem of some very old mother-plants, on one-year-old seedlings with a success rate of 50%.

The habit of these grafted shoots became more juvenile. The number of branches per whorl diminished and often their implantation on the main stem became helical as in one-year-old seedlings. Explants from this rejuvenated material were more morphogeneti-

FIGURE 1. Induction of juvenility:
right = one time grafted plagiotropic branchlet of *Araucaria excelsa* on a seedling.
left = four times successively grafted plagiotropic branchlet of *Araucaria excelsa* on a seedling. This branchlet is more orthotropic than the one shown on the right.

cally active in stage I, the shoot induction phase *in vitro*, than those from the earlier grafts.

Burrows (4) succeeded in improving the rooting percentage of shoots originating from cultures of *A. cunninghamii* by taking as explants coppice shoots from the base of 20-year-old decapitated trees instead of shoots from non-decapitated trees.

2.3. Preparation of stock plants – Stage 0

No information is available on the preparation of mother-plants for tissue culture except when test tube-germinated seedlings were used (11).

With *A. excelsa* we kept the mother-plants (3-year-old rooted cuttings) for a minimum of 6 months, under very dry conditions (6). This reduced infection, after establishment of the cultures, from about 70 to 20%. Another major advantage of the dry pretreatment was that a milder sterilization procedure could be used: 15 minutes in a 0.5% solution of $HgCl_2$ and teepol, without the previously required supplementary disinfection in a 10% solution of commercial bleach (10°C) for 20 min (13). This weaker sterilization procedure improved the viability of the stage I cultures and 20 to 50% of the stem explants showed swelling of one or more axillary buds.

Since stage 0 conditions are created in a greenhouse, they cannot be applied to tall selected genotypes found in the field. However scions of these can be grafted on seedling-rootstocks, and then exposed to stage 0 conditions in the greenhouse.

2.4. Aseptic initiation – Stage I

Most authors (3, 10, 11, 13) have used a piece of the orthotropic stem with 2-3 leaf axils as explant (Fig. 2). With *A. araucaria* and *A. excelsa* leaves were also used as explants (Table 1).

FIGURE 2. Stage I-culture of *Araucaria excelsa*, 1 month after inoculation, arrows show the beginning of axillary shoot development.

FIGURE 3. Culture of *Araucaria excelsa* on elongation medium after 45 days.

With stem pieces as explant the development of the shoots from leaf axils, which is the preferred development, prevails. Most authors use the MS-medium (15) (Table 1). In our experience with *A. excelsa* the composition of the culture medium is not the

TABLE [illegible]. Survey of the published information on tissue culture of *Araucaria* species.

Araucaria spp.	age	explant	salt form medium	environment conditions	hormones (mg/l)	reaction	author
A. angustifolia	seedling 30 d (dark)	stem with 2-3 leaf axils	MS (1962) mod., or Knop	light or dark	K, IAA, 2,4-D, IBA, NAA, (different combinations)	callus, could not be subcultured	(11)
					K 2 + auxin 0,1	shoots could not be rooted	
A. araucaria	25 years	leaves	MS (1962) mod.	light 9 h 3500 lux 22°C	NAA 10, 2iP or K 2	callus could not be subcultured	(1)
	25 years	stem (1 cm)	idem	idem	2,4-D 3 + 2iP 2	callus could be subcultured	
					IAA or 2,4-D	callus could be subcultured	
		megagametophyte	MS		NAA 10, K 1	callus	(5)
A. cunninghamii	seedling 7 days	hypocotyl	MS 1/2		BA 1-10 µM	adventitious buds	(3)
	2 years	stem	MS 1/2		no pronounced cytokinine effect	axillary shoots	
	20 years	coppice shoots	MS 1/2		idem	axillary shoots	
					IBA 10 µM	rooting	(4)
	seedling 18 mo	stem 4 mm upper part	de Fossard (1974)	light 12 h 25°C	different	axillary shoots, rooting 50%	(10)
A. excelsa	70-100 y	stem 2 cm	different	dark 18°C 2 wk, light 16 h, 2000 lux, 24°C	different	shoots	(13)
					0.2% charcoal + 2,4-D 1	elongation	
		leaves	MS	idem	2iP 0.01 + polyamines	callus	
A. balansae *A. birdulii* *A. columnaris* *A. hunsteinii* *A. luxurians* *A. montana* *A. rulei* *A. scopulorum* *Agathis robusta*	seedling 2 y	stem 2 cm	MS 1/2			axillary shoots	(4)

principal factor controlling caulogenesis. On different salt-formulations the best response was obtained when explants were placed horizontally on the culture medium and when the stem pieces were mature enough to have a large diameter. The larger the diameter, the more buds emerged. The basal explants from the stem were more competent than apical ones.

The problem of browning of the explant and/or of the culture medium was mentioned by Handro and Ferreira (11). Browning is more pronounced in the light than in the dark. We encountered the same problem with *A. excelsa* and were able to overcome it by keeping the cultures from the first two weeks in the dark at 18°C. Afterwards the cultures were incubated under a 16 h photoperiod at about 30 μmol m^{-1} s^{-1} at 24°C.

The induction of callus on needles of *A. excelsa* required a mixture of polyamines (spermidine 0.1 mM + spermine 0.025 mM + putrescine 0.025 mM). BA and kinetin (K) (0.01-5 mg/l) were toxic, but 6-(γ,γ-dimethylallylamino)-purine (2iP) was beneficial at 0.01 mg/l in the presence of 4% sucrose. However caulogenesis was never observed.

2.5. Multiplication – Stage II

Burrows (3) obtained a low rate of multiplication after reculturing stem segments of *A. cunninghamii* that were excised from the initial shoot explants after their elongation *in vitro*. At most 60 to 70 shoots, of a size suitable for rooting were produced from a single seedling in four months.

Neither Handro and Ferreira (11) with *A. angustifolia*, nor Haines and De Fossard (10) with *A. cunninghamii*, were able to multiply shoots initiated in stage I.

We also did not succeed to subculture the shoots we obtained from stage I-cultures of *A. excelsa*. When axillary shoots, on the initial explant, reached a length of 0.5-1 cm after about 2.5 months, their base generally turned brown, and the shoots died, even after subculture on various media. The time, which elapsed before the browning appeared, depended on the size of the initial explant. The larger the explant, the longer it took. Necrosis appeared also after transplantation, irrespective of the size of the shoots and the media used.

Therefore, we discontinued the use of multiplication media and tried instead to induce elongation of axillary buds. We tried 1) a sequence of cytokinin containing and cytokinin free media, with or without charcoal as suggested by Boulay (2); 2) antioxidants and complexforming products (ethylenediaminetetraacetic acid, dimethyl sulfoxide); 3) different salt formulations; 4) different growth regulators; 5) manipulation of the water potential of the medium by increasing the sucrose (or mannitol) and/or the agar concentration. In a few cases the axillary buds elongated to about 5 cm-long shoots on a Whites' medium + charcoal 0.2% + 2,4-D 1 mg/l. Charcoal was beneficial but not essential. The addition of Na_2EDTA 20 mg/l to the usual 25 mg NaFeEDTA, and of thiourea 0.1 mg/l and glutamine 100 mg/l improved the elongation and the survival of the shoots. The elongated shoots resembled a seedling with the branches in a helical disposition (Fig. 3).

2.6. Rooting

Rooting of shoots produced *in vitro* has only been reported for *A. cunninghamii*. Haines and De Fossard (10) rooted 3 out of 6 explants on a medium with 6 μ M IBA and Burrows (3) obtained 40% rooting after first placing the shoots for 1 month on a medium with 10 μM IBA and then on a perlite and vermiculite mix (1:1) under high humidity (Table 1).

3. CONCLUSION

Except for *A. cunninghamii*, where rooting of tissue cultured shoots was achieved, micropopagation of *Araucaria* has not yet been possible. The Araucarias obviously are difficult to handle *in vitro*, and more research is required before micropropagation of most species will become reality.

4. REFERENCES

1. BENIEST J, P DEBERGH 1976 Nutritional and hormonal requirements for the growth of *Araucaria araucana* callus *in vitro*. Med Fac Landbouww Rijksuniv Gent 41: 1599-1610
2. BOULAY M 1978 Proc Round Table Conf "In Vitro" Multiplication Woody Species, Gembloux, Belgium, June 6-8 1978
3. BURROWS GE 1981 *In vitro* culture of hoop pine (*Araucaria cunninghamii* Aiton ex Don). Queensl Dep For Res Rep 3: 36-37

4. BURROWS GE 1983 Organ culture of some species of the Araucariaceae. Anatomical aspects of bud development. Proc 2nd Aust Plant Tissue Culture Conf, Sydney, pp18
5. CARDEMIL L, M JORDAN 1982 Light and electron microscopic study of *in vitro* cultured female gametophyte of *Araucaria araucana* (Mol.) Koch. Z Pflanzenphysiol 107: 329-338
6. DEBERGH P, L MAENE 1981 A scheme for commercial propagation of ornamental plants by tissue culture. Sci Hortic 14: 335-345
7. FRANCLET A 1969 Vers une production de masse de Cyprès de forme controlée. Proc 2nd FAO/IUFRO World Consult For Tree Breeding, Washington FO/FTB, 69: 11-18
8. FRANCLET A 1980 Rajeunissement et propagation végétative des ligneux. Ann Rech Sylvicol, AFOCEL France, pp 11-41
9. FRANCLET A 1981 Rajeunissement et micro-propagation des ligneux. *In* Proc IUFRO sect S2 01 5. Int Workshop "In Vitro" Cultivation For Tree Species, Fontainebleau, France, pp 55-64
10. HAINES RJ, RA DE FOSSARD 1977 Propagation of Hoop pine (*Araucaria cunninghamii* Ait.) by organ culture. Acta Hortic 78: 297-302
11. HANDRO W, CM FERREIRA 1980 Preliminary report on tissue culture of *Araucaria angustifolia*. Bol Bot Univ S Paulo 8: 71-84
12. LIBBY WJ, JV HOOD 1976 Juvenility in hedged Radiata Pine. Acta Hortic 56: 91
13. MAENE LJ, PC DEBERGH 1981 Preliminary results on organ culture of *Araucaria excelsa* (Lamb.) R.Br. *In* Proc IUFRO sect S2 01 5. Int Workshop "In Vitro" Cultivation For Tree Species, Fontainebleau, France, pp 155-157
14. MASSART J 1924 La coopération et le conflit des réflexes qui déterminent la forme du corps chez *Araucaria excelsa* R.Br. Acad R Belg, Cl Sc Mém Collect 4° Deux Sér - T V Fasc 8, pp 32
15. MURASHIGE T, F SKOOG 1962 A revised medium for rapid growth and bioassays with tobacco tissue cultures. Physiol Plant 15: 473-497
16. REITZ R, RM KLEIN 1966 Araucariaceas. Flora Illustrada Catarinense, Itajai, Brasil
17. Smith WJ 1980 The wood utilization potential of the Araucarias. *In* Forestry Problems of the Genus *Araucaria*. Proc IUFRO Meeting FUPEF, Curitiba, Brazil, pp 265-285

ACKNOWLEDGEMENTS

The authors thank Dr. Handro for sending an un-edited manuscript, entitled Tissue culture of *Araucaria*: a review.

15. CONIFER MICROPROPAGATION : APPLIED RESEARCH AND COMMERCIAL ASPECTS

M. BOULAY

1. INTRODUCTION

Ten years ago, the main method to produce plants of coniferous species for reforestation was by seedling production. Only Cryptomeria japonica, Thuya sp. and Cupressocyparis leylandii and some other ornamental species were propagated by rooted cuttings. However, evolution of vegetative multiplication techniques of woody plants has altered the situation. Today, numerous species of coniferous trees are propagated by rooted cuttings, among them : Picea abies, Picea glauca, Picea sitchensis, Larix sp., Sequoia sempervirens, Sequoiadendron giganteum, etc. Often the starting material is selected juvenile seedlings, but in some cases (Sequoia sempervirens, Pinus radiata, etc.) selection and vegetative multiplication of adult trees is possible (we consider trees older than 7 or 8 years to be sufficiently adult for selection). Difficulties linked with vegetative propagation of adult genotypes are correlated with maturation of trees. This is especially true for coniferous species. Generally, cuttings taken from seedlings or young plants root easily and manifest a growth habit and vigor characteristic of plants derived from seeds. Nevertheless with some species, even at this very young stage, problems are encountered in vegetative propagation. Classical examples are Pseudotsuga menziesii, Abies sp., and Araucaria sp. When cuttings are taken from adult material and rooted, the vigor of these plants is poor and they are often plagiotropic (see chapter on rejuvenation). In some species with certain treatments (pruning for example) it is possible to regain eventually an orthotropic growth habit.

Because rooting of cuttings is often limited to seedlings or young plants, its industrial application has not been well developed. The lack of knowledge of the genetic value of a clone (often selected in the nursery for only one criterium e.g. growth vigor) demands that a great number of clones are selected at the beginning of the program. Afterwards, clonal tests are necessary to eliminate, from the initial population, genotypes that may express undesirable traits with time. All clones should be kept at a juvenile stage for the mass production of superior genotypes chosen after years of field testing. The experience of forest geneticists in clonal silviculture (Zobel, 1981 ; Libby, 1983) and the demonstration of production gains with eucalyptus species in Congo and Brazil have encouraged some forest companies to develop such programs. Coniferous species have been rooted at a commercial scale in Europe, North America and some other regions.

Since 1974, plants as different as *Fragaria*, *Gerbera*, *Ficus*, *Rosa*, *Orchis*, fruit tree rootstocks and fruit tree scions have been partially or totally propagated by sterile micropropagation methods. In these last ten years, numerous commercial laboratories have been built for these species. Concerning forest tree species, the use of *in vitro* methods for vegetative propagation has taken a slow start. A major reason for this delay is no doubt because forest nurseries are traditionally oriented more towards seedling production than towards rooted cuttings. Furthermore, the low individual price of reforestation plants as compared to the market price of ornamental plants, the lack of selected clones and technical difficulties are also responsible.

Nevertheless, more and more deciduous trees have been propagated on a large scale using *in vitro* methods. In coniferous species, commercial applications are still limited. For that reason, this paper will not be limited to industrial propagation, but will also cover appplied research conducted at university and research institute labs. Coniferous species have a potential market superior to that of deciduous trees, particularly in Europe and North America where they are commonly used for reforestation.

I will discuss species for which commercial *in vitro* production is in progress or projected. I will also describe small scale trials with conifers. Such trials are often necessary to determine such parameters as price, reliability of method, field performance, etc ... which are important limitations in the commercialization of *in vitro* methods for coniferous species. Finally, I will list the work in progress in some research labs and nurseries working in this field.

2. METHODS USED AND COMMON APPLICATIONS

Commercial application of *in vitro* methods is currently restricted to micropropagation by adventitious and/or axillary budding. Other methods such as tissue, cell or protoplast culture, callus budding and somatic embryogenesis, are still at the research level. The main reason why commercial applications are limited to micropropagation is because in most cases the production of true to type genetic copies is intended.

In coniferous species, genetic improvement often starts with provenance selection, elite tree selection and then seed orchard establishment. Many problems can be encountered in this scheme (graft incompatibility, the dependance of fructification on climate, cones being damaged by insects, space requirements, etc). When these problems raise the price of seeds and limit the genetic gain, the *in vitro* cloning of progeny from sexual crosses has great value. However, the most interesting application would be the cloning of trees selected at an adult stage either in natural stands or clonal tests. The possibility of propagation out of season, and some other advantages described below, make micropropagation an important tool for geneticists and nurserymen.

2.1. Embryo or seedling cloning

In coniferous species, adventitious budding on the entire embryo or parts thereof is a reliable technique and often allows a multiplication rate as high as 700 to 2,000 copies per year per embryo. That is about a 10 to 400 fold increase over the rates obtained with rooted cuttings.

Such micropropagation is useful for :

1) cloning selected provenances, which is important in the case of seed harvesting problems (lack of fructification, etc), and where seed germination is poor (e.g. Sequoia sempervirens) ;
2) cloning embryos obtained from controlled pollination (intra -families, inter-clonal, inter-provenances). Because micropropagation reduces the number of seeds required, one can reduce the number of time consuming control crosses ;
3) cloning embryos obtained from interspecific crosses. This could result in hybrid vigour, hybrids with a broader environmental adaptation or hybrids with a good mixing of genetic traits (see Pinus rigida x Pinus taeda, (21) ; Pinus palustris x Pinus elliottii, (36) ;
4) evaluating breeding value (e.g. the capacity of cotyledon fragments of Douglas fir to form adventitious buds is strongly linked to the later vigor of the clone (1)).

Mass production of propagules by micropropagation can also help the breeder to determine microsite effects, competition among trees and intra family variation. Such cloning avoids genotype variation, which often partially masks research results obtained with sexually produced seedlings.

Finally it allows screening for disease and frost or salt tolerance (16, 20).

2.2. Cloning young and adult genotypes

Maturation is one of the most important problems blocking the rooting of cuttings of adult genotypes. Pretreatment of stock trees is often necessary to reactivate portions of a selected genotype (see chapter on rejuvenation). These reactivated ramets are often obtained only in small quantities and, therefore, these techniques are uneconomical unless a system of mass propagation follows. Micropropagation may permit a faster multiplication of this material than rooting of cuttings.

One could also micropropagate plant parts which retain juvenile characteristics in the adult genotype (e.g. basal sprouts of Sequoia sempervirens, epicormic shoots of numerous coniferous species).

Adventitious budding permits the formation of vegetative

meristems on plant organs such as leaves, *Sequoia sempervirens* (3, 18) and leaf primordia, *Picea abies* (37) ; *Pinus sylvestris* (6) ; *Picea pungens* (29). Budding on leaves or dormant buds are rejuvenation methods only possible by *in vitro* techniques. However, it is yet to be established whether the reversion from adult to juvenile is total in adventitious budding.

Micropropagation methods can serve as rejuvenation pretreatments. For example, in 1978, we described (8) the rejuvenation of mature *Sequoia sempervirens* plants by successive subcultures of axillary buds on cytokinin medium. Explants from juvenile parts on an adult tree (basal sprouts), needed three or more subcultures, depending on the genotype, to reach an exponential multiplication rate. Following this work, La Goublaye (24) showed that mean rooting time, growth vigor and plagiotropic habit of *in vitro* plantlets derived from sprouts of a 90 year old *Sequoia sempervirens* tree are correlated with the number of subcultures on cytokinin media. One year after transfer to soil, various dfferences (in mean weight of stems, mean number of orthotropic shoots, rooting ability of cuttings taken from *in vitro* copies) were noted that correlated with the number of subcultures. Fouret (18), using the same species but other criteria, also came to the conclusion that rejuvenation can be achieved by *in vitro* subculture. Using *in vitro* techniques, one could also use plant organs which do not develop into whole plants with other techniques, e.g. meristems (5, 8, 38) or pine brachyblasts (13).

One last application concerns the selection of young genotypes and their conservation under minimal growth conditions until the result of clonal tests are known. With *Sequoia sempervirens*, it is possible to conserve sterile axillary bud cultures at 1°C in the dark for 1 to 2 years. At the end of cold storage, these genotypes can be mass propagated, because the rooting ability of these shoots did not diminish in storage. Other techniques of keeping clones at a juvenile stage (serial propagation, hedging, etc ...) are time consuming and not always effective. Cold storage has been undertaken in New Zealand with *Pinus radiata* micropropagation (see chapter by Aitken in Volume 2). It is intended to store clones at low temperatures *in vitro* for the 5 year duration of field tests of the clones. We are presently investigating if *Sequoia sempervirens* germ plasm can be maintained by cryopreservation (in liquid nitrogen) of meristems of selected trees.

3. SPECIES PROPAGATED ON A SMALL TO LARGE COMMERCIAL SCALE

3.1. *Sequoia sempervirens*

As early as 1950, Ball (3) established continuous cultures of this species. He obtained callus that showed some differentiation. In 1964, Choulet *et al.* (11) obtained callus with bud differentiation. After these preliminary studies, Ball (3) and Boulay (8) regenerated complete plants. At AFOCEL, we are interested in the *in vitro* multiplication of *Sequoia* for the rapid cloning of selected genotypes introduced in France 50 to 100 years ago. Vegetative multiplication by rooted cuttings is possible although there are maturation related problems with this method. We have tried to solve these problems by *in vitro* reactivation.

3.1.1. *Micropropagation techniques used in AFOCEL*. Basal sprouts are the best explant source. If such material is not available, branch segments can be used, although with these the budding rate is low at the beginning of the culture and more time is needed to obtain exponential multiplication. Pieces of stems (10 cm long) are disinfected by a dip in a calcium or sodium hypochlorite solution and then rinsed with sterile water. These stems are cut into 10-20 mm long explants with 1-4 needles. The length of disinfection treatment (5-10 min) depends upon the state of the plant material, i.e. whether it is herbaceous or more lignified. The percentage of cultures that are free of contamination is low if field material is used (10 to 40 %). If explants are taken from rooted cuttings or grafts cultivated in a nursery, the percentage of non-contaminated cultures is from 80 to 90 %. The dissection and introduction of meristems, although time consuming, can result in cultures that are 100 % contaminant free.

In vitro multiplication is based on axillary budding on a Murashige and Skoog (MS) (macro-elements half strength, micro-elements full strength) medium supplemented with a simplified vitamin mixture and N^6-benzylaminopurine (BAP) at 0,5 to 1 mg/l and α-naphthaleneacetic acid (NAA) at 0,01 mg/l (8). The appropriate balance of hormones depends on the genotype and the number of transfers. After one month on this medium *in vitro* formed buds are planted on an elongation medium, singly or in clusters, depending upon their height (buds longer than 10 mm are planted singly). The mineral and vitamin composition of this medium is the same as that of the budding medium except that activated charcoal is added (1 to 10 g/l) and that it contains no hormones. After one month, a five fold or more elongation is obtained. Subculturing is done by the division of the elongated shoots into 10-20 mm long pieces and their transfer to the budding medium. This alternation between budding and elongation is the basis of mass propagation. Cuttings are rooted under *in vivo* conditions as described by Poissonnier *et al* (32). First the growth of the *in vitro* shoots is slowed down and lignification is encouraged before they are transfered to the greenhouse. While the shoots formed *in vitro* are still on sterile elongation medium, they are placed for 8-15 days at 10-15°C under natural light. Subsequently, the base of the 5-8 cm long shoots *in vitro*, are dipped for 24 hours in 5 mg/l of exuberone (commercial solution based on indolbutyric acid (IBA) and 125 mg/l of benlate (commercial fungicide based on benomyl). A quick dip in hormone powder (rhizopon 1 %) can also be used. Rooting takes place in a confined space ; mist is added only during the summer. Heating of the soil (20 to 22°C) and additional light (16 hours of day) are necessary during winter. A mixture of perlite-peat-pine bark in 2:1:2 V/V proportion is used as the substrate. The rooting percentage varies from 75 to 150 % depending upon the season and genotype. Tall shoots can be cut in two before rooting.

After 4 to 6 weeks, well rooted shoots are planted in Melfert containers (Melfert : patented containers filled with a mixture of a peat-pine bark 1:2 V/V, lignite ashes 10 % and osmocote, a 8-9 months slow release fertilizer N.P.K. 18:11:10). The plants are moved to less humid conditions but kept under plastic film for acclimatization. After 10-15 days roots grow through the sides of the containers and the plants can be cultivated under standard nursery conditions.

At the level of the *in vitro* method, there are two problems : contaminants and hyperhydric transformation (waterlogging).

Some contaminants arrive at a late stage of culture. Bacteria are major contaminants (9, 39). Bacteria cause little damage to the shoots of *Sequoia* if they are rooted under *in vivo* conditions. Nevertheless, the possibility of losing cultures in an advanced stage of multiplication to contamination has lead us to use different plant organs as explant source, e.g. meristems (9, 38). In meristem culture, the incidence of contaminants is low and they can be screened easily at the beginning of the multiplication process.

Waterlogging is infrequent if multiplication is alternated with elongation. In cases in which successive transfers on multiplication media are needed (e.g. clone reactivation), waterlogging can be a problem.

Rooting *in vivo* was originally choosen over rooting *in vitro* because the later gave poor results. However, lately the work of Bekkaoui (5) has solved some problems of *in vitro* rooting. *In vivo* rooting is often choosen in commercial propagation because it is more economic and acclimatization is easier.

Plagiotropic growth of transplanted plantlets can be a problem. This plagiotropism depends on the clone and is often only of short duration. Orthotropic growth is generally restored one year after *in vitro* plants are well established, or, for some difficult cases, after cutting back. Plagiotropic growth is common during the *in vivo* rooting process if the *in vitro* shoots are too tall or too succulent.

Nevertheless, our oldest field test (five years old) demonstrates that there are few differences between seedlings, rooted cuttings, and *in vitro* produced plantlets. To date about 50,000 micropropagated copies from 120 selected adult genotypes have been produced. In 1985, another 50,000 *in vitro* plantlets will be produced. These plants, once rooted, are used directly for reforestation. The price of acclimated plants is around 0.50 US $. During the first commercial application of our methods, we found that the quality of *in vitro* shoots produced on the last elongation medium has a determining influence on the rooting percentage. The number of shoots per jar (750 ml) must be no more than 30. Etiolation occurs with a higher shoot density, leading to weak and difficult to root cuttings. Pretreatment at 10-15°C and a shoot length of at least 5 cm, are essential for good rooting.

3.1.2. Other commercial propagation. Georgia Pacific Timber Co., using Dr H.J. Burkhardt's lab in California, has also begun commercial propagation of *Sequoia sempervirens* using micropropagation methods. The program began in 1984, and in 1985 they intend to produce 36,000 plantlets (3,000 plants per month of selected clones). The *in vitro* procedures are combinations of methods described by Ball (3) and Boulay (8) and include modifications made during 2 years of private research. The price of each *in vitro* shoot (in the jar, ready for rooting) is estimated at 0.50 U.S. $. Research is in progress to decrease this price.

Shoots are rooted *in vivo* in a greenhouse with 6 compartments that permit varied mist regulation. Before *in vivo* rooting, shoots are pretreated as described by Poisonnier *et al* (32) and dipped in IBA.

According to Dr H.J. Burkhardt, one of the most important prerequisites for success is humidity control. During the first weeks 90 to 95 % humidity is maintained ; after 3 to 4 weeks, humidity is decreased to 70-75 %. A commercial rooting substrate (Forestry mix) and bottom heat is used. Rooting takes place in 2 to 3 months and the percentage (depending on the clone) is from 50 to 100 %.

One goal is to supply the Company with plants for reforestation and with donor plants that will provide cuttings. Georgia Pacific has, for some years, been using rooted cuttings produced by Dr Libby's techniques (26). However, according to Dr H.J. Burkhardt, these rooted cuttings remain plagiotropic for 3 to 5 years, and therefore, are not acceptable for reforestation. Restitution of orthotropic growth on rooted cuttings requires pruning. In Dr H.J. Burkhardt's lab, meristems of shoots taken from the upper part of the crown of old genotypes, have been cultured *in vitro*. These cultures have produced plants that have orthotropic growth.

Another commercial application has been undertaken by Simpson Timber Co. in California, in collaboration with Dr E. Ball's lab at Santa Cruz.

3.2. Sequoiadendron giganteum

Originating from the Sierra Nevada mountains in California, this species does not regenerate by root sprouts as does *Sequoia sempervirens*. Micropropagation methods designed for this species are used mainly for ornamental propagation.

3.2.1. Work at AFOCEL. One clone of *Sequoiadendron giganteum* cv. *pendula* has been multiplied to nearly 1,200 copies, by methods similar to those used for *Sequoia sempervirens*. The multiplication and elongation media are similar for both species. Compared to *Sequoia*, the multiplication rate is lower mainly because the axillary budding rate is poor. Two to three buds are formed on each stem piece in one month. Elongated *in vitro* shoots are rooted *in vivo*. There is no need for pretreatment because the *in vitro* shoots are sufficiently lignified for good rooting. The shoots are dipped in a rooting powder (rhizopon 1 %) and root within 2 to 3 months. Acclimatization and culture conditions are similar to those of *Sequoia sempervirens*.

After two years of growth, the *in vitro* plantlets resemble seedlings, and not the weeping form of the *pendulum* cultivar that is traditionally propagated by grafting scions onto *Sequoia sempervirens* rootstocks. We are trying to rejuvenate and mass propagate selected adult genotypes by *in vitro* methods (30).

3.2.2. Other commercial propagation. In Clay's nurseries in Canada, Dr M.L. Kurz initiated cultures in 1981. Axillary budding of stem pieces was readily achieved, but rooting of *in vitro* shoots has given problems similar to those with rooted cuttings, i.e. the roots elongate poorly.

The *in vitro* methods include multiplication on a modified Anderson's rhododendron medium with N^6-isopentenyl adenine (2iP) (2 to 5 mg/l) and indoleacetic acid (IAA) (0,1 to 1 mg/l). Cultures are transferred every two months, showing a 4 fold multiplication in that time. Rooting is conducted *in vivo* using the following procedures. *In vitro* shoots are dipped in a commercial hormone mixture (Seradix) and planted in trays. Each rooting tray includes 265 shoots and the rooting substrate is a peat-sand mixture. Trays are put in a greenhouse equiped with a mist system (4 second cycle every 32 minutes during cold days and every 8 minutes during hot days). The soil temperature is maintained at 20°C. After 8 to 10 weeks under mist, the plants are acclimatized to regular greenhouse conditions for 3 months. At the end of this period, the plantlets are potted in small pots and fertilized every week with a 10-52-10 formulation.

In comparison with normal cuttings, Dr M.L. Kurz reports that *in vitro* plantlets have a better form and growth rate. Plants are produced for the market.

Microplants nursery, in England, is also propagating giant sequoia *in vitro* for the ornamental market. Dr Skokes began this production in 1984. He produces microcuttings from axillary bud cultures. The basal medium is similar to Murashige and Skoog with BAP as cytokinin and IBA as auxin. Rooting is done *in vivo* under mist in a rooting substrate that is based on peat. Rooting requires additional fluorescent lighting (16 hours/day) during winter. Several thousands of plants are produced on a year-round basis.

3.3. *Pinus species*.

Micropropagation of numerous pine species has been studied for many years. Pioneer work was done by Dr H. Sommer *et al* (35) on *Pinus palustris*. The system involves adventitious budding on excised embryos or parts of excised embryos and has since been applied to numerous other coniferous species. With this method seeds of high genetic value are propagated . An alternative method developed by David *et al* (13), using brachyblasts (secondary needles) as the explant source, has been used for the propagation of young seedlings or trees.

Three pine species have been commercially propagated through *in vitro* methods : *Pinus radiata* in New Zealand, *Pinus taeda* in North America and *Pinus pinaster* at a small scale in France.

3.3.1. *Pinus radiata*. Research on this species was done first in Canada and North America and later in New Zealand. Multiplication is based on adventitious budding of excised cotyledons (2) (see chapter by Horgan in this volume), and is used commercially for mass propagation of selected crosses.

In 1984, Dr J. Gleed of the Tasman Forestry LTD began to produce *Pinus radiata* plantlets *in vitro*. A production of 200,000 plants is intended by September 1985 to be used for reforestation. In 1986, they plan to reach a production of 2 million plants annually. In brief, the propagation method (2) is as follows : adventitious buds are initiated on excised cotyledons, neoformed buds are elongated and multiplied by axillary budding, shoots are preconditionned for rooting, roots are induced *in vitro*, and plantlets are potted in non-sterile soil and acclimatized. In December 1984, Dr J. Gleed obtained 54,000 bud clusters in culture ; he intends to produce 5,000 plants per clone.

Similarly, the Forest Research Institute of New Zealand has produced since 1981, several thousands of *in vitro* plantlets (2). Dr D. Smith (New Zealand Forest Institute, Rotorua) states that radiata pine propagated *in vitro* appears, after 3 years in the field, to be more mature than normal 1 + 0 seedling planting stock.

Another commercial production has been undertaken at Topline nurseries in New Zealand.

An advantage of micropropagation is that clones can easily be cold stored until field test results are known (see chapter by Aitken in this book).

Multiplication of plantlets from embryos, is not the only way to clone radiata pine *in vitro*. Using a method similar to the one developed by David *et al* (13) for *Pinus pinaster*, Dr J. Aitken has begun the *in vitro* multiplication of 1 year old seedlings, using secondary needles as explants. These explants produce axillary shoots on a modified Lepoivre medium (2) containing 5 mg/l BAP. For elongation, the same medium (without BAP) is used. Rooting will be done *in vivo* by the same method as previously described. In 1984-1985, 200 clones, from controlled pollinations, were multiplied *in vitro* and then transferred to cold storage. Field testing is presently carried out with these same clones. After 6-8 years of testing, selected clones will be taken from cold storage for large scale propagation.

Propagation of mature trees is still a problem. So far 6,000 radiata pine shoots have been produced from 22 clones (4, 8, 12, 20 and 50 years old), and a few of these shoots have rooted.

3.3.2. *Pinus taeda*. In this species micropropagation has been carried out with embryos from seeds. This method (28, 31) is based on adventitious budding on excised cotyledons. In 1980, researchers from North Carolina State produced around 2,000 *in vitro* plantlets. In 1980, The International Paper Company produced several hundreds of plants. Dr Anderson of Buckeye Cellulose also works with loblolly pine tissue cultures.

Leach (25), compared conventional seedlings and *in vitro* plantlets, and found that after 6 months of greenhouse growth *in vitro* plantlets had a greater mortality (38 %) and less shoot growth (50%) than conventional seedlings. Mc Keand *et al* (27), following growth of these two kinds of plants, concluded that differences in height were mainly due to differences in root morphology and not in root physiology. Root systems of *in vitro* plantlets were less branched, and had no tap roots, which leads to a difference in the ion absorption capacity. These data are very important because so few details are known regarding nursery or field growth of *in vitro* plantlets.

Research on *in vitro* propagation of mature loblolly pine has been carried out by the Weyerhaeuser Company in Tacoma, Washington. They patented a method for asexual reproduction in which trees *in vivo* are sprayed with cytokinin, giving rise to juvenile looking shoots. Nevertheless, to my knowledge, the system has not been applied at a commercial level for the mass production *in vitro* of adult genotypes.

3.3.3. *Pinus pinaster*. This important reforestation species for south western France has been studied at University of Bordeaux. Two processes of *in vitro* multiplication have been developed.

The first one is used for the *in vitro* propagation of embryos from seeds germinated under sterile conditions. As opposed to other pine species, adventitious budding, while possible (15) is not very efficient in *Pinus pinaster*. Axillary budding of young stems is easier (13, 14). In addition to the buds from the axils of primary needles on young stem pieces, shoots contained within secondary needle fascicles (brachyblasts) are also used *in vitro*.

In the second method, the brachyblasts are placed on a cytokinin medium and then transfered to an elongation medium. Elongated shoots from the brachyblasts can then be multiplied by axillary budding.

For rhizogenesis, a reliable *in vitro* method has been developed (33). For *in vivo* rooting, we dip shoots, elongated on an activated charcoal medium, in a commercial auxin powder (Rootone). Rooting, acclimatization and later cultivation take place under the same conditions as were used for *Sequoia sempervirens* (see 3.1.). Well rooted plantlets are obtained after two months. So far 2,000 *in vitro* plantlets have been produced by this method. Field tests show that three years after planting, the growth of *in vitro* plantlets is comparable to that of normal seedlings or rooted cuttings of young donor plants.

A small number of plants have been regenerated from 3 to 4 year old plants and adult genotypes (11 years old) (19). After one year of growth, these *in vitro* plantlets look more like rooted cuttings of adult trees (poor vigor, rythmic growth, early appearance of secondary needles), than like normal seedlings.

3.4. *Pseudotsuga menziesii*

Cheng (10) and Reilly and Brown (34) were the first to propagate Douglas fir through adventitious budding on excised embryos or cotyledons. Since then, Weyerhaeuser Company has started micropropagation of that species. Seeds from selected crosses were used as starting material. In 1980, 3,000 *in vitro* plantlets were produced and 5,800 in 1982. The present production is not known. Some problems have been encountered with the *in vitro* plantlets of Douglas fir. In 1981, Weyerhaeuser reported the results from a 2-year-old field test (personal communication). Some *in vitro* plantlets showed growth and vigor comparable to that of normal seedlings while others were plagiotropic, slow growing, and resembling shrubs.

In our institute, using another method (axillary budding on young stems from sterile germinated seeds) (7), the same results were obtained. Nevertheless, 3 or 4 years after field planting, most of the plantlets showed normal growth and vigor.

With regards to the propagation of old genotypes, research continues. At AFOCEL, F. Bekkaoui (personal communication) using pretreated mother trees (23) failed to propagate adult material at an exponential rate. Differences in hormone and Ca/K ratios, in reactivation of meristems, and in peroxidase between young, old and pretreated trees are presently under investigation.

At Clay's nurseries, in Canada, Dr M.L. Kurz is also doing research on old material. Using dormant buds in *in vitro* culture, he is trying to determine the best time of the year for harvesting buds, and the best procedures for transfering cultures from a medium with hormones to a hormone free one. On modified Anderson's and Murashige and Skoog's medium, a high percentage of buds, collected in March just before flushing, elongated or gave rise to adventitious shoots. Before transfer to hormone free medium, cytokinin was supplied for 6 weeks. Shoot induction was obtained without callusing.

Whenever the buds *in vitro* developed normally, they had one growth flush followed by axillary budding. The shoots grew for about 120 days and reached a height of 5 to 13 cm. They became dormant at the end of the summer. Maintaining the photoperiod at 18 hours during the dormant period prevented browning of needle tissue at the base of the shoots. Active growth resumed at the end of the fall.

Dr P. Evers in Wageningen (Netherlands) is studying similar problems.

3.5. *Thuya plicata and occidentalis*

Coleman and Thorpe (12) have described *in vitro* plant generation from juvenile and mature tissues of western red cedar (*Thuya plicata* Donn.) by adventitious budding.

In Clay's nurseries, Dr M.L. Kurz has produced 300 *in vitro* plantlets of _Thuya occidentalis_ var. (Smaragd). _In vitro_ multiplication is obtained by alternating modified Anderson's medium without hormones and modified Murashige and Skoog's medium (1/4 strength) with IBA (0.1 -1 mg/l). Explants are stored in tubes and are only mass propagated when there is a demand. _In vivo_ rooting, acclimatization and cultivation procedures are the same as those described for _Sequoiadendron giganteum_ (see 3.2.). The potting medium is a peat-perlite mixture (40:60 V/V). These plants are sold mostly for ornamental use. According to Dr M.L. Kurz, _in vitro_ plantlets have a better form and look more juvenile than rooted cuttings.

3.6. _Cunninghamia lanceolata_

Starting from experiments first carried out at AFOCEL (Boulay, unpublished data), Engelmann (17) has propagated two clones, one selected from a one year old seedling and the other from an adult genotype. The clones are multiplied by axillary budding of stem pieces (for more details see chapter by Bigot and Engelmann in this book).

In a test of 1,000 shoots, the rooting percentage _in vitro_ was 70 % for the younger clone and 53 % for the older one. For rooting _in vivo_, shoots were dipped for 5 hours in an IBA (1 mg/l) solution and were then placed for 24 hours in a humid chamber. This resulted in 25 % rooting. After 4 weeks in a rooting medium under mist, rooted plantlets were acclimatized, repotted in containers with a pine bark-peat mixture, and placed in a greenhouse at 24°C.

Engelmann and Bigot observed that the juvenile clones produced two kinds of copies. The first kind looked like seedlings, the second kind was heavily branched and lacked a leader. All copies of the older clone were plagiotropic. The rooting percentage improved during successive _in vitro_ transfers. None of the treatments applied to the mature clone gave rise to orthotropic growth, but some propagules of the mature clone displayed morphological characteristics typical of the juvenile form.

4. OTHER ATTEMPS AT MASS PROPAGATION

Many research labs have attempted to mass propagate numerous species.

4.1. European countries

In Sweden, Von Arnold is working with Picea abies, Pinus contorta and Pinus sylvestris. Rooted plantlets were produced from excised embryos or part of embryos by adventitious budding. Plantlets were not produced on a large scale but their growth was comparable to nursery seedlings. Buds taken from adult trees gave rise to adventitious buds that elongated, rooted and regenerated plantlets.

The Hillesög Company, is also attempting to mass propagate Picea abies and Pinus sylvestris through in vitro culture.

In Umea, at the Forest Genetics Institute, Pinus sylvestris and Pinus contorta are being propagated by David's system (i.e. : brachyblast culture, shoot elongation, axillary budding, elongation, rooting). Explants are taken from young seedlings 1 or 2 years old. Multiplication takes place on a modified MS medium with BAP and IBA. After elongation, in vitro shoots are transferred to a half strength MS medium with IBA. The shoots are then transferred to an inorganic nutrient solution. Once rooted, the plantlets are potted in peat and cultivated under short day and low temperature to induce bud dormancy. After this resting period, the plantlets are cultivated as seedlings.

Field testing has shown that in vitro plantlets have a root system less symetric and less stable than that of seedlings. Nevertheless, plagiotropic growth has not occured.

In Belgium, Debergh (see chapter in this volume) is attempting to produce donor plants of Araucaria heterophylla that have juvenile characteristics and are in a good phytosanitary state. Picea glauca, Pinus pinaster, Picea abies, Abies and Larix species are also being studied.

In Italy, research on micropropagation of coniferous species is conducted by Dr P. Grossoni (Cupressus sempervirens), Dr S. Biondi (Pseudotsuga menziesii) and Dr L. Lubrano. The latter is working on Pinus pinaster and Pinus radiata using methods described by David

(13) and Aitken (2). The main goal is to clone selected, genetically improved seeds.

In Czechoslovakia, Dr V. Chalupa is using axillary bud and *in vitro* rooting techniques to propagate *Larix decidua*, *Picea abies*, *Pinus sylvestris* and *Pseudotsuga menziesii*. One hundred to one thousand *in vitro* plantlets per clone per year are produced and used as donor plants for cuttings.

In West Germany, Dr R. Ahuja intends to propagate mature genotypes of *Picea abies*, *Larix decidua* and *Pseudotsuga menziesii*. In Escherode, Dr H. Meier-Dinkle has started research to mass propagate seeds obtained from intra or interspecific crosses. Rejuvenation of old genotypes, or old clones propagated by cuttings for several years, is also intended.

In Yugoslavia, Drs B. Kolevska and S. Jelaska have developed a method of bud and shoot formation in juvenile tissue culture of *Pinus nigra*(22).

In England, *Picea sitchensis* is being studied (see chapter by John and Webb in this book). Shell Oil Company in their laboratory in Kent England, are researching *in vitro* propagation techniques for *Pinus carabaea* and *Pinus oocarpa* using axillary budding and brachyblast culture (Dr J. Stead).

Lastly in France, *Larix* sp. are being studied at an INRA lab (Dr D. Cornu) and *Pinus caribaea* and *Araucaria* sp. at the CTFT (Technical Center of Tropical Forestry).

4.2. North America and Canada

Much work has been carried out in this part of the world on *in vitro* propagation of coniferous species and has been cited above or in different chapters in this book. We want just to mention Dr H. Sommer's works with *Pinus palustris*, *Pinus elliotii* and their hybrids, and Dr T. Thorpe's work with *Picea glauca*, *Picea mariana* and some other species. In Clay's nurseries, Dr M.L. Kurz cultures *Picea glauca* and *Chamaecyparis nootkatensis*. This list is incomplete; there are other scientists working on coniferous species on a small scale in these countries.

4.3. South Korea

A mass clonal propagation method of *Pinus rigida* x *Pinus taeda* F1 has been developed using embryo culture (21). The interspecific hybrids are promising and *in vitro* propagation could provide the means to plant on a large scale the small number of successful hybrid crosses. The level of production through *in vitro* culture is not known.

5. CONCLUSION

This overview of mass propagation research with different coniferous species is not all inclusive. Nevertheless, it shows the great importance of *in vitro* systems for the mass propagation of coniferous species. Many conifer plantations are presently regenerated with improved seed but there is a trend to clonal forestry.

The commercial propagation of coniferous species through *in vitro* culture is still limited to 2 or 3 species, first *Pinus radiata* in New Zealand (see 3.3.1.), second *Sequoia sempervirens* in North America and France (see 3.4.), and third *Pinus taeda* in North America (see 3.3.2.). Some other species are on the way (*Pseudotsuga menziesii*, *Pinus rigida* x *taeda*). In most cases mass propagation starts from excised embryos or parts of embryos, followed by adventitious budding. Propagation of mature genotypes is for most species still in the research stage. Maturation problems, as found in rooted cuttings, also seem to be a barrier in micropropagation. Rejuvenating treatments and donor plant pretreatments are often not complete (see others chapters in this book). More research is required to determine what the differences are between adult and juvenile plant material before juvenile copies from adult selected genotypes can be expected.

In my opinion, the recently achieved somatic embryogenesis of *Picea abies* and a number of other conifers (Tulecke, Durzan in volume 2) is one of the most important results of recent years. If similar methods could be developed for other coniferous species and for mature genotypes, we would have an elegant method to resolve most of the problems of *in vitro* propagation of difficult coniferous species. However, somatic embryogenesis may introduce a level of genetic instability, that is unacceptable for forest regeneration purposes. So far however, no evidence of lethal or semilethal development has been observed on sugar pine (Gupta and Durzan unpublished).

6. REFERENCES

1. ABOEL NIL M, ZS WOCKOK 1978 Morphogenetic heterogeneity of cultured Douglas fir cotyledons and its correlation with seed weight. Proc 4th Int Congr Plant Tissue Cell Cul. Univ Calgary, Alberta Canada
2. AITKEN-CHRISTIE J, TA THORPE 1984 Clonal propagation : gymnosperms in cell culture and somatic cell genetics of plants. In Cell Culture and Somatic Cell Genetics of Plants, Vol 1, Academic Press, pp 82-95
3. BALL E, DM MORRIS, JA RYDELIUS 1978 Cloning of Sequoia sempervirens from mature trees through tissue culture. In Proc Round Table Conf "In vitro" Multiplication Woody Species, Gembloux, Belgium, June 6-8, 1978, pp 181-226
4. BEKKAOUI F 1983 Microbouturage in vitro du Sequoia sempervirens (Endl). Stade de la rhizogenèse considéré comme critère de juvénilité chez deux clones d'âges différents. DEA Paris VI, 43 p
5. BEKKAOUI F, A FRANCLET, N WALKER 1985 Culture in vitro de méristèmes de douglas âgés et juvéniles. Ann AFOCEL 1984, pp 5-72
6. BORNMAN CH, E JANSSON 1980 Organogenesis in cultured Pinus sylvestris tissue. Z Pflanzenphysiol 96 : 1-6
7. BOULAY M 1976 Recherches sur la propagation du douglas par culture in vitro. Ann AFOCEL 1975, pp 84-145
8. BOULAY M 1978 Multiplication rapide du Sequoia sempervirens en culture in vitro. Ann AFOCEL 1977, pp 38-65
9. BOULAY M 1985 Aspects pratiques de la multiplication in vitro des essences foretières. Ann AFOCEL 1984, pp 7-43
10. CHENG TY 1975 Adventitious bud formation in culture of douglas fir (Pseudotsuga menziesii Mirb.) Franco. Plant Sci Lett 5 : 97-103
11. CHOULET B, A ROBERT, T HIGUGHI, F BARNOUD 1964 Quelques aspects chimiques et biochimiques des ligno-celluloses synthétisées dans une souche de Sequoia sempervirens. Actes Symp Int Grenoble, pp 481-490
12. COLEMAN WK, TA THORPE 1977 In vitro culture of western red cedar (Thuya plicata Don) I - Plantlet formation. Bot Gaz 138 : 298-304
13. DAVID H, K ISEMUKALI, A DAVID 1978 Obtention de pin maritime (Pinus pinaster Sol.) à partir de brachyblastes ou d'apex caulinaires de très jeunes sujets cultivés in vitro. C R Acad Sci, Paris 287 : 245-247
14. DAVID A, H DAVID, M FAYE, K ISEMUKALI 1979 Culture in vitro et micropropagation du pin maritime (Pinus pinaster Sol.). AFOCEL Etud Rech 12 : 33-40

15. DAVID A, H DAVID, T MATEILLE 1982 *In vitro* adventitious budding on *Pinus pinaster* cotyledons and needles. Physiol Plant 56 : 102 -107
16. DINER A, R MOTT 1982 A rapid axenic assay for hypersensitive resistance of *Pinus lambertiana* to *Cronartium ribicola*. Phytopathology 72 : 864-865
17. ENGELMANN F 1982 Multiplication végétative *in vitro* de clones juvéniles et matures de *Cunninghamia lanceolata*. DEA Univ Pierre et Marie Curie, Paris VI, pp 43
18. FOURET Y 1984 Micropropagation *in vitro* du *Sequoia sempervirens* (Endl.) : effet de la fréquence des repiquages sur le rajeunissement de trois clones issus de matériel d'âge différent et recherche d'un critère précoce de juvénilité. DEA Univ Pierre et Marie Curie, Paris VI, pp 43
19. FRANCLET A, A DAVID, H DAVID, M BOULAY 1980 Première mise en évidence morphologique d'un rajeunissement de méristèmes primaires caulinaires de pin maritime âgé (*Pinus pinaster* Sol.) C R Acad Sci, Paris 290 : 927-930
20. HRIB J, V RYPACEK 1983 *In vitro* testing for the resistance of conifers to fungus *Phaseolus schweinitzii* (Fr.) Pat. on callus cultures. Eur J For Pathol 13 : 86-91
21. KIM JH, SK HYUN, SK LEE, KJ LEE, KC KIM 1982 Mass clonal propagation of *Pinus rigida* x *taeda* F1 by embryo culture. Res Rep Inst For Genet, Off For, Suweon-Korea, V 18, pp 74-79
22. KOLEVSKA-PLETIKAPIC B, S JELASKA, J BERLJAK, M VIDAKOVIK 1983 Bud and shoot formation in juvenile tissue culture of *Pinus nigra*. Silvae Genet 32 : 115-119
23. La GOUBLAYE de NANTOIS T de 1980 Rajeunissement chez le douglas (*Pseudotsuga menziesii*) en vue de la propagation végétative. Etude sur la plagiotropie des parties aériennes et racinaires. DEA Univ Paris VI, pp 44
24. La GOUBLAYE de NANTOIS H de 1981 Vieillissement et rajeunissement chez le *Sequoia sempervirens* (Endl.) en relation avec la propagation végétative. Thèse 3ème Cycle Univ Paris VI, pp 170
25. LEACH GN 1979 Growth in soil of plantlets produced by tissue cultured loblolly pine. TAPPI 62 : 59-61
26. LIBBY WJ 1983 The clonal option. Norsk Inst Skogforskning 1432, AS.NLH Norway, pp 3-32
27. McKEAND SE, HL ALLEN 1984 Nutritional and root development factors affecting growth of tissue cultured plantlets of loblolly pine. Physiol Plant 61 : 523-528
28. MEHRA-PALTA A, RH SMELTZER, R MOTT 1978 Hormonal control of induced organogenesis. Experiments with excised plant parts of loblolly pine. TAPPI 61 : 37-40
29. MISSON JP, M COUMANS, P GIOT-WIRGOT, Th GASPAR 1982 Induction de bourgeons adventifs de *Picea pungens* en culture *in vitro*. Z Pflanzenphysiol 107 : 161-167
30. MONTEUUIS O 1984 Etude de la multiplication végétative de *Sequoiadendron giganteum* (Lindl.) en vue du clonage. DEA Clermont II, pp 30
31. MOTT RL, HV AMERSON 1981 A tissue culture process for the clonal production of loblolly pine plantlets. Tech Bull n° 271, North Carolina Agric Res Serv, pp 1-14

32. POISSONNIER M, A FRANCLET, MJ DUMANT, JY GAUTRY 1980 Enracinement de tigelles issues _in vitro_ de _Sequoia sempervirens_. Ann AFOCEL 1979, pp 232-253

33. RANCILLAC M, M FAYE, A DAVID 1982 _In vitro_ rooting of cloned shoots in _Pinus pinaster_. Physiol Plant 56 : 97-101

34. REILLY K, CH BROWN 1976 _In vitro_ studies of bud and shoot formation in _Pinus radiata_ and _Pseudotsuga menziesii_. Georgia For Res Pap 86, pp 1-9

35. SOMMER HE, CH BROWN, P KORMANIK 1975 Differentiation of plantlets in long leaf pine (_Pinus palustris_) Mill. tissue cultured _in vitro_. Bot Gaz 136 : 196-200

36. STINE M, H SOMMER 1983 Adventitious shoot formation on embryos of _Pinus palustris_ Mill., _Pinus elliottii_ Engelm. and _Pinus palustris_ x _elliottii_. _In_ 17th Southern For Tree Improv Conf, pp 349-350

37. VON ARNOLD S, T ERIKSSON 1979 Induction of adventitious buds of Norway spruce (_Picea abies_) grown _in vitro_. Physiol Plant 45 : 29-34

38. WALKER N, E DUMAS, A FRANCLET, F BEKKAOUI 1985 Techniques de culture _in vitro_ des méristèmes de _Sequoia sempervirens_ et _Pinus pinaster_. Ann AFOCEL 1984 : pp 87-109

39. YOUNG PM, AS HUTCHINS, ML CANFIELD 1984 Use of antibiotics to control bacteria in shoot cultures of woody plants. Plant Sci Lett 34 : 203-209

40. ZOBEL B 1981 Vegetative propagation in forest management operations. _In_ Proc 16th Southern For Tree Improv Conf Blacksburg VA, pp 149-159

16. IN VITRO PROPAGATION OF POPLAR AND ASPEN [1]

M.R. AHUJA

1. INTRODUCTION

1.1. Biology and distribution

Poplar and aspen belong to the genus Populus, and family Salicaceae. They are dioecious, with male and female flowers (catkins) occurring on separate trees. The diploid chromosome number in poplar and aspen is 2n = 38. The genus *Populus* is comprised of five sections, consisting of more than 30 species widely distributed in the Northern Hemisphere (20, 22). The five sections of the genus *Populus* are: 1) Turanga, consisting of a single species, *P. euphratica* Oliv., which is native to western and Central Asia, as well as North and Central Africa; 2) Leucoides, contains four species, none of which are of economic importance; 3) Leuce, a large group further divided into two subsections: Albidae, consisting of white poplars, *P. alba* L. and *P. monticola* T. Brand, and Trepidae, consisting of aspens (several species); 4) Tacamahaca, collectively known as balsam poplars, is the largest section of the genus, and contains several species of economic importance; and 5) Aigeiros, the "true poplars", to which more than 90% of the cultivated poplars (mainly cottonwoods and black poplars) of the world belong. Three main species are included in the section Aigeiros: eastern cottonwood (*P. deltoides* Marsh.), distributed over most of the United States of America and southern Canada; arid-zone cottonwood (*P. fremontii* Wats.), distributed over south-western portion of the United States; and black poplar (*P. nigra* L.), a native of Eurasia.

The subsection Trepidae of the section Leuce contains six species of aspens. European aspen (*P. tremula* L.) is widely distributed in Europe, western Asia and North Africa.

---

[1] Dedicated to Dr. G.H. Melchior on his 60th birthday.

Quaking aspen (*P. tremuloides* Michx.) is the most widely distributed aspen in North America. Bigtooth aspen (*P. grandidentata* Michx.), a close relative of quaking aspen, has a limited range of distribution within the Northeast of the United States and Southeast Canada. A few species akin to European aspen occur in the Far East. These include Chinese aspen (*P. adenopoda* Maxim.), Korean aspen (*P. davidiana* Schneid.), and Japanese aspen (*P. sieboldii* Miq.).

Hybridization is frequent between certain species of the genus *Populus*, and consequently a number of natural hybrids exist. Only two will be mentioned here. Euramerican poplar (*P.* x *euramericana* (Dode) Guinier), occurring in North America and Europe, is derived from hybridization between *P. deltoides* and *P. nigra* following exchange of these species across the Atlantic Ocean. Grey poplar (*P.* x *canescens* Sm.) is a hybrid between *P. tremula* and *P. alba*, and covers the whole range of overlap between the parental species in central and northern Europe, as well as western Asia.

Poplars and aspens occur interspersed throughout all the forests of temperate and cold regions of the Northern Hemisphere. There is enormous genetic variation in the genus *Populus*, which confers on its genotypes a wide range of adaptability. Because of their fast growth rates, short rotation cycle, and ability to grow on marginal soils, poplars and aspens are gaining importance as forest trees.

1.2. Economic importance

Poplar and aspen wood has varied uses. It has been employed in paper, matchstick, and plywood industries. Poplar and aspen wood has also been used as light packing material, hardboards, and for making pelleted animal feed. Because of rapid growth rates, poplar and aspen are potentially valuable for the production of biomass and chemical feedstock. It has been proposed that aspen and poplar can be employed in the so-called 'energy plantations' (45), the biomass of which could be used for industrial or commercial energy either through direct combustion or via conversion into chemical fuels, such as alcohol, or gaseous fuels. Because of extensive suckering habit, aspens can be employed in self-perpetuating 'energy plantations'. They can be used until there is a decline in the regeneration of healthy shoots from the root suckers.

2. CONVENTIONAL METHODS OF VEGETATIVE PROPAGATION

Poplars can be propagated by stem cuttings (hardwood cuttings), which root readily in soil or peat-perlite substrate. Leuce poplars (aspens), on the other hand, are difficult to propagate by woody stem cuttings, because of lack of preformed root primordia (13). However, aspens can be vegetatively propagated by root suckers, graftings (27, 36), and green shoots (8). Because of cost factors, graft incompatibilities (36), and limited availability of a specific genotype as planting material, vegetative propagation by root suckers, graftings, and green shoots has only limited commercial applicability in the aspen regeneration programs. For this reason aspens have so far been mostly propagated by seed. However, this situation is likely to change in the future in view of the fact that *in vitro* techniques are becoming available for rapid and reliable clonal propagation of selected aspen genotypes (1, 4, 24).

3. REVIEW OF TISSUE CULTURE WORK

3.1. Callus culture

More than 50 years ago Gautheret (25) successfully cultured cambial tissues from a number of tree species, including a poplar (*P. nigra*). Some 30 years after Gautheret's pioneering work, Mathes (34) established long-term callus cultures from stem explants of tripolid quaking aspen (*P. tremuloides*). Mathes obtained both roots and shoots on the callus of the triploid aspen. However, it was not certain whether these roots were directly connected to shoots to give rise to plantlets. A few years later, Wolter (44) demonstrated that shoots could be induced on callus cultures of *P. tremuloides* by inclusion of 0.2 - 0.5 mg/l benzylaminopurine (BA) to the basal medium, and that shoots could be rooted on Wolter's basal medium. In the same year, Winton (41) also produced shoots on callus of triploid aspen (*P. tremuloides*) by employing Wolter's basal medium containing 0.5 mg/l BA. These shoots were rooted on Wolter's medium containing 0.04 mg/l 2,4-dichlorophenoxyacetic acid (2,4-D) and 1.0 mg/l kinetin.

These earlier studies by Mathes, Wolter, and Winton on aspen culture, although of historical importance, left some doubt about the normality and genetic stability of the regenerants. Later, however, Winton (42, 43) demonstrated that normal aspen plantlets, although few in numbers, could be regenerated from callus cultures of triploid quaking aspen (*P. tremuloides*) and tetraploid European aspen (*P. tremula*). Chalupa (15) extended the studies of Winton and Wolter on several species of *Populus*, and regenerated plantlets from callus cultures. A few years later, Chalupa (16) compared the growth rates, morphology and genetic constitution of the callus derived plantlets in *Populus* species, and claimed that clonal regenerants were essentially similar to the parent plants, both morphologically and genetically. Lester and Berbee (30), on the other hand, observed a wide range of variation in height, number of branches, leaf traits and chromosome numbers in the callus derived plants in several clones of black poplar (*P. nigra*) and Euramerican poplar (*P. x euramericana*).

By employing an Aspen Culture Medium (ACM) (1), which is a modified version of Woody Plant Medium (WPM) (31), calli were induced on young stem segments of leaf petioles by supplementing ACM with 1.0 - 2.0 mg/l naphthaleneacetic acid (NAA) or 2,4-D. Callus growth was enhanced by inclusion of 0.01 mg/l BA in the medium. Multiple shoots could be induced following culture of callus segments on ACM fortified by 0.4 - 0.5 mg/l BA. Roots were induced on excised microshoots cultured on ACM supplemented with 0.5 mg/l indolbutyric acid (IBA) and 0.1 mg/l NAA. However, callus cultures were not employed by us for clonal propagation of aspen because: 1) only a limited number of shoots are formed on callus cultures, and 2) occurence of genetic variation in long-term callus cultures. We intend to employ callus cultures for searching for somaclonal variation in aspen. However, for large scale clonal propagation of selected parental genotypes, somaclonal variation should be avoided. Organ cultures, particularly bud meristems, have been shown to provide reasonably reliable material for clonal propagation on a large scale, with minimal risks of somaclonal variation.

3.2. Organ culture

Whitehead and Giles (40) investigated the potential of axillary bud explants of poplars for clonal propagation. Bud explants from three *Populus* species, namely *P. nigra* "Italica", *P.* "*Flevo*" (*P. deltoides* x *P. nigra*), and *P. yunnanensis* were cultured on modified Murashige and Skoog's medium (MS) containing 0.2 mg/l BA. Bud break occurred within 2-3 weeks, and within 4 weeks initial shoots had lengthened enough to be cut into 5 mm sections and placed again on the same medium. Once proliferation had started, the cultures were transfered to modified MS medium containing 0.1 mg/l BA and 0.02 mg/l NAA. After 6-8 weeks 120-220 shoots had formed on each original bud explant. These shoots could be either rooted or used for production of more shoots. Within three months after rooting, the tissue culture derived poplar plants were 1-1.5 meters tall. They estimated that more than 1 million plantlets per year could be produced from one bud.

By employing essentially the same bud explant method of Whitehead and Giles (40), Christie (18) regenerated plantlets from several Populus species: *P. alba*, *P. tremula*, *P. tremuloides*, *P.* x *canescens*, and *P. alba* x *P. glandulosa*. A similar program of micropropagation was carried out by Barocka *et al.* (7) for aspen multiplication on a commercial scale.

While previous studies in the genus *Populus* were carried out on bud explants of juvenile or young plants, later studies were extended to older trees ranging in age from 10 - 20 years (17), or even older and mature trees between 17 to 40 years (1, 4). In addition to bud explants, Ahuja (1) carried out regenerative studies by employing leaf discs and young leaves and root explants of *P. tremula*, *P. tremuloides*, and their hybrids (hybrid aspen). Microshoots that developed on leaf or root explants could be later rooted on a rooting medium. However, for large scale clonal propagation of aspen and hybrid aspen, Ahuja (1, 4), employed meristem explants from dormant buds. Table 1 summarises the results of plantlet regeneration studies from callus and organ cultures in poplars and aspens.

4. MICROPROPAGATION

In order to extend tissue culture technology to large scale

Table 1. Plantlet regeneration in *Populus* by tissue culture

Species/hybrid	Age	Explant	Reference
P. alba	Y	Axillary buds	(17)
P. alba x *P. glandulosa*	Y	Apical and axillary buds	(18), (28)
P. alba x *P. tremula*	Y	Apical and axillary buds	(17)
P. x *canadensis*	S	Stem tip	(9)
P. x *canescens*	Y	Apical and axillary buds	(17)
P. x *euramericana*	S	Stem, Shoot tip	(15), (32)
P. "Flevo" (*P. deltoides* x *P. nigra*)	Y	Axillary buds	(40)
P. glandulosa	R	Anther	(29)
P. grandidentata	M	Apical and axillary buds	(A)[a]
P. nigra	R	Anther	(38)
P. nigra "Italica"	Y	Axillary buds	(40)
P. tremula 2n	S	Stem	(15)
P. tremula 4n	S	Stem	(43)
P. tremula 2n	Y	Apical and axillary buds	(18), (7)
P. tremula 2n	Y,M	Apical and axillary buds, stem leaf, root	(1), (4)
P. tremuloides 3n	S	Stem	(41), (42)
P. tremuloides 2n	S	Cotyledon, hypocotyl	(39)
P. tremuloides 2n	Y	Apical and axillary buds	(18)
P. tremuloides 2n	Y,M	Apical and axillary buds, root	(1), (4)
P. tremula x *P. tremuloides* 2n, 3n	Y	Apical and axillary buds	(7)
P. tremula x *P. tremuloides* 2n, 3n	Y,M	Apical and axillary buds stem, leaf, root	(1), (4)
P. yunnanensis	Y	Axillary buds	(40)

S = seedlings; Y = young trees; R = reproductive organ; M = mature trees

[a](A) = Ahuja, unpublished

clonal propagation, it is necessary to develop a method that: 1) is relatively simple, 2) has a high multiplication rate, 3) has a high degree of reproducibility, 4) is applicable to juvenile and mature tissues, 5) induces direct differentiation of microshoots with minimal callus formation, 6) allows rooting of microshoots in agar medium or soil-free substrates or soil, 7) gives high survival rates of microshoots or plantlets upon transfer to rooting or potting conditions, 8) is cost effective, that is, plantlet vs. seedling costs are comparable, 9) can be easily adapted to nursery practices, and 10) ensures, to a large extent, the genetic stability of propagules under *in vitro*, nursery, and field conditions, and in progeny tests. Although such an ideal method(s) of micropropagation is not yet available for any forest tree species, at least some aspects of such a procedure have been accomplished in the genus *Populus*.

We have been interested in rapid clonal propagation of selected trees of European aspen (*P. tremula*), quaking aspen (*P. tremuloides*) and their hybrids (hybrid aspen). Initially bud explants were cultured from young plants established from root suckers of selected trees in the nursery. Subsequently, however, meristem cultures were established from primary buds directly obtained from mature trees. Bud explants from more than 100 aspen and hybrid aspen trees ranging in age from 17 to 40 years were investigated for their *in vitro* regenerative potential (1, 4). To start with, a 4-step micropropagation procedure was employed (Fig. 1). This procedure consisted of: 1) bud break and conditioning of meristems on ACM-1 (ACM + 0.5 mg/l BA), 2) growth and proliferation of microshoots on ACM-2 (ACM + 0.5 mg/l BA + 0.02 mg/l NAA), (Fig. 2), 3) root formation on microshoots on ACM-3 (ACM + 0.5 mg/l IBA + 0.1 mg/l NAA), (Fig. 3), and 4) transfer of plantlets to pots, and after hardening in controlled chambers and greenhouse, transplantation to the field (Fig. 4). This 4-step procedure has been simplified to reduce the number of steps and to make the micropropagation method cost-effective. From four, first it was reduced to a 3-step method (Fig. 1) by combining steps 1 and 2 to achieve meristem conditioning and microshoot proliferation on ACM-2. Later step 3 (rooting medium, ACM-3) was eliminated from the scheme, and excised microshoots were directly rooted in

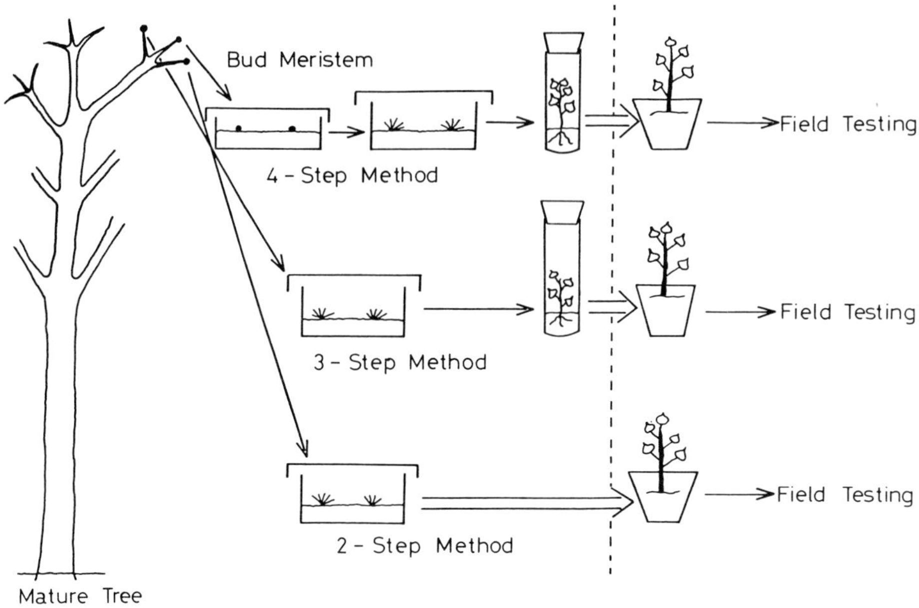

FIGURE 1. Diagram showing 4-step, 3-step, and 2-step micropropagation procedures. Initially, a 4-step method was employed for plantlet regeneration. Subsequently, it was simplified to a 2-step micropropagation method. This 2-step method has been successfully employed for rapid and large scale clonal propagation of elite aspen and hybrid aspen trees.

peat-perlite substrate. Roots differentiated in 70 - 90% of the microshoots within two weeks on this substrate, which is about the same time it took to root on the ACM-3 medium. Thus it was possible to evolve an effective 2-step micropropagation method from the original 4-step method (Fig. 1). By employing this 2-step micropropagation method (4), it has been possible to reduce costs and minimize the exposure of tissues to unnatural *in vitro* conditions. In these experiments microshoots developed on the bud meristems with minimal prior callus formation. A few regenerants with morphological variation were detected when the 4-step micropropagation method was employed. However, with this 2-step micropropagation procedure, we have not observed morphological variants in the *in vitro* produced propagules.

FIGURE 2. Bud explant from a mature hybrid aspen tree showing proliferation of microshoots on ACM-2.

FIGURE 3. Plantlet regeneration from an excised microshoot of a hybrid aspen on ACM-3.

FIGURE 4. Tissue culture derived plants from a number of aspen and hybrid aspen clones under field conditions.

Cytological examination of randomly selected aspen and hybrid aspen plants derived by the 2-step procedure revealed a normal genetic constitution (4). That does not preclude the presence of recessive gene mutations, if they were indeed induced under the cultural conditions. However, by minimizing the exposure of tissues to in vitro environment, along with minimal callus formation, the genetic risk of instability may have been reduced by our 2-step micropropagation method. By this method we have regenerated more than 5000 plants from mature aspen and hybrid aspen trees. Of the about 100 trees tested, 50% differentiated microshoots on ACM-2, and thus could be clonally propagated. The regenerative potential of so far nonresponsive trees is being further tested by broadening the hormone spectrum (0-5 mg/l of BA or zeatin, along with low concentrations of NAA) of ACM, and variation in cultural conditions (Ahuja, unpublished). Preliminary results from our laboratory indicate that, besides the genotype, cytokinin levels, inclusion of a few juvenile leaves along with the dormant bud meristem, time of the year (early spring), and storage of dormant buds at low temperatures before culture influence the morphogenetic response.

5. FIELD TESTING AND COSTS

Tissue culture derived plants in poplars and aspens have been planted under field conditions (1, 4, 7, 18, 32). In the first growing season plants reach a height of 1 to 2.5 metres. These are healthy, vigorous, and free of any morphological abnormalities.

The costs of tissue culture derived plants are rather difficult to calculate, because the techniques are still evolving. Only rough estimates can be worked out. Brown and Sommer (14) estimated that 1000 containerized plantlets of sweet gum (*Liquidambar styraciflua*) would cost US$ 123 from shoot tip culture, and 1000 containerized seedlings from improved seed would cost $ 75. McKeand and Weir (35) have estimated that loblolly pine (*Pinus taeda*) plantlets could be produced for $ 50 to $ 100 per thousand by the cotyledon culture method, whereas 1000 bare-root seedlings cost about $ 20. Therefore, the cost of a plantlet is still generally higher than that of a seedling. However, the higher price may be worth it because of the relative genetic uniformity and the genetic gain in the propagules. After improvement, simplification, and automation of tissue culture techniques, the costs could be further reduced, to make plantlet production for reforestation economically attractive. Our 2-step micropropagation method (4) for apsen is a step in that direction.

6. JUVENILITY, MATURITY, AND REJUVENATION

In a number of forest tree species explants derived from juvenile tissues, such as embryos, cotyledons, seedlings, etc. have been employed for clonal propagation. Micropropagation from juvenile material, although useful for certain studies, has the disadvantage that it is difficult to predict how a seedling will perform in later years. Clonal propagation of most mature hardwoods and conifers by tissue culture is still very difficult (10). Of course, it would be ideal if genetic, biochemical and molecular markers were available in the juvenile stages, such as seedlings, that are correlated with yield traits. However, that not being the case at present, we still have to probe into basic processes involved in juvenility, maturation, and rejuvenation. Rejuvenation treatments (10, 23) currently under investigation include: 1) severe pruning of trees to stimulate production of lateral juvenile meristems, 2) serial grafting of buds from older trees onto juvenile rootstocks, 3) spraying of plants with cytokinin before culture, 4) serial subculturing of explants *in vitro*, 5) cytokinin stimulated induction of adventitous buds, and 6) treatment of resting buds with chemicals that may leach out substances that might be inhibitory to the rejuvenation process. Another approach might be to store juvenile material under low temperature conditions or cryogenically for a number of years until the germplasm has been evaluated following maturity and progeny tests. At that stage the cryopreserved material could be retrieved and recultured for micropropgation. Such tissue 'gene banks' could also be valuable for preservation of forest tree germplasm already being depleted by pathogens or unfavourable environmental conditions. It is, however, not entirely known whether these rejuvenation and cryopreservation procedures carry a risk of genetic instability.

A degree of rejuvenation has been accomplished in the mature tissues of some forest tree species. Boulay (12) has developed *in vitro* methods for large scale propagation of 100 years old *Sequoia sempervirens*. By culture of bud explants, clonal propagation has been accomplished in 100 years *Tectona grandis* (26), 60 years old *Prunus avium* (19), 10 - 20 years old *Eucalyptus* spp (33), and 17 - 40 years old *Populus tremula*, *P. tremuloides*, and their hybrids (1, 4). Bonga (11) has reported on plantlet formation (few in number) from the female cone tissues of 25 - 30 years old *Larix decidua*. Rejuvenation potential of several other tree species was investigated in our tissue culture laboratory at Grosshansdorf. On a variety of different media bud explants were cultured from more than 50 years old softwoods, *Picea abies*, *Pseudotsuga menziesii*, and *Larix decidua*, and hardwoods, *Fagus sylvatica*, *Quercus robur*, and *Q. petraea*. Excepting slight differentiation on bud explants of a cultivar (*f. purpurea*) of *Fagus sylvatica* (5), and callusing on *Larix decidua*, there was hardly any growth and differentiation on the bud explants of the other forest tree species (Ahuja, unpublished). We are continuing morphogenetic investigations on the juvenile and mature tissues of these forest tree species.

7. REGENERATIVE POTENTIAL OF PROTOPLASTS AND CELLS

Protoplasts have been isolated from cell suspension cultures of a *Populus* hybrid TT32 (21), seedlings of quaking aspen (*P. tremuloides*) (37), and young leaves from mature trees of European aspen (*P. tremula*), quaking aspen (*P. tremuloides*) and their hybrids (2, 3). Recently, we have also isolated protoplasts (Fig. 5) from juvenile leaves derived from bud meristem cultures of aspen and hybrid aspen (Ahuja, unpublished). Budding of individual protoplasts has been observed in the protoplast cultures of *Populus* spp. (2, 3, 21). Sustained cell divisions have not, so far, been observed. Consequently, it has not been possible to regenerate plantlets from the protoplasts cultures of Populus (6).

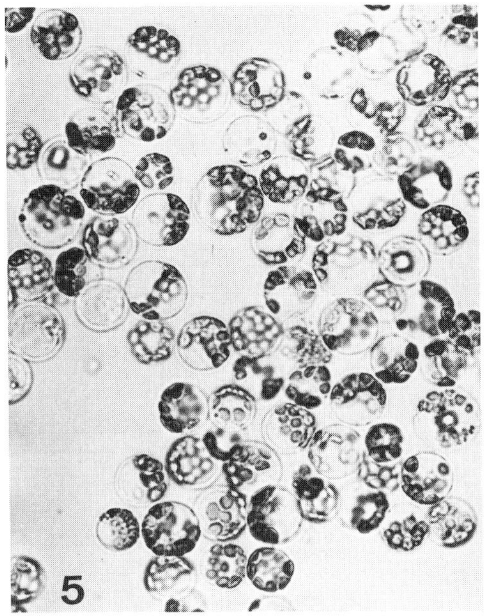

FIGURE 5. Freshly isolated mesophyll protoplasts from tissue culture derived juvenile leaves of a hybrid aspen clone.

Cell suspension cultures have been established from callus of the Populus hybrid TT32 (21), and from callus as well as young leaves and meristems of European aspen (P. tremula) (Ahuja, unpublished). Initial cell divisions have been observed in the cell suspension cultures of aspen. However, somatic embryogenesis has not, so far, been observed in these cultures.

8. FUTURE RESEARCH AND PROSPECTS

Poplars and aspens are gaining importance as forest tree species. They exhibit fast growth rates, and have a short rotation cycle. Besides, they can grow on marginal sites, and on agricultural wastelands that are lying fallow. Furthermore, they can improve the ecological stability of coniferous stands. In addition to various uses of poplar and aspen wood, these tree species are good candidates for 'energy plantations'. Tissue culture technology offers prospects for mass cloning of superior genotypes for reforestation programs, storage of tissues under super low temperatures, isolation of new genotypes through protoplast fusion and gene transfers, and utilization of genetic variability that has been detected in the protoclones or somaclones. In this direction, complete protoclones have to be developed for regeneration of plantlets from protoplasts and cells in poplar and aspen.

Since aspen tissues, obtained from juvenile and mature material, can be induced to undergo organogenesis _in vitro_, this species can serve as a model system for understanding problems of growth, differentiation, juvenility, maturity, rejuvenation, and gene transfers in the forest tree species.

9. ACKNOWLEDGEMENTS

This research was supported by a grant C080 from the Federal Ministry of Research and Technology. The author thanks Drs. G. H. Melchior and Hans J. Muhs for discussions, E. Wenk and R. Amenda for technical assistance, and H. Seuthe for organizing greenhouse and field planting of tissue culture derived plants.

10. REFERENCES

1. AHUJA MR 1983 Somatic cell differentiation and rapid clonal propagation of aspen. Silvae Genet 32: 131-135
2. AHUJA MR 1983 Developmental potantial of mega and normal protoplasts in Populus. In I Potrykus, CT Harms, A Hinnen, R Hütter, PJ King, RD Shillito, eds, Protoplasts 1983 Poster Proceedings, Birkhäuser Verlag, Basel, pp 28-29
3. AHUJA MR 1983 Isolation and culture of mega and normal protoplasts in aspen. Silvae Genet 32: 225-227
4. AHUJA MR 1984 A commercially feasible micropropagation method for aspen. Silvae Genet 33: 174-176
5. AHUJA MR 1984 In vitro induction of organogenesis in juvenile and mature beech. Silvae Genet 33: 241-242
6. AHUJA MR 1984 Protoplast research in woody plants. Silvae Genet 33: 32-37
7. BAROCKA KH, M BAUS, E LONTKE, F SIEVERT 1985 Tissue culture as a tool for in vitro-mass-propagation of aspen. Z Pflanzenzüchtung 94: 340-343
8. BEHRENS V, GH MELCHIOR 1978 Hinweise zur Verbesserung des Verfahrens der vegetativen Vermehrung von Aspen und Aspenhybriden durch Grünstecklinge aus jungen Wurzelschösslingen. Holzzucht 32: 17-21
9. BERBEE FM, JG BERBEE, AC HILDEBRANDT 1972 Induction of callus and trees from stem tip cultures of a hybrid poplar. In vitro 7: 269
10. BONGA JM 1982 Vegetative propagation in relation to juvenility, maturity, and rejuvenation. In JM Bonga, DJ Durzan, eds, Tissue Culture in Forestry, Martinus Nijhoff/Junk Publ, The Hague, pp 387-412
11. BONGA JM 1984 Adventitious shoot and root formation in tissue cultures of mature Larix decidua. In Proc Int Symp Recent Adv For Biotechnol, Traverse City, Michigan, pp 64-68
12. BOULAY M 1979 Multiplication et clonage rapide du Sequoia sempervirens par la culture in vitro. In "Micropropagation d'Arbres Forestieres", Ann AFOCEL, Nangis, No 12, 6/79, pp 49-56
13. BRAUN HJ, G SCHLENCKER 1964 Das Vorkommen von Wurzelkeimen in den Sprossachsen verschiedener Populus-Sorten. Mitt des Vereins f forstl Standortkunde und Forstpflanzenzüchtung Nr 14: 65-68
14. BROWN CL, HE SOMMER 1982 Vegetative propagation of dicotyledonous trees. In JM Bonga, DJ Durzan, eds, Tissue Culture in Forestry, Martinus Nijhoff/Junk Publ, The Hague, pp 109-149
15. CHALUPA V 1974 Control of root and shoot formation and production of trees from poplar callus. Biol Plant 16: 316-320
16. CHALUPA V 1977 The use of regenerants from tissue culture of forest trees in tree breeding. In Use of Tissue Culture in Plant Breeding. Czech Acad Sci, Prague pp 183-193

17. CHALUPA V 1979 In vitro propagation of some broad-leaved forest trees. Commun Inst For Czech 11: 159-170

18. CHRISTIE CB 1978 Rapid propagation of aspen and silver poplar using tissue culture techniques. Proc Int Plant Propagators Soc 28: 255-260

19. CORNU C, C CHAIX 1981 Multiplication par culture in vitro de merisiers adultes (*Prunus avium*) application a un large eventail de clones. *In* Proc IUFRO sect S 2 01 5. Int Workshop "*In vitro*" cultivation For Tree Species, Fontainebleau, France, pp 71-79

20. DICKMAN DI, K STUART 1983 The Culture of Poplars in Eastern North America. Michigan State Univ Press, East Lansing, Michigan

21. DOUGLAS G 1982 Protoplast isolation from totipotent cell-cultures of *Populus* hybrid TT32. *In* A Fujiwara, ed, Plant Tissue Culture 1982, Proc 5th Int Cong Plant Tissue Cell Cult, Tokyo, pp 605-606

22. FAO 1980 Poplars and Willows in Wood Production and Land Use. FAO Forestry Ser No 10, Rome

23. FARNUM P, R TIMMIS, JL KULP 1983 Biotechnology of forest yield. Science 219: 694-702

24. FRÖHLICH HJ 1982 Fortschritte bei der vegetativen Vermehrung. Forstarchiv 53: 3-9

25. GAUTHERET RJ 1934 Culture du tissu cambial. CR Acad Sci (Paris) 198: 2195-2196

26. GUPTA PK, AL NADGIR, AF MASCARENHAS, V JAGANNATHAN 1980 Tissue culture of forest tees: clonal multiplication of *Tectona grandis* L. (teak) by tissue culture. Plant Sci Lett 17: 259-268

27. HERRMANN S, H SEUTHE 1982 Vegetative Vermehrung von Hybrid-Aspen durch Okulation. Holzzucht 36: 17-22

28. KIM HK, SK LEE, YW CHUN 1981 Mass propagation of tree species through in vitro culture. I. Bud culture of *Populus alba* X *P. glandulosa* F1. Res Rep Inst For Genet Korea 17: 57-63

29. KIM HK, EW NOH, JI PARK 1983 Haploid plant formation through anther culture of *Populus glandulosa*.

30. LESTER DT, JG BERBEE 1977 Within-clone variation among black poplar poplar trees derived from callus culture. For Sci 23: 122-131

31. LLOYD G, BH MCCOWN 1981 Commercially feasible micropropagation of mountain laurel (Kalmia latiflora) by use of shoot tip culture. Proc Int Plant Propagators Soc 30: 421-427

32. LUBRANO L 1981 Micropropagation du peuplier. *In* IUFRO Sect S2 01 5. Int Workshop "In vitro" Cultivation For Tree Species, Fontainbleau, France, pp 81-86

33. MASCARENHAS AF, S HAZARA, U POTDAR, DK KULKARNI PK GUPTA 1982 Rapid clonal multiplication of mature forest trees through tissue culture. *In* A Fujiwara, ed, Plant Tissue Culture 1982, Proc 5th Int Cong Plant Tissue Cell Cult, Tokyo, pp 719-720

34. MATHES MC 1964 The *in vitro* formation of plantlets from isolated aspen tissues. Phyton 21: 137-141

35. MCKEAND SE, RJ WEIR 1984 Tissue culture and forest productivity. J For 82: 212-218

36. MELCHIOR GH, H SEUTHE 1985 Kopulationsunverträglichkeit zwischen Aspenarten und Aspen- und Hybridaspenunterlagen. Gartenwissenschaft (in press)

37. VERMA DC, SR WANN 1983 Isolation of high yield of viable protoplasts from quaking aspen seedlings and cultured loblolly pine cell suspensions. *In* I Potrykus, CT Harms, A Hinnen, R Hütter, PJ King, RD Shillito, eds, Protoplasts 1983 Poster Proceedings, Birkenhäuser Verlag, Basel, pp 10-11

38. WANG C, C SUN, T CHU 1975 The induction of *Populus* pollen-plants. Acta Bot Sin 17: 56-59

39. WANN SR, DW EINSPAHR 1985 Reliable plantlet formation from seedling explants of *Populus tremuloides* (Michx.). Silvae Genet (in press)

40. WHITEHEAD HCM, KL GILES 1977 Rapid propagation of poplars by tissue culture methods. NZ J For Sci 7: 40-43

41. WINTON LL 1968 Plantlet formation from aspen tissue culture. Science 160: 1234-1235

42. WINTON LL 1970 Shoot and tree production from aspen tissue cultures. Am J Bot 57: 904-909

43. WINTON LL 1971 Tissue culture propagation of European aspen. For Sci 17: 348-350

44. WOLTER KE 1968 Root and shoot initiation in aspen callus cultures. Science 219: 509-510

45. ZAVITKOVSKI J 1979 Energy production in irrigated, intensively cultured plantations of Populus 'Tristis No. 1' and jack pine. For Sci 25: 383-392

17. EUROPEAN HARDWOODS

V. CHALUPA

1. INTRODUCTION

Hardwood forests cover about 40% of the forested area in Europe. The proportion of hardwoods in the forest increases from the north to the south of Europe. Whereas in the northern European countries (Norway, Sweden, Finland) hardwoods cover only 10-14% of the total forested area, in central and southern Europe hardwoods cover about one third (Federal Republic of Germany 32%, Czechoslovakia 36%), and in some southeast European countries more than one half (Bulgaria 74%, Rumania 76%) of the forested area.

The most important European hardwoods include various species of the genus Quercus, Fagus, Betula, Tilia, Fraxinus, Acer, Carpinus, Sorbus, Castanea, Prunus, Ulmus, Salix. Hardwoods are propagated mainly by seeds (with the exception of willows). However, propagation by seeds is not without problems. For example, because of air pollution damage many forest stands have ceased to produce seeds. To complicate matters, vegetative propagation of hardwoods by cuttings is not satisfactory for most hardwood tree species either (48).

Micropropagation of hardwood trees should speed up hardwood tree improvement programs (5, 6, 22, 23, 37). Breeding of hardwoods is a slow and difficult process, because the juvenile non-flowering stage often lasts a long time. On the other hand micropropagation of selected rapidly growing and resistant trees could bring significant genetic gains in a short time. Rapid vegetative propagation of selected trees with valuable forms of wood, highly valued in the furniture industry, would have important economic effects.

Experiments with tissue cultures of European hardwood species have been carried out for a long period. Gautheret (26, 27) was the first to achieve callus proliferation and adventitious bud differentiation in tissue cultures of hardwoods. For a long time the main objective was to establish hardwood callus cultures. Until 1975 most efforts were focussed on regeneration of trees from such callus cultures. Experiments with organ cultures are of more recent origin.

2. PROPAGATION OF HARDWOODS BY ORGAN CULTURES

The method of axillary bud proliferation is applicable to most European hardwood species. The main advantages of this method are that the multiplication rate is high, that organ cultures are genetically stable, that only small pieces of tree are required for culture establishment, and that virus-free trees can be produced. In some cases, new shoots can also be produced from adventitious buds formed on the stem or callus.

The main stages of hardwood micropropagation by organ culture are: culture establishment, shoot multiplication, rooting of shoots, and transfer of rooted plantlets to the field. Each stage is critical for the successful regeneration of trees. The greatest difficulties are experienced with shoot multiplication, transfer of hardwood plantlets from the agar medium to soil, and hardening off of the plantlets.

2.1. Culture establishment and shoot multiplication

Shoot tips and short nodal segments were most commonly used as explants for the establishment of hardwood cultures (9, 10, 11, 13, 14, 15, 16, 17, 39, 42, 53). Embryos dissected from seeds are another source of explants (2). Actively growing shoots are often better as explants than dormant shoots for hardwood culture establishment (9, 10, 11, 13). The appropriate time to collect actively growing shoots is in the spring and early summer, when shoots are not yet severely contaminated with bacteria and fungi. Excellent sources of actively growing shoots are young seedlings (1-12 months old) grown from seeds in a greenhouse or under laboratory conditions (11, 13). Hardwood shoots taken in the winter can be used for culture establishment if the explants are taken at the end of bud dormancy. The rapidity of axillary bud sprouting is closely connected with the end of dormancy. Shoots taken in the winter period are often severely contaminated and disinfection of winter explants is often difficult.

Before surface sterilization of the explants, all leaves are removed from the actively growing shoots and the stems are cut into 1-2 cm long nodal segments with 1-3 axillary buds. The most commonly used sterilizing solutions contain sodium or calcium hypochlorite (2, 9, 43) or mercuric chloride (10, 13, 14, 15, 43). For sterilization of actively growing shoots, which are more sensitive to the sterilizing process, a weak solution of mercuric chloride (0.1-0.2%) has been employed (10, 13, 14, 15). If the duration of sterilization is correctly chosen, microorganisms are destroyed and plant tissue is not damaged. The duration of effective surface sterilization of actively growing hardwood shoots usually ranges from 18 to 25 min (10, 13, 14, 15). After sterilization with mercuric chloride, the explants are washed three times in sterile distilled water.

Some hardwood species (Quercus, Castanea, Fraxinus, Betula, Carpinus) contain polyphenolic compounds, which are oxidized after explant excision. After placement on the agar media these substances are released and darken the agar medium. This inhibits growth of the explants and often results in their death. Browning occurs more frequently if explants from older plants are used. To prevent browning of hardwood explants several methods are recommended: a) explants are transplanted into fresh medium whenever browning occurs, b)newly excised explants are soaked for several hours in sterile distilled water or in a solution of antioxidants, c) explants are grown on media containing antioxidants, d) explants are incubated in low light intensities or in darkness.

The growth and proliferation of shoots in hardwood cultures was stimulated by cytokinin. Of the tested cytokinins (kinetin, benzylaminopurine (BAP), N^6-isopentenyladenine (2iP), zeatin) BAP was the most effective cytokinin. For most hardwood species the effective concentration of BAP was low (0.2-1.0 mg l^{-1}). Shoot proliferation of some species required a higher concentration of BAP. Nutrient media used in some hardwood species contained besides cytokinin a low concentration of auxin. Naphthaleneacetic acid (NAA) and indolebutyric acid (IBA) were the most frequently used auxins. The addition of auxin to the nutrient medium often stimulated growth of axillary buds and shoot elongation. The concentration of auxin used in most nutrient media was low (0.02-0.1 mg l^{-1}).

2.2. Induction of roots

Shoots excised from proliferating cultures were rooted either *in vitro* or in a conventional potting substrate (7, 13, 42). The advantage of the *in vitro* method is that shoots are rooted under controlled environmental conditions and that root induction can be stimulated by an appropriate concentration of auxin in the nutrient medium. The danger of root deformation and root damage by transplanting from the agar medium to soil are disadvantages of *in vitro* rooting. Therefore, rooting is often carried out *in vivo*. Excised shoots that are rooted directly in the potting mixture form roots that are well adapted to growth in soil. Furthermore, *in vivo* rooting is less laborious and less costly.

Direct rooting in a potting mixture is suitable mainly for species, which form roots easily, such as willows, birches, and elms. Microshoots were inserted into the potting mixture (peat and perlite, peat and sand) and were subsequently maintained in a warm (21-25°C) and humid (80-90% RH) atmosphere (7, 42). The substrate temperature was maintained by electric heating cables, the relative humidity by a humidistatically controlled vaporiser.

For induction of root formation on agar nutrient media, a reduced concentration of salts and organic compounds was used. When the salt concentration in the medium was lowered to one-half or to one-third of the normal strength, the rooting percentages increased (10, 13, 14, 43). Low salt media such as Gressoff and Doy (29) or woody plant medium (36) stimulated root formation of hardwoods more than the high salt medium of Murashige and Skoog (38).

Root formation was stimulated by auxins. The most frequently used auxins were IBA and NAA, which were used either separately, or together (10, 13, 14, 15). The nutrient media used for root formation did not contain cytokinin, because cytokinins stimulate callus formation and inhibit root formation. A low concentration of auxin (0.1-0.5 mg l^{-1}) in the medium stimulated root induction and supported root elongation on the same medium. A two step process of root formation, whereby roots are induced by a high auxin concentration and then elongate on an auxin-free medium, was employed for a few hardwood species (43, 53).

Shoots placed on the agar medium formed roots within 10-14 days; after another 1-2 weeks the roots elongated. For transplantation into soil plantlets with short roots are preferred, because the danger of root damage and root deformation is higher when plantlets with long roots are used.

2.3. Propagation of selected species by organ culture

A large number of commercially important hardwood species are already propagated by organ culture, and the number of species responding positively is continually increasing.

The media used most frequently and with greatest success are Murashige and Skoog (MS) medium (38), woody plant medium (WPM) (36), and broadleaved tree medium (BTM) (10, 13, 15). Gresshoff and Doy (GD) medium (29), Wolter and Skoog (WS) medium (55), and White medium were used less frequently.

For propagation of hardwood species the age of the tree providing the initial explants is significant. Explants taken from young plants are usually best for the establishment of cultures. Some hardwood species were propagated using explants from mature trees.

To date a large number of hardwood species has been propagated *in vitro* by organ culture.

2.3.1. *Acer platanoides* L. (16). Shoot tips were cultured on a modified MS medium supplemented with BAP (5 μM) and with IBA (0.5-5.0 μM). After approximately six weeks in culture, formation of multiple shoots was observed. *A. platanoides* cv. Crimson Sentry produced less than 10 shoots per explant per culture period. Excised shoots were rooted on agar medium supplemented with IBA.

2.3.2. *Acer pseudoplatanus* L. (13, 44). Seedling shoot tips and short nodal segments were grown on a modified MS medium and on WPM supplemented with cytokinin. A low concentration of BAP (0.2-1.0 mg l^{-1}) and IBA (0.1 mg l^{-1}) stimulated shoot proliferation. Microshoots, 1.5-3 cm long, were excised and were rooted on GD medium and on WPM (half-strength). Rohr (44) propagated adult trees with a wavy grain pattern, using stump sprouts for the initial explants. Shoot tips were grown and rooted on diluted MS medium (1/3 of original concentration). Single node explants produced axillary shoots which were excised and rooted *in vitro*. Shoots and roots developed simultaneously. Activated charcoal (5 g l^{-1}) added to the nutrient medium promoted formation of roots and growth of the explants. The cultures were maintained in darkness at 23°C during the first week and were subsequently cultured under long day (16 h) at the same temperature. Roots developed within six weeks.

2.3.3. *Betula pendula Roth. (10, 13, 28), and Betula pubescens Ehrh. (13)*. A high multiplication rate of both species was achieved on BTM and WPM containing a low level of cytokinin. BAP (0.2-1.0 mg l^{-1}) was the most effective cytokinin for the promotion of shoot proliferation and multiple shoot formation. Shoot tips and short nodal segments, used as initial explants, produced new shoots within 4 weeks. After the second or third transfer multiple shoots developed (Fig. 1). Every 4 weeks 10-30 shoots were harvested from each culture and used for rooting (Fig. 2). The basal portion of the cultured tissue was divided into 5-10 pieces which were placed on a fresh multiplication medium, where proliferation continued.

Formation of multiple shoots was achieved from multiple explants of young seedlings and from mature trees. Shoots taken from the lower part of crowns of mature trees were used as initial explants. Explants taken from mature trees began to proliferate and form multiple shoots after a longer period than explants of seedlings.

Microshoots 1-3 cm long, excised from proliferating cultures, were rooted either on agar nutrient media or as cuttings in a conventional potting mixture. For rooting of microshoots on agar media, GD medium, BTM and WPM (half-strength) were used. Media were supplemented with a low concentration of auxin (IBA 0.1-0.4 mg l^{-1}). Good root development occurred in 95-100% of the microshoots within 2-3 weeks.

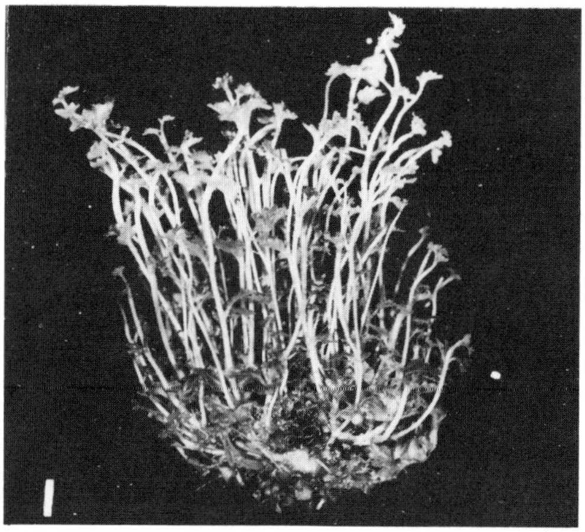

FIGURE 1. Multiple shoots of Betula pendula.
FIGURE 2. Rooted plantlet of Betula pendula.

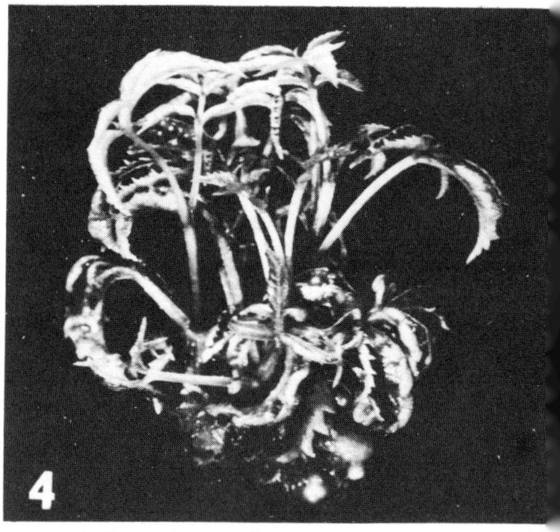

FIGURE 3. Formation of multiple shoots of Carpinus betulus.
FIGURE 4. Multiple shoots of Fraxinus excelsior.

2.3.4. Carpinus betulus L. (10, 13). Nodal segments were placed on BTM or WPM. Each segment produced 1-3 new shoots within 5-6 weeks. A low concentration of BAP (0.2-1.0 mg l^{-1}) stimulated the proliferation of shoots (Fig. 3). Rooting of a high percentage (80-90%) of excised microshoots was obtained on WPM (half-strength) containing both IBA 0.2 mg l^{-1} and NAA 0.2 mg l^{-1}. After

3-4 weeks the rooted plantlets were transplanted to a potting mixture.

2.3.5. Castanea sativa Mill. (4, 51, 52, 53, 54). Shoots were multiplied on a modified MS medium (half-strength of nitrates) and on Lepoivre (40) medium. On medium without BAP shoot growth was slight and shoot multiplication was near zero. The maximum number of shoots was obtained with 1-2 mg l^{-1} of BAP, but these shoots failed to elongate. Concentrations of BAP between 0.1 and 0.5 mg l^{-1} were most suitable for optimum proliferation and elongation of shoots. Biondi *et al*. (4) induced formation of multiple shoots on explants taken from stump sprouts of mature chestnut trees. Newly formed summer buds responded best in culture. MS medium and Schenk and Hildebrandt medium promoted the growth of axillary buds. Strong intervarietal differences were found; some varieties responded significantly better in culture than others.

Shoots longer than 2 cm were used for rooting. The highest rooting percentages were obtained when excised microshoots were placed on MS medium (half-strength) containing auxin (IBA 3 mg l^{-1}) for 12 days. Subsequently the shoots were transferred to an auxin-free medium for root elongation. In the best treatment 72% of shoots rooted. The rooting percentage was low if the cultures were kept in darkness for 2 weeks prior to rooting. The highest rooting percentage was obtained when microshoots excised from cultures were treated as cuttings: i.e., the basal part of the shoots was immersed in a concentrated IBA solution (0.5-1.0 mg l^{-1}) for a period of 2 to 15 min, after which the shoots were placed on an auxin-free agar nutrient medium. Immersion in IBA (1 mg l^{-1}) for 2 minutes was the most effective treatment. Roots started to develop after 12-14 days.

2.3.6. Fagus sylvatica L. (1, 9). Shoot tips and nodal segments from actively growing seedlings and embryonal explants were used as initial material. Modified MS medium and WPM supplemented with a low level of BAP (0.2-1.0 mg l^{-1}) and with auxin (IBA or NAA 0.02-0.1 mg l^{-1}) stimulated shoot growth. The number of shoots produced was usually limited to one or two per explant. In few cases 5-9 shoots per explant were produced. Bud explants from mature trees remained mostly unresponsive under the experimental conditions. Plantlet regeneration was achieved from juvenile explants.

2.3.7. Fraxinus excelsior L. (13). Shoot growth from axillary buds was best on WPM and on modified MS medium supplemented with cytokinin. Multiple shoot formation (Fig. 4) was obtained on WPM and on MS medium supplemented with higher concentrations of BAP (2.0-5.0 mg l^{-1}). Within 4 weeks each explant produced 2-6 new shoots which were divided and placed on a fresh medium of the same composition where proliferation of shoots continued.

Excised microshoots 2-3 cm long were rooted on low salt, agar nutrient media. GD medium and WPM (half-strength) supplemented with an auxin (IBA 0.2-0.5 mg l^{-1}) promoted root formation. Roots developed within 2-4 weeks and the rooted plantlets were transplanted to a potting substrate.

2.3.8. Prunus avium L. (17, 18, 21, 43). Shoots were multiplied on modified MS medium or on Lepoivre medium (40) supplemented with an appropriate balance of cytokinin and auxin. Shoots taken from the crown of mature trees or from root suckers were used as a source for initial explants. Nutrient media containing BAP as cytokinin (1 mg l^{-1}) and IBA as auxin (0.1 mg l^{-1}) stimulated rapid shoot proliferation. Each explant produced 4 to 15 shoots (depending on the clone) in 4-5 weeks. For elongation of short shoots cultures were transferred to media supplemented with a high concentration of GA_3 (1-10 mg l^{-1}). Cornu and Chaix (17) found that six month storage of cultures at 2°C without light doubled the proliferation rate of cultures after their return to normal temperature. Some clones survived more than one year in cold storage.

Microshoots, 2-3 cm long, were excised and rooted on an auxin-free modified MS medium (1/5 of original salt concentration). Prior to placing on this auxin-free agar medium, the basal part of the shoots was immersed in concentrated IBA solution (5 x 10^{-3} M) for several seconds. High rooting percentages (67-100%) were obtained under low light intensities and at relatively low temperature (19°C). For rooting large numbers of microshoots of mature wild cherry trees, another method was used (17). Elongated microshoots were rooted on a diltuted agar MS medium supplemented with an auxin. IBA was more effective than IAA or NAA. For most clones the optimum concentration of IBA was 1 mg l^{-1}. Clones which showed poor rooting, could be rooted on media containing a high level of auxin (IBA 10 mg l^{-1}) but with some callus formation. Low light intensities and lower temperature (21°C) were most effective in promoting rooting of excised shoots. Druart et al. (21) rooted wild cherry shoots longer than 1.5 cm on agar media containing IBA (2 mg l^{-1}). High rooting percentages (90%) were obtained when the cultures stayed for 8-10 days in darkness at the beginning of the rooting stage. More roots were formed at 25 than at 28°C.

2.3.9. Prunus padus L. (13). MS medium and WPS supplemented with cytokinin and auxin promoted the proliferation of shoots. Shoot tips and nodal segments used as initial explants produced new shoots within 4-5 weeks. Multiple shoots developed after the third transfer (Fig. 5). Every 4 weeks cultures were divided into 4-8 pieces which were transferred to a fresh multiplication medium. A high multiplication rate was achieved on MS medium containing a low level of BAP (0.6-1.0 mg l^{-1}) and auxin (IBA 0.1-0.2 mg l^{-1}).

Shoots, 2-4 cm long, excised from proliferating cultures were rooted. High rooting percentages (75-90%) were obtained on WPM containing a low concentration of auxins (IBA 0.5 mg l^{-1} and NAA 0.5 mg l^{-1}). After 3-4 weeks the rooted plantlets were transplanted to a potting mixture.

2.3.10. Quercus robur L. (9, 10, 13, 15) and Quercus petraea (Matt.) Liebl. (13). Shoot growth from axillary buds was most rapid on BTM and WPM supplemented with a low concentration of cytokinin. On MS medium only a few explants formed long shoots. On WPM shoots attained a length of 3-5 cm and well developed leaves within four weeks when supplied with 0.2-2.0 mg l^{-1} BAP. These shoots were cut into short nodal segments and placed on a fresh medium. Within 3-4 weeks each segment produced 1-9 new shoots (Fig. 6) that were divided and subcultured for further proliferation.

Microshoots, 2-4 cm long, showed 75-90% rooting in 2-3 weeks on GD medium, BTM and WPM (all media half strength) containing a low

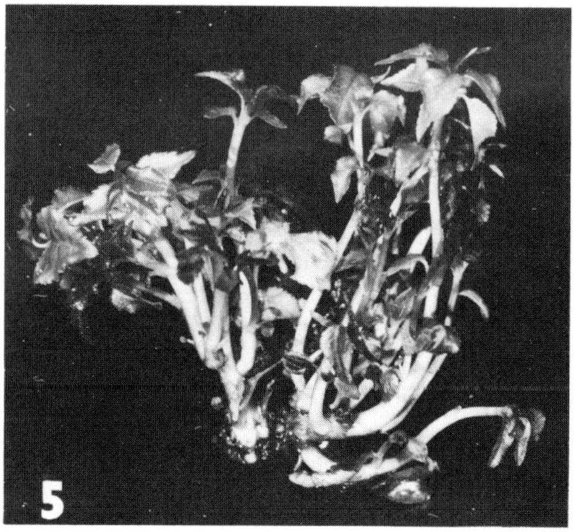

FIGURE 5. Multiple shoots of Prunus padus.
FIGURE 6. Formation of multiple shoots of Quercus robur.

FIGURE 7. Multiple shoots of Silix alba.
FIGURE 8. Multiple shoots of Sorbus aucuparia.

concentration of sucrose (10-15 g l^{-1}) and auxins (IBA 0.3 mg l^{-1} and NAA 0.1-0.3 mg l^{-1}).

2.3.11. Quercus suber L. (2, 39). Embryos cultured on an agar medium produced shoots (2). Pardos (39) used nodal segments of seedlings to obtain new shoots from axillary buds. His cultures were grown on MS medium containing cytokinin (BAP) and auxin (NAA). High concentrations of BAP (10^{-6} and 10^{-5} M) induced several buds (up to eight) from each node, while low concentrations only produced one bud per node. Shoots produced on media with a high concentration of BAP had very short internodes and twisted leaf edges. The single shoots that formed at low BAP concentrations were longer and their leaves showed a normal morphology.

Microshoots were excised and rooted on MS agar medium. They were first placed on media containing auxin (IBA or NAA) for 7 and 15 days, and then transferred to auxin-free basal MS medium. Rooting ability was related to shoot length, the minimum shoot length for rooting being 10 mm. Shoots placed for 7 days on agar medium with 1 mg l^{-1} IBA formed roots the easiest. Activated charcoal improved rooting when added to the basal MS medium after the hormonal treatment. Plants transferred from test tubes to soil did not survive longer than three months.

2.3.12. Salix alba L. (10, 13, 14, 25, 42), Salix fragilis L. (13, 14), Salix viminalis L. (10, 13, 14, 42). WPM supplemented with a low concentration of BAP (0.2-1.0 mg l^{-1}) stimulated rapid growth of shoots, with formation of multiple shoots occurring after the third transfer (Fig. 7). Every 4 weeks 5-10 shoots were harvested from each culture and used for rooting. The production of microshoots varied greatly between clones and species. Salix viminalis clones produced multiple shoots the fastest, whereas Salix alba clones were the least productive.

Rapid micropropagation of shoots was also achieved by repeated division of stems into a large number of segments. Nodal segments placed on GD medium or on WPM containing a low concentration of sucrose (10-15 g l^{-1}) and of auxin (IBA 0.1 mg l^{-1}) rooted within 5-10 days while at the same time producing a shoot. Within 50-60 days these shoots elongated to 40-60 mm. Each shoot was then again cut into 3-5 nodal segments and the whole cycle was repeated.

Some microshoots were rooted on low salt media (GD, BTM, WPM) containing a low concentration of sucrose (10-15 mg l^{-1}) and auxin (IBA 0.1-02 mq l^{-1}). Most microshoots were rooted in a conventional potting mixture (peat and perlite 1:1, or peat and vermiculite

FIGURE 9. Multiple shoots of *Sorbus torminalis*.
FIGURE 10. Multiple shoots of *Ulmus campestris*.

FIGURE 11. Multiple shoots of *Tilia cordata*.
FIGURE 12. Rooted plantlet of *Tilia cordata*.

1:1) at high relative humidity (85-95%) and temperature (24-25°C). High rooting percentages (95-100%) were obtained.

2.3.13. *Sorbus aucuparia L. (10, 13, 14) and Sorbus torminalis (L.) Cr. (13)*. MS medium with low concentrations of BAP (0.6-1.0 mg l^{-1}) stimulated multiple shoot formation (Fig. 8,9). A high shoot multiplication rate was obtained on MS medium with a low concentration of BAP (0.6-1.0 mg l^{-1}) and IBA (0.05-0.1 mg l^{-1}). Shoot tips and short nodal segments were used as initial explants. Four weeks after the second transfer the first multiple shoots appeared. Every 4 weeks 3 to 15 shoots were harvested from each culture and used for rooting. The basal portion of the cultured tissue was divided into 3-10 pieces which were placed on a fresh medium where shoot muliplication continued. Multiplication rates of _S. aucuparia_ cultures were higher than those of _S. torminalis_ cultures.

Microshoots, 2-3 cm long, were rooted on GD medium and WPM (both media half-strength) with a low concentration of sucrose (5-10 g l^{-1}) and auxins (IBA 0.3 mg l^{-1} and NAA 0.3 mg l^{-1}). Most shoots rooted within 3-4 weeks.

2.3.14. _Tilia cordata Mill. (10,13,15), and Tilia platyphyllos Scop. (13)_. Multiple shoot formation (Fig. 11) occurred on a modified MS medium and on WPM containing BAP (0.2-2.0 mg.l^{-1}) and IBA (0.1 mg l^{-1}). Each nodal segment produced 1-8 new shoots within 3-4 weeks. The mean number of shoots produced ranged from 2 to 6 shoots in each 4 week culture period, depending on the clone.

Microshoots were excised and rooted on GD medium, BTM or WPM (half-strength) with a low concentration of auxins (IBA 0.3 mg l^{-1} and NAA 0.1 mg l^{-1}). Within 2-3 weeks 80-100% of the microshoots formed roots (Fig. 12).

2.3.15. _Ulmus campestris L. (3,9,13), Ulmus laevis Pall. (9), Ulmus glabra Hud. (9)_. Proliferation of axillary shoots of elms was achieved on a modified MS medium and on WPM supplemented with BAP. Higher concentrations of BAP (0.7-2.0 mg l^{-1}) stimulated formation of multiple shoots (Fig. 10). For micropropagation of mature trees of _U. campestris_, Biondi _et al_. (3) used bud explants from root suckers. These buds proliferated on a medium supplemented with 0.7 mg l^{-1} of BAP.

Microshoots, 2-4 cm long, were excised and rooted either on agar nutrient media or as cuttings in a potting mixture. High rooting percentages (80-90%) were obtained within 2-3 weeks on GD medium and on WPM (half-strength) with a low concentration of IBA (0.3-0.5 mg l^{-1}). Good rooting was also obtained by immersing the basal part of the shoots in a concentrated IBA solution (80 mg l^{-1}) for a period of 8 hours followed by planting in a peat and perlite 1:1 v/v mix. The shoots rooted at high air humidity (80-95%) and temperature (25°C) within 2-3 weeks.

3. PROPAGATION OF HARDWOODS BY CALLUS CULTURES

Callus tissues of various hardwood species were first established a long time ago (26, 27, 33, 34). However, the formation of shoots in callus cultures is still impossible for many hardwood species.

Roots were induced in callus cultures of Ulmus campestris, Betula pendula, and Tilia cordata on media supplemented with a low concentration of NAA or IBA (7, 33, 34, 35). Wright and Northcote (56) induced short roots in callus cultures of Acer pseudoplatanus on media containing a low concentration of kinetin. Huhtinen and Yahyaoglu (32) obtained roots in callus cultures of Betula pendula grown on a medium containing IAA and kinetin. Chalupa (7) induced roots in callus cultures of Salix alba and Ulmus campestris on media with a low concentration of NAA.

Shoot formation and regeneration of plantlets from callus cultures was achieved only with a few hardwood species. Huhtinen and Yahyaoglu (30, 32) regenerated early flowering trees of Betula pendula from a cambial callus that was initiated on MS medium with kinetin (0.5 mg l^{-1}) and IAA (25 mg l^{-1}). After several weeks roots differentiated first and green shoots later. These shoots were excised and rooted on MS medium containing 2,4-D (0.1 mg l^{-1}). The newly regenerated plants formed male flowers at a young age. In another report Huhtinen (31) described the process of plantlet regeneration from anther cultures of Betula pendula. Callus tissues derived from anthers were cultured on MS medium containing casein hydrolysate (1 g l^{-1}), kinetin (0.1-1 mg l^{-1}) and 2,4-D (1-10 mg l^{-1}). After transfer of this callus to basal medium containing BAP and NAA shoots with small green leaves were formed. Radojevic (41) cultured anthers of Aesculus hippocastanum on MS medium containing casein hydrolysate (200 mg l^{-1}), kinetin (1 mg l^{-1}) and 2,4-D (1 mg l^{-1}). After 5-8 weeks in culture embryos developed which grew into plantlets. After transfer of these plantlets to a medium lacking growth hormones, long roots developed. Haploid karyotypes were observed in the root cells of three plantlets.

Druart (19, 20) obtained adventitious shoots and plantlets from callus of four Prunus species, Ulmus campestris and Aesculus hippocastanum. Chalupa (7) induced shoot formation in callus cultures of Ulmus campestris grown on modified MS medium with BAP (0.5-1.0 mg l^{-1}). Excised shoots were rooted on WS medium containing IBA (0.5 mg l^{-1}). All propagules regenerated from the same mother tree showed uniformity in growth rate and leaf form (8). In another study, Chalupa (11) cultured shoot tips and short nodal and internodal segments of Betula pendula on modified MS medium containing BAP (0.4-0.6 mg l^{-1}) and IBA (0.05 mg l^{-1}). Within 4-6 weeks callus tissue with numerous buds was formed on the lower part of the shoot tips and on the nodal and internodal segments. Callus tissue with shoot buds was also formed on new, developing leaves which were in contact with the agar medium. Callus with buds was transferred to fresh medium and continued to form new buds which gradually developed into shoots. Excised shoots were rooted on a modified MS medium (half-strength) containing IBA (0.1 mg l^{-1}). The regenerated trees showed genetic uniformity.

Srivastava and Steinhauer (49) regenerated birch plants from callus tissue derived from catkins of Betula pendula. Catkins were cultured on White medium supplemented with IAA, kinetin, casein hydrolysate, and adenine. The cut ends of the axis and scales of catkins in contact with the medium, produced callus. Many roots and shoots differentiated from the callus. Shoots were excised and rooted on a fresh medium. The callus maintained its organogenetic capacity for more than 10 months. Histological studies of the callus revealed cells of different shape; cell clusters simulating early stages in embryogeny were also noticed. Srivastava and Steinhauer also regnerated new plants of Betula pendula from callus derived from shoot buds (50). Buds were cultured on modified White medium supplemented with 2 mg l^{-1} each of GA_3, IAA and BAP. These buds produced callus in 2 weeks, roots within 4 weeks and buds after 6 weeks.

Simola (47) regenerated birch plants from callus derived from young leaf blades of Betula pendula f. purpurea. Callus was formed on a medium supplemented with casein hydrolysate (1 g l^{-1}). Shoot differentiation was stimulated when casein hydrolysate was omitted and zeatin or zeatin riboside (5 or 10 mg l^{-1}), and a low concentration of auxin (NAA or NOA, 0.1 or 0.2 mg l^{-1}) or no auxin were added. Excised shoots were rooted on media containing a low level of auxin or lacking auxin. Most rooted plantlets (95%) survived the transfer from aseptic culture to soil.

In spite of numerous experiments, the induction of shoots in callus cultures of such main hardwood species as *Fagus*, *Fraxinus*, *Acer*, *Tilia* has not yet been achieved. Simola (45, 46) found that the poor capacity for organogenesis of forest tree callus cultures is not due to abnormalities in cell organelles. For example, callus cultures of *Betula pendula*, *Sorbus aucuparia*, and *Alnus incana* showed a normal fine structure. The callus cells accumulated starch and had many mitochondria. Therefore, the basic prerequisites for differentiation were present.

4. TRANSPLANTING AND HARDENING OFF OF PLANTLETS

Hardwood plantlets produced *in vitro* require an acclimatization period to adapt to the natural environment, i.e., to lower humidity, higher temperatures and higher light intensities. Plantlets grown *in vitro* desiccate more rapidly than greenhouse grown plants, mainly because of reduced levels of epicuticular wax and by the slowness of stomatal response to water stress. Plants grown *in vitro* on media containing sugar have a limited capacity for photosynthesis, and their dark respiration is higher. After transplanting it takes one or two weeks before they achieve a positive carbon balance.

Plantlets lifted from the agar medium were first washed to remove agar from the roots. The rooted plantlets were then transplanted to a potting mixture and were covered by clear plastic or by glass jars to maintain a high relative humidity. When a larger number of plantlets was transplanted, they were grown in a chamber under intermittent mist. After one or two weeks the high relative humidity (80-95%) was gradually reduced to normal values. During this process the plantlets were protected against direct sun radiation (10, 13, 14, 15, 16, 17, 43, 47).

A rapid stem elongation and formation of new leaves that are anatomically adapted to low relative humidity, was important for survival of the hardwood plantlets. After properly adapted leaves had formed, the hardwood plants were placed outdoors and grown for 1-2 months in partial shade. Survival rates of 65-90% were obtained for *Quercus*, *Betula*, *Ulmus*, *Sorbus*, *Prunus*, *Tilia*, *Fraxinus* (10, 13, 15, 17, 47).

5. FIELD PERFORMANCE OF MICROPROPAGATED TREES

Trees produced *in vitro* should be phenotypically identical and genetically stable. Genetic stability of cultures grown *in vitro* depends on the type of culture. Organ cultures are generally genetically stable. Callus cultures of hardwoods may be more stable than callus cultures of herbaceous plants. For example, Simola (45) reported that callus from *Alnus*, *Betula* and *Sorbus* was cytologically more stable than callus from herbs.

Experiments with hardwood trees indicate that trees produced by organ cultures are phenotypically uniform. Propagules of one tree each of *Betula pendula*, *Ulmus campestris*, *Sorbus aucuparia* and *Salix alba* were grown in a controlled environment in growth cabinets (9, 10, 13, 14). The growth rates, the color and the form and size of the leaves varied little within each clone. Simola (47) found that trees regenerated from callus cultures of *Betula pendula* f. *purpurea* are phenotypically uniform. A red color was characteristically prominent in all young plants and later in the youngest parts of the shoots. For the confirmation of genetic stability of micropropagated hardwood species long-term field tests are necessary.

Survival of trees propagated by organ culture and planted after hardening off in the field was high. For example, in *Quercus*, *Betula*, *Ulmus*, *Sorbus*, and *Salix* it reached 90-100% (13). Losses during the first winter period were low. The planted trees withstood even the severe frosts (-25 to -30°C) of the winter of 1984/85 without significant losses. This is an important practical advantage of micropropagation because with traditional techniques, i.e., rooting of softwood cuttings, a large percentage of propagules generally dies in the first winter (48). Simola (47) found that all birch trees regenerated from callus cultures withstood the winter frosts of central Finland.

6. PROPAGATION OF MATURE TREES OF HARDWOODS

The capacity of trees to be propagated vegetatively decreases with increasing age. Juvenile explants grow and proliferate easily while explants taken from adult parts of the tree often either do not grow at all or they start to proliferate with a delay (5). Some parts of mature hardwood trees, usually the roots and the base of the trunk, retain juvenile characteristics for a long time. For micropropagation of mature hardwood trees it is advantageous to use these juvenile parts of the tree as a source of explants.

Roots are used as explants when dealing with hardwood species that easily produce shoots on root segments naturally (*Ulmus*, *Sorbus*, *Prunus*). Cornu and Chaix (17), Cornu *et al*. (18), and Riffaud and Cornu (43) propagated mature tree of *Prunus avium in vitro* using suckers initiated on root segments in a greenhouse. Chalupa (13) and Biondi *et al*. (3) used root suckers to propagate mature trees of *Ulmus campestris*.

Stump sprouts are often used for micropropagation of mature trees. Chalupa (13,15) propagated mature trees of *Quercus robur*, *Tilia cordata* and *Salix alba* using stump sprouts as a source of primary explants. Biondi *et al*. (4) propagated mature trees of *Castanea sativa* either by grafting buds on stump sprouts or by using stump sprouts directly. Rohr (44) propagated mature trees of *Acer pseudoplatanus* by rooting shoot tips taken from stump sprouts.

For rapid propagation of mature trees *in vitro*, their rejuvenation is sometimes necessary. As main methods of rejuvenation of mature trees Franclet (24) mentioned repeated cutting back, pruning or hedging, micrografting and sprays with BAP solution. *In vitro* culture is also a rejuvenating technique, and is more efficient than the others. Franclet (24) was able to completely rejuvenate mature clones of wild cherry, selected for timber value, by *in vitro* methods.

Micropropagation of mature trees of some hardwood species is possible even if shoots are taken from branches in the crown. For example, Cornu and Chaix (17) and Cornu et al. (18) propagated mature trees of Prunus avium and Chalupa (13) mature trees of Betula pendula by culturing shoots taken from the crown.

7. REFERENCES

1. AHUJA MR 1984 In vitro induction of organogenesis in juvenile and mature beech. Silvae Genet 33: 241-242
2. BELLAROSA R 1981 In vitro culture of Quercus suber L. embryos. In Proc IUFRO Sect S2 01 5. Int Workshop "In Vitro" Cultivation For Tree Species, Fontainebleau, France, pp 119- 125
3. BIONDI S, L CANCIANI, N BAGNI 1984 Uptake and translocation of benzyladenine by elm shoots cultured in vitro. Can J Bot 62: 2385-2390
4. BIONDI S, L CANCIANI, G DE PAOLI, N BAGNI 1981 Shoot formation from bud cultures of mature chestnut. In Proc IUFRO Sect S2 01 5. Int Workshop "In Vitro" Cultivation For Tree Species, Fontainebleau, France, pp 181-185
5. BONGA JM 1982 Vegetative propagation in relation to juvenility, maturity, and rejuvenation. In JM Bonga, DJ Durzan, eds, Tissue Culture in Forestry. Martinus Nijhoff, Junk Publ, The Hague, pp 387-412
6. BROWN CL, HE SOMMER 1982 Vegetative propagation of dicotyledonous trees. In JM Bonga, DJ Durzan, eds, Tissue Culture in Forestry. Martinus Nijhoff, Junk Publ, The Hague, pp 109-149
7. CHALUPA V 1975 Induction of organogenesis in forest tree tissue cultures. Commun Inst For Cech 9: 39-50
8. CHALUPA V 1977 The use of regenerants from tissue culture of forest trees in tree breeding. In FJ Novák, ed, Use of Tissue Cultures in Plant Breeding. Proc Int Symp, Sept 6-11, 1976, Olomouc, Inst Exp Bot, Praha, pp 183-193
9. CHALUPA V 1979 In vitro propagation of some broad-leaved forest trees. Commun Inst For Cech 11: 159-170
10. CHALUPA V 1981 Clonal propagation of broad-leaved forest trees in vitro. Commun Inst For Cech 12: 255-271
11. CHALUPA V 1981 In vitro propagation of birch (Betula verrucosa Ehrh.). Biol Plant 23: 472-474
12. CHALUPA V 1981 Tissue culture of forest trees. Colloq Pflanzenphysiol Humbold-Univ, Berlin Nr 4: 101-109
13. CHALUPA V 1983 Micropropagation of conifer and broadleaved forest trees. Commun Inst For Cech 13: 7-39
14. CHALUPA V 1983 In vitro propagation of willows (Salix spp), European mountain-ash (Sorbus aucuparia L.) and black locust (Robinia pseudoacacia L.). Biol Plant 25: 305-307
15. CHALUPA V 1984 In vitro propagation of oak (Quercus robur L.) and linden (Tilia cordata Mill.). Biol Plant 26: 374-377
16. CHENG TY 1978 Clonal propagation of woody plant species through tissue culture techniques. Proc Int Plant Propagators Soc 28: 139-155
17. CORNU D, C CHAIX 1981 Multiplication par culture in vitro de merisiers adultes (Prunus avium): application à un large

éventail de clones. In Proc IUFRO Sect S2 01 5. Int Workshop "In Vitro" Cultivation For Tree Species, Fontainebleau, France, pp 71-79

18. CORNU D, JL RIFFAUD, P CAPELLI 1981 In vitro propagation of wild cherry tree (Prunus avium L.). In Proc IUFRO Sect S2 01 5. Int Workshop "In Vitro" Cultivation For Tree Species, Fontainebleau, France, pp 133-134
19. DRUART P 1980 Plantlet regeneration from root callus of different Prunus species. Sci Hortic 12: 339-342
20. DRUART P 1981 Technique de régénération de plantules sur racines de ligneux cultivés in vitro. In Proc IUFRO Sect S2 01 5. Int Workshop "In Vitro" Cultivation For Tree Species, Fontainebleau, France, pp 251-253
21. DRUART P, P BOXUS, O LIARD, B DELAITE 1981 La micropropagation du merisier à partir de la culture du méristème. In Proc IUFRO Sect S2 01 5. Int Workshop "In Vitro" Cultivation For Tree Species, Fontainebleau, France, pp 101-108
22. DURZAN DJ 1980 Progress and promise in forest genetics. In Proc 50th Anniv Conf Pap Sci Technol: The Cutting Edge. Appleton, WI, May 8-10, 1978 Inst Pap Chem, Appleton pp 31-60
23. DURZAN DJ, RA CAMPBELL 1974 Prospects for the mass production of improved stock of forest trees by cell and tissue culture. Can J For Res 4: 151-174
24. FRANCLET A 1981 Rajeunissement et micropropagation des ligneux. In Proc IUFRO Sect S2 01 5. Int Workshop "In Vitro" Cultivation For Tree Species, Fontainebleau, France, pp 55-64
25. GARTON S, PE READ, RS FARNHAM 1983 Effect of stock plant nutrition on macro- and micropropagability of Salix. Acta Hortic 131: 141-151
26. GAUTHERET R 1934 Culture du tissu cambial. CR Acad Sci Paris 198: 2195-2196
27. GAUTHERET R 1940 Nouvelles recherches sur le bourgeonnement du tissu cambial d'Ulmus campestris cultivé in vitro. CR Acad Sci Paris 210: 744-746
28. GRELLIER B, R LETOUZE, DG STRULLU 1984 Micropropagation of birch and micorrhizal formation in vitro. New Phytol 97: 591-599
29. GRESSHOFF PM, CH DOY 1972 Development and differentiation of haploid Lycopersicon esculentum /tomato/. Planta 107: 473-497
30. HUHTINEN O 1976 Early flowering of birch and its maintenance in plants regenerated through tissue cultures. Acta Hortic 56: 243-249
31. HUHTINEN O 1978 Callus and plantlet regeneration from anther culture of Betula pendula Roth. In TA Thorpe 4th Int Cong Plant Tissue Cell Cult, Calgary, Abstr No 1740, pp 169
32. HUHTINEN O, Z YAHYAOGLU 1974 Das frühe Blühen von aus Kalluskulturen herangezogenen Pflänzchen bei der Birke (Betula pendula Roth.). Silvae Genet 23: 32-34
33. JACQUIOT C 1949 Observations sur la néoformation de bourgeons chez le tissu cambiel d'Ulmus campestris cultivé in vitro. CR Acad Sci Paris 229: 529-530
34. JACQUIOT C 1955 Formation d'organes par le tissu cambial d'Ulmus campestris L. et de Betula verrucosa Gaert. cultivé in vitro. CR Acad Sci Paris 240: 557-558

35. JACQUIOT C 1966 Plant tissues and excised organ cultures and their significance in forest research. J Inst Wood Sci 16: 22-34
36. LLOYD G, B McCOWN 1980 Commercially-feasible micropropagation of mountain-laurel, Kalmia latifolia, by use of shoot tip culture. Proc Int Plant Propagators Soc 30: 412-427
37. MURASHIGE T 1974 Plant propagation through tissue cultures. Annu Rev Plant Physiol 23: 135-166
38. MURASHIGE T, F SKOOG 1962 A revised medium for rapid growth and bioassay with tobacco tissue cultures. Physiol Plant 15: 473-497
39. PARDOS JA 1981 In vitro plants formation from stem pieces of Quercus suber L. In Proc IUFRO Sect S2 01 5. Int Workshop "In Vitro" Cultivation For Tree Species, Fontainebleau, France, pp 186-190
40. QUOIRIN M, P LEPOIVRE 1977 Étude de milieux adaptés aux cultures in vitro de Prunus. Acta Hortic 78: 437-442
41. RADOJEVIC L 1978 In vitro induction of androgenic plantlets in Aesculum hippocastanum. Protoplasma 96: 369-374
42. READ PE, S GARTON, KA LOUIS, ES ZIMMERMANN 1982 In vitro propagation of species for bioenergy plantations. In A Fujiwara, ed, Plant Tissue Culture 1982, Proc 5th Int Cong Plant Tissue Cell Cult, Tokyo, pp 757-758
43. RIFFAUD JL, D CORNU 1981 Utilisation de la culture in vitro pour la multiplication de merisiers adultes (Prunus avium L.) sélectionnés en forêt. Agronomie 1: 633-640
44. ROHR R 1985 Vegetative propagation of wavy grain sycamore maple (Acer pseudoplatanus L.). (personal communication)
45. SIMOLA LK 1981 Ultrastructure of callus cultures from trees. In Proc IUFRO Sect S2 01 5. Int Workshop "In Vitro" Cultivation For Tree Species, Fontainebleau, France, pp 201-210
46. SIMOLA LK 1982 Ultrastructure of callus cultures from Betula pendula and Picea abies. In A Fujiwara, ed, Plant Tissue Culture 1982, Proc 5th Int Cong Plant Tissue Cell Cult, Tokyo, pp 173-174
47. SIMOLA LK 1985 Propagation of plantlets from leaf callus of Betula pendula f. purpurea. Sci Hortic (in press)
48. SPETHMANN W 1982 Stecklingsvermehrung von Laubbaumarten. I. Versuche mit Ahorn, Esche, Eiche, Buche, Kirsche, Linde, Birke. Allg Forst Jagdztg 153: 13-24
49. SRIVASTAVA PS, A STEINHAUER 1981 Regeneration of birch plants from catkin tissue culture. Plant Sci Lett 22: 379-386
50. SRIVASTAVA PS, A STEINHAUER 1981 Isozymes in differentiating shoot bud cultures of Betula pendula Roth. Z Pflanzenphysiol 103: 341-346
51. VIEITEZ AM, ML VIEITEZ 1980 Culture of chestnut shoots from buds in vitro. J Hortic Sci 55: 83-84
52. VIEITEZ AM, E VIEITEZ 1980 Plantlet formation from embryonic tissue of chestnut grown in vitro. Physiol Plant 50: 127-130
53. VIEITEZ AM, ML VIEITEZ 1983 Castanea sativa plantlets proliferated from axillary buds cultivated in vitro. Sci Hortic 18: 343-351
54. VIEITEZ AM, ML VIEITEZ, A BALLESTER 1981 In vitro chestnut regeneration: anatomical and chemical changes during the rooting process. In Proc IUFRO Sect S2 01 5. Int Workshop "In Vitro" Cultivation For Tree Species, Fontainebleau, France, pp 149-152

55. WOLTER KE, F SKOOG 1966 Nutritional requirements of Fraxinus callus cultures. Am J Bot 53: 263-269
56. WRIGHT K, DH NORTHCOTE 1972 Induced root differentiation in sycamore callus. J Cell Sci 11: 319-337

18. NORTH AMERICAN HARDWOODS

D.D. MCCOWN AND B.H. MCCOWN

1. INTRODUCTION

1.1. Scope and perspective of chapter

The number of deciduous woody plants being successfully microcultured (i.e. in commercial numbers) has slowly increased in the last 5 years. However, many more woody crops are in the research and development phase or have not yet proven successful. Much of the developmental work has shifted from the public research sector to the commercial laboratories. Details of the research in the latter is often difficult to obtain. Thus a cataloguing of information on all the North American hardwoods currently being cultured is necessarily incomplete and inaccurate. We all take a more generalized approach and use the experience gained with different groups of hardwood trees as examples of responses commonly observed. The genera with which we have the most experience are *Amelanchier*, *Betula*, *Populus*, *Quercus*, and *Ulmus*.

We know of no commercial application of a micropropagation system based on adventitious (*de novo*) shoot generation of North American hardwoods. Instead, non-adventitious shoot cultures are used preferentially. Careful manipulation of *preformed* terminal and axillary vegetative buds in both the isolation and multiplication phases has made it possible to avoid the complexities of using callus or organ culture and the subsequent formation of adventitious buds. Since commercial practice must guarantee the micropropagated material as having the same characteristics as the parent used for isolation, avoidance of reorganization of a meristem via *de novo* cell differentiation and the potential resultant genetic changes (somaclonal or epigenetic mutants) is an important consideration. In addition, shoot cultures are relatively simple to manipulate and the same standard procedures are applicable across a wide range of species. Thus, only procedures employing shoot culture generated and maintained by non-adventitious means will be discussed in this chapter.

1.2. Purpose of micropropagation of deciduous trees

The primary objective of the microculture of deciduous woody plants has been to generate large numbers of propagules with predictable and desirable characteristics. In many cases this has been impossible or at best marginally successful with classical clonal propagation techniques. Many of these plants cannot be readily rooted from cuttings and problems associated with grafting, e.g. incompatibility, delayed graft failure, and expense, make micropropagation a desirable alternative. With most of the plants currently being commercially micropropagated, sexual reproduction is not a viable alternative because of the frequent seed crop failure, low germination rates, or genetically non-uniform offspring.

Micropropagation has another very significant advantage in that it allows integration into modern production systems. Micropropagation is generally performed in controlled environment facilities which generate large numbers of vigorous, uniform plants. Many stages are or have the potential of being automated including the harvesting of the culture, the sticking of the microcuttings into rooting plugs, and the transplanting of the plant/plugs to the field for continued growth.

Microculture does have a definite place in programs designed to genetically improve trees. Not only can the cultures be the source tissue used in molecular manipulation, but cultures can act as a germplasm storage mechanism; and as a technique for rapidly increasing germplasm for testing and evaluation (1). These aspects will be discussed in other chapters of these volumes.

2. MICROPROPAGATION TECHNIQUES AND APPLICATIONS

2.1. Isolation and stabilization

While it is possible to detail the source plant characteristics, specific tissues and timing for isolation of a plant in culture, there is still a certain measure of serendipity involved. Each plant needs to be individually evaluated as it will have a unique growth stage, degree of juvenility/maturity and a host of other factors that will impinge on success. The ideal scenario is to have a very juvenile plant that can be maintained in a "clean" environment and from which tissue can be harvested at the appropriate developmental stage. This is not always possible since the plant to be isolated in culture may be a mature specimen in the field. Forcing of detached shoots may be possible using resting buds but some woody plants do not respond well to forcing. If the mature specimen can be propagated using classical techniques (cutting, grafting), then this can provide a more easily-manipulated source material.

Our experience has been that the most successful tissue for isolation is buds that are just beginning to expand. The entire bud can be sterilized and the outer scales carefully removed before cutting. Shoots that are just finishing a flush of growth also can be readily isolated. We have experienced only marginal success with late summer/fall isolations since the seasonal growth cycle response is so deeply ingrained in northern woody plants. The approach using newly elongated vegetative shoots has been successful with 8 clones of *Amelanchier*, 5 clones of *Populus*, 10 clones of *Betula*, 2 clones of *Quercus*, and 5 clones of *Ulmus*. The newly formed leaves were carefully removed leaving the intact terminal and axillary buds. The tissue was surface sterilized in commercial bleach (5.3% sodium hypochlorite) in distilled water 15:85 v/v) plus 0.05% liquid detergent. The optimum time is generally 10-15 minutes. The tissue is then rinsed twice in sterile water. Various pretreatments like alcohol rinses have been ineffective or disadvantageous.

The sterile tissue is placed directly on solid medium. An intermediate liquid stage was evaluated and seemed to give little advantage. The most generally efficacious medium for initial isolates is WPM (2) modified with 0.1 μM to 1.0 μM benzyladenine (BA) and 0.2% gelrite (Kelco, San Diego, CA) as the gelling agent. Actively growing tissues are grown in baby food jars with Magenta B-caps (Magenta Crop., Chicago, IL). Cultures are grown under 25 hours cool white flourescent light (200 $\mu Em^{-2}s^{-1}$) at 24 to 26°. The frequency that newly isolated tissue needs subculturing varies. A few plants e.g. _Cornus_, _Magnolia_, _Quercus_, produce phenols and may require fresh medium weekly. Others e.g. _Amelanchier_, _Betula_, _Ulmus_ are not subcultured until a newly formed shoot can be removed from the isolated tissue.

In many cases the original tissue, although sterile and apparently uninjured by the insult of the sterilization procedure, will remain quiescent for many weeks or months. This also occurs with newly-formed shoots. Excising them from the explant and moving to fresh medium does not necessarily produce continued growth and a protracted quiescent stage can ensue. The theories to explain such behavior are several although a primary one relates to rejuvenation of the parent material. Work on _Betula_ clones using seedlings or fully mature specimen trees demonstrated an increased period of quiescence with increased source tissue maturity.

In addition to problems with shoot quiescence is the phenomenon of episodic growth. Tissue from fully mature plants will often produce one or a few new shoots, become quiescent for a period, then repeat the cycle. As the tissue becomes more juvenile the episodic flushes/quiescent periods dampen in frequency and amplitude. However, this change may span a period of years. Some plants e.g. _Quercus_, have never stopped episodic flushing _in vitro_. On occasion the tissue never produces normal growth and slowly deteriorates on each subculture.

The above observations are represented diagrammatically in Figure 1 where the establishment in culture is divided into 3 phases. These 3 phases constitute the microculture period.

During the isolation phase, one is deceived into thinking that all will be successful because shoot growth is rapid and of good quality. However this phase usually is of short duration

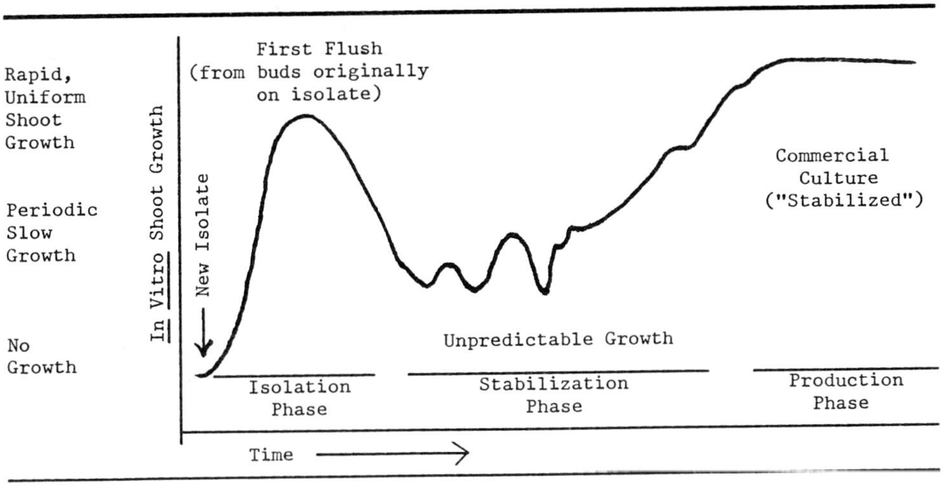

FIGURE 1. A generalized scheme of the 3 important phases in the microculture period through which a shoot must progress to be successfully microcultured. The second period is rooting and acclimation of shoots.

since the growth merely consists of the expression (growth) of buds isolated with the original explant. Once these buds have fully elongated, the rapid growth stops and the cultures enter what we call the stabilization phase. During this phase, growth is completely dependent on meristems generated in culture, and thus the culture environment has a dominant effect. Whether the culture will ever leave this phase is difficult to predict. Juvenile explants will progress through this phase most rapidly and consistently, thus rejuvenation may be an important aspect of "stabilization". The third and final phase is the production phase where the shoot cultures are now stabilized, that is the growth and quality of shoots do not change with subculturing and growth is uniform and continuous. Only cultures that have fully entered the production phase are commercially economical and can be used in subsequent acclimation and rooting procedures.

Some predictive information can be gained as to how (or if) a plant will progress through the phases of the microculture period by studying its normal growth patterns in the field (Fig. 2).

Plant Type	Ease of stabilization *in vitro* easy, short period required	moderately difficult but possible	long period, very difficult
Herbaceous	XXXXXXXXXXXXXXXXXXXXX		
Woody Shrubs	XXXXXXXXXXXXXXXXXX		
Hardwood Trees (examples)			
Group I	XXXXXXXXXXXXXXXXXXXXXXX		
Amelanchier			
Betula			
Populus			
Salix			
Ulmus			
Group II		XXXXXXXXXXXXXXXXXXXXX	
Acer			
Fraxinus			
Prunus			
Group III			XXXXXXXXXXXXXXXXXXXX
Carya			
Fagus			
Quercus			

FIGURE 2. The relative ease of stabilizing shoot cultures of various plants. The ranges shown are due to differences in the original plant physiology and in genotype responsiveness.

In general, plants that have a growth pattern marked by continuous shoot growth (growth occurs whenever the environment permits) will progress through these phases most rapidly. Thus herbaceous, seedling (juvenile) and non-dormant tissues stabilize best. Plants that have seasonal growth characterized by definite growth flushes (episodic growth) will be more difficult to fully stabilize. With those plants that have episodic growth that is dominated by internal controls (that is all growth occurs from extension of preformed buds, occurs only once or twice a season, and cannot be modified by environment) progression through stabilization has generally not been successful.

2.2. Maintenance of stabilized cultures

Once the tissue is in constant growth it is frequently subcultured on a medium different from the isolation medium. Generally vitreous tissue is produced on gelrite-solidified medium. Although the gelrite medium has the advantage of supporting fast growth, the resultant vitreous microcuttings cannot be successfully rooted. Agar-solidified medium will yield non-vitreous tissue, but the growth may be significantly inhibited. One solution is to use a combination of agar and gelrite to solidify the medium. At KHN Laboratory (Knight Hollow Nursery, Inc., Madison, WI), we find medium prepared with 0.3% agar plus 0.1% gelrite supports satisfactory growth of most clones (Fig. 3).

FIGURE 3. Actively-growing and stabilized shoot cultures of Betula (grown on agar-based medium), of Amelanchier (grown on agar/gelrite-based medium), and of Ulmus (grown on gelrite-based medium).

It is only after stabilization has occurred that any optimization studies should be initiated. Conditions used before this period are chosen to support continuous but not necessarily the best quality growth. Since the sensitivities of the tissue appear to change as stabilization progresses, the medium and environment combination that yields the best multiplication and shoot quality can only be determined after stabilization.

One of the major problems observed during this stage is over-callusing. In our experience, there is a direct inhibition of shoot growth by callusing that occurs at the bases of shoot cultures. In some cases (Acer), the problem is so serious that the genus is considered difficult to micropropagate even though it can be stabilized. Anti-auxins (e.g. TIBA) generally do not solve the problem and may compound the problem by further inhibiting shoot growth. With some plants (e.g. Ulmus), callusing is stimulated by high cytokinin levels (above 1.0 μM BA). Good quality shoots can only be produced by using a low cytokinin level (below 0.1 μM BA) and sacrificing multiplication rates.

2.3. Acclimation, rooting and field establishment

There has been no demonstrated advantage to rooting in culture (Stage III) for the hardwoods we micropropagate. Cultures with shoots for rooting are destructively-harvested and the shoots planted in either peat/plug flats (Techniculture, Castel & Cooke, Salinas, CA) or a horticultural mix of peat/perlite. Various hormone treatments have been explored but generally found to have no effect or be inhibitory to rooting. Some researchers have found a pretreatment beneficial with some woody plants (3). Shoots should be approximately 3 cm long with well-developed leaves. Shoots with poorly developed leaves generally deteriorate.

It is critical to rooting success that shoots continue to active growth over the entire rooting period. With Ulmus (also observed with Amelanchier), shoots which cease growth during the rooting phase often do not resume normal growth despite an adequate root system. Observations of death of the terminal meristem in culture coupled with this post-rooting meristem problem led to speculation about a calcium deficiency developed in culture that had carry-over effects on rooted shoots. Similar shoot-tip necrosis had been observed in cultures of herbaceous plants and the benefits of increased calcium levels have been documented (4). A multiplication medium of WPM with the calcium increased to 15 mM plus 0.02 μM BA solidified with 0.2% gelrite increased the number of Ulmus microcuttings that remained in growth from 50% to 85% (Table 1). Of the remaining 15% of the shoots, about half will eventually break into growth and the other half never show shoot growth.

Table 1. The performance in actual commercial nursery practice of Ulmus micropropagules obtained from 2 in vitro sources. Media for both sources was WPM. Data from Bernard Fourrier, McKay Nursery, Waterloo, WI.

	Number of microcuttings observed	% Rooted in greenhouse	% Transplanted to field: of total m-cuttings observed	% Transplanted to field: of rooted m-cuttings
Source 1 -agar -normal Ca	11,000	51%	25%	49%
Source 2 -gelrite -high Ca	3,100	84%	78%	93%

Betula and Populus microcuttings generally root easily directly from culture and rooting of 98% in 4 weeks is routine. The problem of cessation of growth discussed above with Ulmus has not been observed with Betula or Populus.

We have conducted or observed field plantings of micropropagated hardwood trees including Amelanchier, Betula, Populus, and Ulmus. In general, the plants respond like seedlings of the species. We find that all of the precautions necessary for handling seedling-transplants must be applied to micropropagules. Of special importance for the maintenance of quality of the transplant is the prevention of root circling and deformation. The use of modern plug systems that rely on air-pruning of the roots has been very effective in controlling this problem. With the above precautions and with good field conditions just after planting (e.g. irrigation for the first week), high field survival (greater than 90%) of the transplants is practical.

One interesting study of McCown and Amos (5) comparing the performance of field grown _Betula platyphylla szechuanica_ produced both by micropropagation and from a selected, high performing seed source demonstrates several important considerations in woody plant microculture. The initial study involved a single seedling population (Fig. 4). The micropropagated clone showed very uniform growth however its total annual growth was in general less than the growth of the seedling population. The important question was whether the smaller size of the micropropagated plants was due to the genetic potential of the original plant used for isolation or some feature inherent in the micropropagation technique.

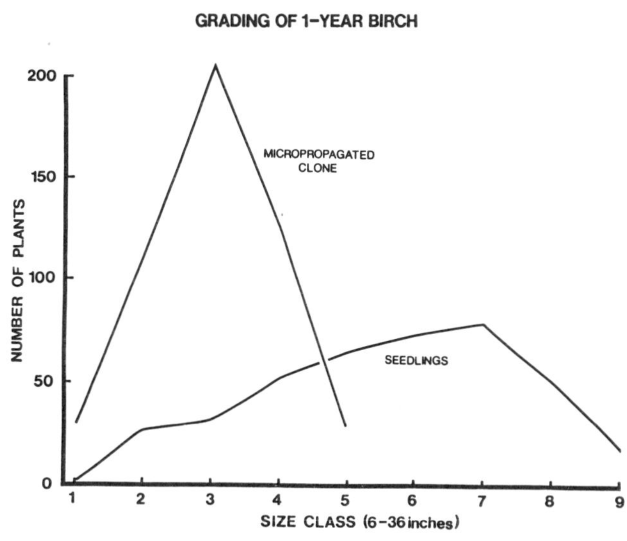

FIGURE 4. The grading by size (height) of _Betula platyphylla szechuanica_ liners, propagated by two techniques and grown in the same field for one season. The size classes were 3 inch intervals except the last class which included all plants greater than 30 inches in height. The data are for 600 plants from each source (From McCown and Amos, 5).

The second study (unpublished, McCown and Amos) involved six micropropagated clones which were selected from the seedling population in the first study. Of the 6 clones, 3 were isolated from seedlings with a small size grade and 3 from plants in the large grade. The data in the graph (Fig. 5) clearly demonstrates the importance of parental genetic potential as the average height (and node number) of the "small parent" clones performed in the small growth range while the average height (and node number) of the 3 "large parent" clones graded in the large range. The performance of the original (first study) micropropagated plant is also shown.

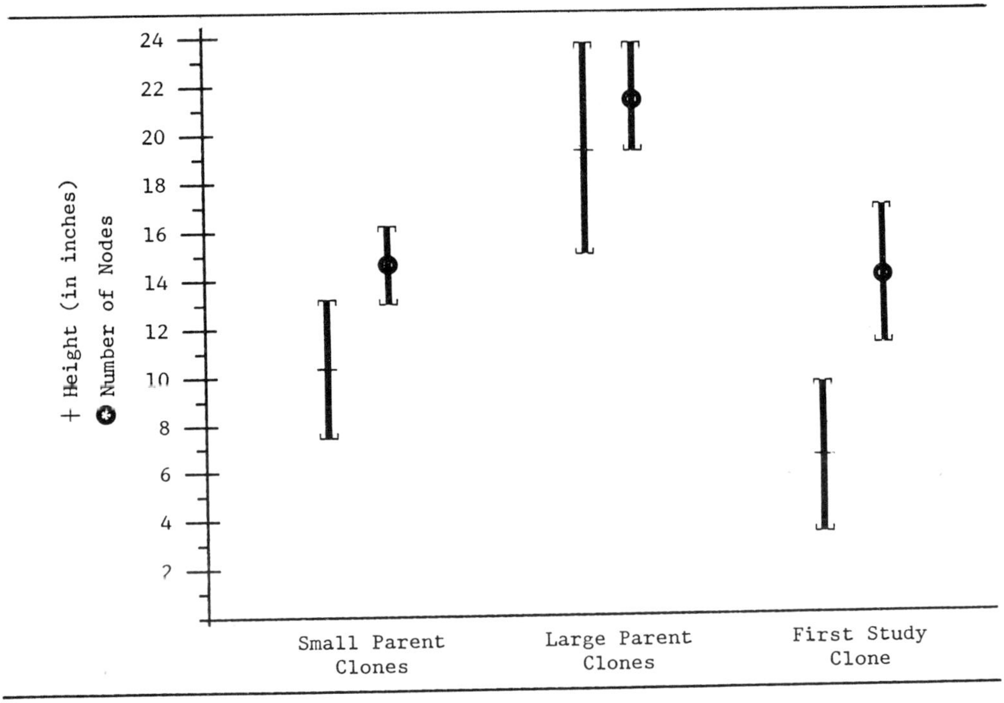

FIGURE 5. The grading by height and number of nodes of <u>Betula platyphylla szechuanica</u> plants, propagated by microculture and grown in the greenhouse. Parental plants were selected from a seedling population that had been grown in the field the previous year and were graded for height at the end of the season. The growth in the greenhouse of the original microcultured clone (first study) is shown for comparison. Small Parent and Large Parent data are 3 clones of 10 plants each. First study is data for 16 individuals. Data presented with $\pm$ one standard deviation.

2.4. Production specifics

The time to get a particular plant in production (and for KHN Laboratory this is defined as an annual minimum production of 5000 microcuttings) is directly related to the time required for acclimation and stabilization. Once the tissue is in constant growth it is simple arithmetic to predict the number of microcuttings that will be produced. For example with *Ulmus* it may take 12 to 18 months to get the tissue in constant growth. For a production culture of elm, KHN uses 10 to 15 pieces of explant in a container and when a mature culture is subcultured the shoots are divided in half, terminal portion in one vessel and the basal portion (with a fresh cut on the base of the shoot) in another vessel. The axillary buds on the shoot base will break and produce multiple shoots while tips generally do not produce axillary shoots on the required low BA medium. One stock culture of *Ulmus* will yield 2.5 to 3 new cultures. Shoot production rates for *Betula* are higher because they can be grown on a higher cytokinin medium, thus stock cultures can yield from 5 to 10 new cultures.

The time between subculturing most trees is approximately 4 weeks. If cultures are intended for harvesting and rooting, they are allowed to grow for 6 weeks to enhance shoot length and leaf development.

The critical variable in determining cost is the number of shoots produced per culture. Most of the costs can be calculated on a per culture basis (media preparation time, chemicals transfer technician time, overhead) and the advantages of getting 40 shoots rather than 10 per culture are obvious. There is an optimum number of shoots per culture that is species specific and this can be manipulated to a certain extent by hormone concentration in the medium. By increasing the cytokinin it is possible to increase the number of shoots produced but they will be smaller and may not be as easy to handle. KHN Laboratory tries to maintain a minimum 15 shoots per culture but this is not always possible. For example, *Magnolia* and *Viburnum* tend to have relatively large leaves in culture and 6 to 10 rootable shoots per culture is optimum. These are both expensive crops to produce. *Betula* and *Amelanchier* can have upwards of 40 rootable shoots per culture.

3. CONCLUDING REMARKS

The number of woody plants being microcultured successfully has increased in the last decade and we remain enthusiastic. For North American hardwoods, however, the Group III plants and the plants that show serious over-callusing pose a great challenge. Unfortunately, very little serious basic research is being done to illuminate the basis of these problems. Without more information much of the current work on these plants falls to the most primitive level of research, trial and error.

A second problem that is increasing in importance is the cost associated with microculture. The labor intensive nature of the technique has up to now limited micropropagation to plants with high individual value. This is why many of the plants currently micropropagated are woody ornamentals. If we are to consider micropropagating most forest species seriously, then microculture will have to become more automated so labor costs can be reduced. Again, the research effort on this front is minimal. We need to combine the talents of plant biologists and engineers to effectively address automation technology.

Other important problems are associated with micropropagules after they are removed from culture. Nonuniformity in rooting, slowness to root, shoot meristem problems (we discussed only one example), and root deterioration are common. Many of the growers who have the responsibility of producing a crop from rooted microcuttings are intimidated by the small size and "high tech" history of the plants. Most of these problems will be overcome as more species are micropropagated and experience increases.

4. REFERENCES

1. MCCOWN BH 1985 From gene manipulation to forest establishment: Shoot cultures of woody plants can be a central tool. TAPPI J 68:116-119
2. LLOYD G, B MCCOWN 1980 Commercially-feasible micropropagation of Mountain Laurel, *Kalmia latifolia*, by use of shoot-tip culture. Comb Proc In the Plant Propagators Soc 30:421-427

3. ZIMMERMAN RH, I FORDHAM 1985 Simplified method for rooting apple cultivars _in vitro_. J Am Soc Hortic Sci 110:34-38
4. SHA L, BH MCCOWN, LA PETERSON 1985 Occurrence and cause of shoot-tip necrosis in shoot cultures. J Am Soc Hortic Sci 110:631-634
5. MCCOWN B, R AMOS 1979 Initial trials with commercial micropropagation with birch. Comb Proc Int Plant Propagators Soc 29:387-393

19. TISSUE CULTURE OF *JUGLANS*

G.H. MCGRANAHAN, J.A. DRIVER and W. TULECKE

1. INTRODUCTION

The English or Persian walnut, *Juglans regia* L., and the eastern black walnut, *J. nigra* L., are the commercially important species within the genus *Juglans*. *J. regia* is known for its edible nuts. Worldwide production has been estimated at between 700,000 and 800,000 metric tons/year (11) making it one of the most important nut crops in the world today. *J. nigra* has long been prized as a timber species. Its wood is used for veneer, furniture, and other items where its characteristic luster and quality are needed (2).

For commercial plantings, cultivars of *J. regia* are propagated clonally by grafting on to seedling rootstock. The preferred rootstocks in California are *J. hindsii* Jeps. ex R.E. Smith (northern California black walnut) and 'Paradox', a hybrid between *J. hindsii* and *J. regia*. *J. nigra*, *J. regia* and their hybrid are used in Europe. In contrast, commercial plantations of *J. nigra* are usually entirely of sexually produced, seedling origin. However, grafted plantations of *J. nigra* have been recommended as a means to increase vigor, form, and uniformity within plantations (2).

For both species, the next step towards improved performance and uniformity is the development of clonal rootstock or "own-rooted" cultivars.

In the past, this step could not be taken because of difficulties in vegetatively propagating *Juglans* by any means other than grafting (13). Thus, the major impetus behind the tissue culture (i.e. micropropagation) of walnuts has been the need to propagate rootstocks, selections, and elite trees on their own roots both for research and for commercial applications. Recent advances in biotechnology have provided additional stimulus. New genotypes derived through somaclonal variation, endosperm culture, *in ovulo* culture, or transformation may be developed in the future (see several chapters in Vol. 1 and 2).

The purpose of this chapter is to review the progress that has been made in tissue culture of *Juglans*. Methods that have been developed in the authors' laboratories are described in detail.

2. HISTORY

Difficulties in conventional propagation efforts, specifically rooting of cuttings, led to the first studies of tissue culture of walnut. In 1969 Cummins and Ashby (9) using debarked shoots of J. nigra as the primary explant source obtained callus growth on a modified White's medium (26) supplemented with α-naphthalene acetic acid (NAA) (5.0 mg/l) and kinetin (0.2 mg/l). Callus cultures survived 4 weeks after excision from explants that had been cultured for 15 weeks. No organogenesis was observed in this callus. Several years later a meristem tip program was initiated by Lenartowicz and Millikan (16) to produce pathogen-free clones of black walnut. Meristems (0.8 to 1.0 mm) increased 8-fold in size before a "lethal browning" occurred.

In 1981, the first reports of morphogenesis in walnut tissue culture were published. Rodriquez and Sanchez-Tames (19) initiated cultures from stem segments, leaf disks, whole peeled cotyledons and root segments of J. regia seedlings and exposed them to different concentrations of Cheng's basal medium (6,7), auxins (β-indolebutyric acid [IBA], NAA, and 2,4-dichlorophenoxyacetic acid [2,4-D]) and cytokinins (kinetin, N^6-Benzylaminopurine [BAP]), with or without a 12% coconut milk supplement. Roots were obtained from callus derived from cotyledons and root segments. The best results were obtained from cotyledons grown in the presence of NAA (5 mg/l) and kinetin (0.5 mg/l). The leaf disks also produced a few weak roots. Rodriquez and Sanchez-Tames concluded that stem segments were non-morphogenic, leaf portions were potentially morphogenic, and cotyledons and root segments were clearly morphogenic. In 1982, Rodriquez (20) published a more detailed description of his work with cotyledons cultured on half-strength Cheng's basal medium (6,7) supplemented with various concentrations of the cytokinins and auxins mentioned above. The cytokinins were not required for callusing or rooting but at low concentrations (2μM) they improved rooting percentages when combined with IBA or NAA at 40μM. Root differentiation was often found associated with globular formations and anthocyanin production.

The first paper reporting in vitro clonal propagation of Juglans was also published in 1981. Chalupa (4) using nodal explants from seedlings of J. regia obtained shoot multiplication on a modified Murashige Skoog (MS) medium (18) containing BAP (0.1-0.6 mg/l) and NAA (0.1-0.3 mg/l). Roots also developed and rooted explants were transferred to a mixture of peat and perlite.

Additional papers reporting in vitro shoot multiplication in Juglans were published in 1982. Working with J. nigra, Somers et al. (22) evaluated explant source (dormant buds, shoot tips, axillary buds and embryos), method of sterilization (period of time in 0.5% NaOCl), medium [MS (18), WPM (15), Vieitez and Vieitez (25), and White's (26)] as well as levels of IBA, BAP, gibberellic acid (GA_3). A rapid transfer schedule (days 1, 3, 5, and 42) and the addition of polyvinylpyrrolidone (PVP) were also tested in an attempt to overcome problems associated with explant exudate. They found that dormant walnut buds could not be aseptically cultured but that embryos from ungerminated seeds were frequently free of microorganisms. Between 10 and 20 minutes of 0.5% NaOCl were usually required for adequate surface sterilization. Optimum levels of hormones for proliferation of axillary shoots from shoot tips were 0.2 to 1.0 mg/l BAP with 0.0 to 0.1 mg/l IBA. Rapid transfer was more satisfactory than addition of PVP to the medium and GA_3 was not beneficial in their system. The best media for the embryo cultures were WPM and MS supplemented with 5.0 mg/l BAP. Attempts to root axillary shoots were not successful. Rodriquez (21) in 1982, reported shoot multiplication in _J. regia_ on half-strength Cheng's basal medium (6,7), supplemented with BAP (0.4μM) and IBA (0.4μM).

In a preliminary account, also in 1982, of morphogenesis in _J. regia_ endosperm cultures, Cheema and Mehra (5) compared endosperm cultured with and without embryos on MS media (18) supplemented with various hormones and complex growth substances. Indirect rooting (from callus) occurred whether or not the embryo was included in the culture. Direct rooting (without intervening callus) occurred only in cultures grown without embryos. These cultures also produced "shoot-bud-like structures" after callus cultured on MS with casein hydrolysate (500 mg/l) plus 2,4-D and kinetin (1 mg/l) was transferred to basal MS.

Two additional reports on embryo culture were published in 1983, one on _J. nigra_ (3) and one on _J. regia_ (8). The latter study focused on testing the salts of a variety of published media (MS (18), Cheng's (6,7), White's (26), Gamborg _et al._ B5 (12) and several others) at one-half and full strength. Growth was assessed after 45 days by a leaf growth index (% explants with at least one leaf) x (number of leaves per explant) x (length of longest leaf), stem and root length, and fresh weight. Overall, MS (18) and B5 (12) were best for growth. BAP even at low levels (0.1 mg/l) was considered detrimental.

MS was the basal medium for a _J. nigra_ embryo culture study (3) designed to assess the effect of β-indoleacetylphenylalanine (IAAPhe) (2 mg/l) with BAP (2 mg/l) on axillary shoot formation. The maximum number of axillary shoots were formed when the medium contained both IAAPhe and BAP. IAAPhe alone in basal medium promoted elongation but not axillary shoot formation. Included in this study was a successful attempt to root axillary shoots. Over 50% of the shoots rooted when excised and transferred to one-half strength MS medium supplemented with IBA (1 mg/l). Filter paper bridges were used but the authors concluded that bridges were not necessary. Rooting occurred both in the dark and under a light/dark regime.

Steffan and Millikan (23) attempted to propagate mature J. nigra trees by meristem tip culture but growth has not been sustained beyond 8 weeks. A lethal browning presumably due to the oxidation of endogenous phenolic compounds was a major obstacle to culturing. An overnight soak in 0.01 M sodium diethyldithiocarbamate (Na-DIECA), filter paper bridges, and supplements of PVP have been suggested to overcome it. Two percent walnut endosperm extract was also reported to improve establishment and growth of cultures. Other work (14,22) suggests that rapid transfer is an adequate means for overcoming browning.

Efforts to micropropagate Juglans have also been underway in France for several years (A. Franclet, Association Foret-Cellulose, personal communication). Meynier, (17) has succeeded in micropropagating, rooting and acclimatizing a clone of J. regia x nigra using meristem tip culture to overcome problems of contamination. The original explants were derived from a clone which had been cut back annually in the field. D. Cornu and his colleague C. Jay-Allemanel (Cornu personal communication) have also been successful in micropropagation of J. regia x nigra. Their work has been focused on rejuvenation and should be published shortly.

Two research programs in California are now actively engaged in walnut tissue culture research. Driver, at the Plant Research Laboratory in Modesto, developed an improved medium for micropropagation of 'Paradox' walnut rootstock (10) and also developed techniques for rooting and acclimatization (see Chapter by Driver and Suttle, Vol. I). Micropropagated 'Paradox' rootstocks now have been planted in the field in large numbers and field trials are underway in several locations. At the United States Department of Agriculture/Agriculture Research Service laboratory in the Department of Pomology, University of California, Tulecke and McGranahan developed methods for obtaining somatic embryos and repetitively embryogenic lines from cotyledon tissue. Plants derived from these somatic embryos have also been grown in soil (24). The methods currently used for these culture systems are described in detail below.

3. MICROPROPAGATION

The methods described here have been successful for 'Paradox' (J. hindsii x J. regia), J. hindsii and Pterocarya stenoptera C. DC. The medium described is the corrected version of DKW (10) Table 1, which was a modification of WPM (15). WPM has also been used for J. nigra (14). (Note: In the original publication of DKW (10) several typographical errors occurred in the medium formulation, specifically myo-inositol and K_2SO_4. The corrected version (DKW-C) is presented in Table 1.)

3.1. Plant material

One to three month old greenhouse or field-grown seedlings can be directly initiated into culture using the techniques described. Mature selections are more difficult to culture due to excessive callusing, slow growth and contamination in culture. Although decisive studies on the effect of in vivo pretreatments on success in culture have not been undertaken, the following protocol is generally used to initiate mature selections. Dormant scionwood from the desired selection is grafted onto seedling rootstock. After the first flush in the spring has grown 30-60 cm it is cut back to 3-4 buds. These (and all future growth) are sprayed every 3-4 days with a solution of BAP (100 mg/l) and GA_3 (50 mg/l) to induce vigorous growth. Shoots are cut back successively to 3-4 buds each time they regain 15-20 cm in length. Systemic fungicide sprays are sometimes used to control contamination. Driver has observed that shoots taken from the 5th or 6th flush perform best in culture. Suckers from mature trees have also been cultured after using the same repeated cut and BAP and GA spray program but without grafting.

3.2. Initiation and multiplication

Vigorously growing stems or suckers are collected from plants in the field or greenhouse. Leaves are removed leaving 5 mm of the petiole intact to aid in distinguishing orientation. Stems are rinsed well with tap water and cut into nodal segments 3-4 cm in length, with the bud usually centered on the explants. Segments are surface-disinfected in covered jars in 20% commercial bleach with 3% liquid soap in distilled H_2O (final sodium hypochlorite concentration 1%) on a shaker or magnetic stirrer for 15-45 minutes depending on the hardiness of the tissue. They are then removed and rinsed at least 3 times in sterile distilled water. The base is recut and explants are placed on DKW-C medium (Table 1). They are transferred to fresh media if discoloration in the medium occurs. Transfers may be required daily for the first week and then weekly until the length of the bud reaches at least 2 cm. The new shoot is then excised and placed on fresh medium. Biweekly transfers are recommended unless a discoloration appears in the medium in which case more rapid transfers are desirable. When the new shoot is 4-5 cm long it is cut into nodal segments which are placed on fresh medium. Each node carries a new bud that will grow into a new shoot. With repeated transfers and cuttings the growth rate usually increases until it reaches a point of stabilization. This is considered the rapid multiplication phase.

Environmental conditions for initiation and multiplication vary in our labs. Both 24 hr light and 16/8 light/dark regimes have been used. Light is provided by a 1:1 combination of cool white and plant growth fluorescent lamps which give a photo flux density of 45-60 $\mu E\ m^{-2}\ s^{-1}$ Temperatures range between 24 and 30°C in both light regimes.

3.3. Rooting

Thousands of micropropagated cuttings of 'Paradox' have been rooted with a high (>90) percent success. Investigations on rooting of other species are underway. Preliminary results suggest that both genotype and juvenility influence walnut rooting. The rooting method described here follows that used by Driver at Plant Research Laboratory (patent pending). Rooting trials are still in progress at the USDA/ARS University of California, Davis laboratory.

Vigorous shoots (3-10 cm long) from multiplication phase cultures are cut at the base and placed in DKW-C rooting pretreatment medium (Table 1). Light intensity is increased from 50 to 66 $\mu E\ m^{-2}\ s^{-1}$, the light period is decreased from 24 to 17 hours and temperature is decreased from 28 to 19°C. After 7 days pretreated cuttings are washed in a slurry of systemic fungicides and the bases are dipped in a rooting powder composed of 2 g IBA potassium salt in 100 g talc. They are then placed in peat plugs in high density flats (Castle and Cook, Techniculture, Salinas, CA) enclosed in plastic to maintain high humidity. These enclosed flats are placed in the same environment as was used for pretreatment.

Within a week the bases swell and at this point the cuttings may be removed from plugs and placed directly in the soil in the field as described by Driver and Suttle in Vol. 2. Rooted walnuts have been difficult to transplant when conventional techniques are used. Walnut roots in the early phases of growth outpace top growth by a factor of 5. Thus, potting container size is a critical factor. Young roots are also very sensitive to injury and if the root tips are damaged, the shoot tips also die-back. As roots age they change from white to brown. The darker roots are much better able to withstand changes in environment. The method developed by Driver allows for direct field planting of plantlets before roots become visible and is commercially feasible. If few plants are needed, plantlets can be carefully transplanted to soil and acclimatized as described in the next section.

4. SOMATIC EMBRYOGENESIS

The methods described here for obtaining somatic embryos have been successful with seed from 5 cultivars of J. regia, J. hindsii and Pterocarya sp. (24).

4.1. Initiation and propagation of plants

Immature walnut fruits (6-11 weeks after pollination) are surface sterilized and dissected to remove cotyledon tissue. The lobes of the cotyledon growing into the endosperm are excised (0.5-1.5 cm pieces) and placed on a conditioning medium. This medium is made by adding IBA (0.01 mg/l), BAP (1.0 mg/l), kinetin (2.0 mg/l) and L-glutamine (250 mg/l) to DKW-C medium, Table 1. The cultures are grown in the dark at ambient room temperature (20-22°C). After 2-3 weeks the tissue is transferred to basal DKW-C without hormones or glutamine and transferred to fresh medium every 2-3 weeks. Embryogenic tissue appears after 6-24 weeks in culture and is removed to new basal medium. Somatic embryos develop and also form new somatic embryos on their root tips, cotyledons and hypocotyls. Brown tissue masses derived from hypertrophied somatic embryos also produce globular and other stages of somatic embryos. Both of these sources of somatic embryos are used to maintain embryogenic lines.

Vigorous somatic embryos with well-developed cotyledons and radicle are selected and placed on basal medium in the cold (2-6°C) for 2 months to break dormancy. Germination occurs after removal from the cold when the cultures are placed in light (16/8 photoperiod with cool white and Grolux, $\sim$ 135 $\mu Em^{-2}s^{-1}$). Root growth is improved by transfer to a simple White's medium (26) with lowered sucrose (0.5% w/v) and added charcoal (0.5% w/v). Cotyledon growth, greening and apical development takes place in the next 2-4 weeks. These germinating somatic embyros with small emerging primary leaves and elongating roots are transferred to peat plugs in vials with the above White's medium. One month later they are washed thoroughly, placed in sterilized soil in 8 ounce styrofoam cups and covered with polyethylene bags to ensure high humidity. These plantlets are watered twice weekly with nutrient solution with a systemic fungicide added to prevent wilt. When the plantlets are 5-8 cm high they are transplanted without disturbing the roots to 10 inch pots and acclimatized by gradually perforating the plastic bag. Once fully acclimatized to ambient temperature and humidity fluctuations, the plants are taken to the greenhouse or lathe house for the first year's growth.

5. PROBLEMS AND FUTURE RESEARCH NEEDS

Inspite of recent advances, the methodology for tissue culture and micropropagation of *Juglans* is still in the early stage of development. Many questions remain to be answered and many techniques are yet to be developed. The following section includes areas which are in need of further investigation.

5.1. Micropropagation of related species

Techniques have been developed to micropropagate 'Paradox' rootstock. The same techniques have been applied with varying levels of success to *J. regia* and *J. hindsii* but further improvements such as medium optimization are needed. Preliminary studies suggest that mature *J. regia* may require a higher salt concentration than 'Paradox'. Additional work on *J. nigra* is needed as well, although it appears that DKW-C may be an adequate medium for it.

5.2. Micropropagation of mature trees

Explants taken from mature trees present unique problems for growth in culture (see Vol. 1). Suckers from rootstock of mature trees of *J. hindsii* and vigorous shoots of mature 'Paradox' have been cultured but mature cultivars of *J. regia* have been difficult to establish in culture inspite of treatments described earlier. Latent contamination and slow elongation of shoots prevent rapid multiplication. The conditions under which the source plant is grown and the treatments given the explant in culture are equally important. More work is needed on both of these aspects. Micropropagation of mature trees will be essential for multiplication and evaluation of elite trees on their own roots.

5.3. Somatic embryogenesis

Somatic embryogenesis from immature cotyledons and repetitive embryogenesis from somatic embryos have been demonstrated for *J. regia*, *J. hindsii* and *Pterocarya* sp., a related genus. The methodology provides a means to make use of "somaclonal variation" in *Juglans* and provides a basis for further work on somatic embryogenesis. Other results suggest that somatic embryogenesis can be used to produce triploid plants from endosperm cultures (Tulecke, unpublished). Still needed however, is the methodology to produce somatic embryos from established cultivars.

Little work has been done on plant regeneration from protoplasts, cell suspension, callus, and leaf disks. These areas need to be investigated if techniques involving recombinant DNA are to be used for *Juglans*. Recent advances in somatic embryogenesis and micropropagation have demonstrated that the regenerative potential of walnut is good. Prospects for achieving regeneration in other tissues are promising.

5.4. *In vitro screening*

An important area which is only beginning to be approached in tissue culture is *in vitro* screening for tree improvement (1). Conventional methods of screening for response to nutrients, salts, pests and diseases in trees require large amounts of space and time. This often limits the numbers of genotypes which can be tested and therefore the rate of progress of genetic improvement. *In vitro* screening systems are also essential for progress in gene transfer technology. Tissue culture provides a compact system which can be environmentally controlled. Nodal explants, detached leaves, roots, shoots, embryos, callus and cells are all potential target tissues for screening programs. Investigations in this area are needed to increase the rate of genetic improvement in *Juglans*.

6. CONCLUSION

The impact of tissue culture will be wide-ranging in both research and production of walnuts. The genetic make-up of orchards could change drastically. The combination of a seedling rootstock with a clonal scion, a partial safe-guard against the genetic vulnerability recognized in monoculture systems, could be abandoned in the future in favor of clonal rootstock with clonal scions. Within the next 5 years clonal rootstock for *J. regia* is expected to be commercially available. This is only the beginning of a new trend in walnut production. During this period, all researchers and growers of walnut will need to be alert not only to the advantages of monoculture, but also to its genetic vulnerability.

Breeding strategies will also change. Individual selection followed by clonal propagation and release will provide a rapid means of harnessing the genetic potential of superior trees. Changes in breeding methodologies are also expected. Triploids from endosperm, somaclonal variants, inter- and intra-generic hybrids derived through embryo rescue may be developed in the near future.

In the more distant future, recombinant DNA technology will probably be applied to *Juglans* but it is still to early to predict the impact of these novel techniques.

7. REFERENCES

1. AMERSON HV, RL MOTT 1982 Phytopathology and tissue culture alliance. *In* JM Bonga, DJ Durzan, eds., Tissue Culture in Forestry. Martinus Nijhoff, The Hague, pp 208-230
2. BEINEKE WF 1983 The genetic improvement of black walnut for timber production. Plant Breed Rev 1:236-266
3. CARUSO JL 1983 *In vitro* axillary shoot formation and rooting in black walnut mature embryos. Proc 3rd North Central Tree Improv Conf, pp 144-149
4. CHALUPA V 1981 Clonal propagation of broad-leaved forest trees *in vitro*. Commun Inst For Cech 12:255-271
5. CHEEMA GS, PN MEHRA 1982 Morphogenesis in endosperm cultures. *In* A Fujiwara, ed, Plant Tissue Culture 1982. Proc 5th Int Cong Plant Tissue Cell Cult, Tokyo, pp 111-112
6. CHENG TY 1975 Adventitious bud formation in culture of Douglas fir (*Pseudotsuga menziesii* (Mirb.) Franco). Plant Sci Lett 5:97-102
7. CHENG TY, TH VOQUI-DINH 1977 Regeneration of Douglas fir plantlets through tissue culture. Science 198:306-307
8. COSSIO F, G MINOLTA 1983 Prove preliminari di coltura "in vitro" di embrioni isolati di noce (*Juglans regia* L.) e confronto tra differenti combinazioni di sali minerali. Riv Ortoflorofruttic Ital 67:287-298
9. CUMMINS JN, WC ASHBY 1969 Aseptic culture of *Juglans nigra* stem tissues. For Sci 15:102-103
10. DRIVER JA, AH KUNIYUKI 1984 *In vitro* propagation of Paradox walnut rootstock. HortScience 19:507-509
11. Food and Agriculture Organization of the United Nations. 1983. FAO Production Yearbook 36, FAO Rome, Italy
12. GAMBORG OL, RA MILLER, K OJIMA 1968 Nutrient requirements of suspension cultures of soybean root cells. Exp Cell Res 50:151-158
13. HARTMANN HT, DE KESTER 1983 Plant Propagation Principals and Practices. 4th ed, Prentice-Hall, Englewood Cliffs, NJ
14. HEILE C, G GAFFNEY, J PREECE, O MEYERS 1984 *In vitro* culture of *Juglans nigra* seedling shoot tips. HortScience 19:100
15. LLOYD G, B MCCOWN 1980 Commercially-feasible micropropagation of mountain laurel, *Kalmia latifolia*, by use of shoot tip culture. Proc Int Plant Propagators Soc 30:421-427
16. LENARTOWICZ A, DF MILLIKAN 1977 Meristem culture of *Juglans nigra* L. Trans Mo Acad Sci 10 & 11:279
17. MEYNIER V 1985 Mise en culture *in vitro* de meristemes de Noyers hybrides. CR Acad SC Paris 301:261-264
18. MURASHIGE T, SKOOG F 1962 A revised medium for rapid growth and bioassays with tobacco tissue cultures. Physiol Plant 15:473-497
19. RODRIQUEZ R, R SANCHEZ-TAMES 1981 Cultivo de tejidos y diferenciacion en nogal. Rev Fac Cienc Univ Orviedo (Ser Biologia) 22:21-28
20. RODRIQUEZ, R 1982 Callus initiation and root formation from *in vitro* culture of walnut cotyledons. HortScience 17:195-196
21. RODRIQUEZ R 1982 Stimulation of multiple shoot buds formation in walnut seeds. HortScience 17:592

22. SOMERS PW, JW SAMBECK, JE PREECE, G GAFFNEY, O MYERS Jr 1982 In vitro micropropagation of black walnut (Juglans nigra L.). Proc 7th North Am For Biol Workshop, pp 224-230
23. STEFFAN SJ, DF MILLIKAN 1983 Micropropagation of mature black walnut trees. Annu Rept North Nut Growers Assoc 74:26-29
24. TULECKE W, GH MCGRANAHAN 1985 Somatic embryogenesis and plant regeneration from cotyledons of walnut. Plant Sci 40:57-63
25. VIEITEZ AM, D VIEITEZ 1980 Plantlet formation from embryonic tissue of chestnut grown in vitro. Physiol Plant 50:127-130
26. WHITE PR 1940 Vitamin B6, nicotinic acid, pyridine, glycine and thiamine in the nutrition of excised tomato roots. Am J Bot 27:811-821

Table 1. Basal medium (DKW-C) used for walnut tissue culture.[1]

Component	mM	mg/l
NH_4NO_3	17.7	1416.0
$Ca(NO_3)_2 \cdot 4H_2O$	8.3	1968.0
K_2SO_4	8.9	1559.0
$MgSO_4 \cdot 7H_2O$	3.0	740.0
$CaCl_2 \cdot 2H_2O$	1.0	149.0
KH_2PO_4	1.9	265.0
m-Inositol	0.55	100.0
Sucrose	87.6	30,000.0
	µM	
$Zn(NO_3)_2 \cdot 6H_2O$	57.2	17.0
$MnSO_4 \cdot H_2O$	198.2	33.5
$CuSO_4 \cdot 5H_2O$	1.0	0.25
H_3BO_3	77.6	4.8
$Na_2MoO_4 \cdot 2H_2O$	1.6	0.39
$FeSO_4 \cdot 7H_2O$	121.5	33.8
Na_2EDTA	135.0	45.4
$NiSO_4 \cdot 6H_2O$	.02	0.005
Thiamin·HCl	5.9	2.0
Nicotinic acid	8.1	1.0
Glycine	26.6	2.0

[1]The following modifications are used for:

a) Initiation and multiplication (micropropagation): addition of BAP (4.4 µM = 1 mg/l) and IBA (0.05 µM = 0.01 mg/l).

b) Rooting pretreatment: adjustment of NH_4NO_3 (5.7 mM = 456.2 mg/l), $Ca(NO_3)_2 \cdot 4H_2O$ (2.7 mM = 634.0 mg/l), and sucrose (153.8 µM = 53 g/l) and addition of IBA (0.74 µM = 0.15 mg/l).

c) Conditioning for somatic embryogenesis: addition of l-glutamine (1.7 mM = 250.0 mg/l), BAP (4.4 µM = 1 mg/l), IBA (0.05 µM = 0.0 1 mg/l), and kinetin (9.3 µM = 2.0 mg/l).

Media are adjusted to pH 5.4-5.7, solidified with Gelrite (Merck and Co.) (0.21-0.24%) and autoclaved for 20 min. at 121°C.

20. MULBERRY

K. OHYAMA and S. OKA

1. INTRODUCTION

The mulberries (*Morus alba*, *M. bombysis* and *M. multicaulis*) are important plants in the sericultural industry. They are valued for their foliage which constitutes the chief feed for the silkworm (*Bombyx mori* L.). Some species of mulberry are grown for their edible fruit and useful timber.

Wild mulberry species grow in mountainous areas, in temperate and tropical climates, in Asia, the Middle East, and Central and South America. However, it has been commercially cultivated mainly in China, Japan, Korea and the U.S.S.R., where most of the silk cocoons of the world have been produced. In other countries such as Thailand, Iran, Indonesia and Brazil, etc., sericulture is being progressively introduced to share an important role in agricultural development.

The genus *Morus* is usually classified into thirty species based on the morphological characteristics of flower organs and leaves as well as geological distribution (11). However, it seems difficult to apply a strict concept of "biological species" to *Morus*, as most *Morus* species can be easily hybridized by pollination. Japan has three recognized species, *M. alba*, *M. bombysis* Koiz. and *M. multicaulis* P. and most breeding of mulberry is carried out by interspecific crossing of these species.

Conventional methods of vegetative propagation of mulberry are grafting and rooting of cuttings. In Japan, 70% of nursery plants are produced by means of grafting. Grafting of mulberry is rather simple compared to grafting of fruit and forest trees. Rootstocks are usually prepared from 1-year-old seedlings and scions are taken from branches of mature trees before bud break. The grafts are then grown in the field for another year. Therefore, it takes two years to obtain nursery plants, which poses economic problems.

Rooting of softwood and hardwood cuttings of selected mature trees has the advantage of obtaining nursery plants in one year. However, they have the following drawbacks. In softwood cuttings, successful rooting depends on so many factors with regard to the physiological state of the cutting and environmental conditions that the results tend to be greatly variable. In hardwood cuttings, only a limited number of varieties have enough rooting ability. However, the story is different in tropical regions, where this is often the easiest method of propagation. Because most of the tropical species display a strong rooting ability, planting stem cuttings directly in the field is good enough for establishing a mulberry plantation (30).

There has been increasing interest in using tissue culture as a means for clonal propagation of woody plants. This is true in those plants whose cuttings are difficult to root, particularly when they are taken from the mature tree (3, 12). There are two pathways for obtaining propagules through tissue culture (14). One is adventitious initiation of shoots or embryoids from callus or explants. The other is enhanced axillary branching by means of buds or meristem culture. The latter system is useful in species where plant regneration from cultured tissues is difficult to attain.

In mulberry bud cultures multiple shoots and subsequent plantlets are easily formed. Although the technique remains a laboratory-scale one, it has potential for high multiplication rates and for obtaining rooted plants even with varieties which show poor rooting by conventional methods.

2. ORGAN CULTURE

2.1. Bud culture and plantlet regeneration

Bud meristems are good material for producing multiple shoots in both gymosperm and angiosperm plants. Pre-existing meristems easily develop into shoots (2), and are likely to be genetically stable.

Ohyama (16) and Ohyama and Oka (19) have attempted the *in vitro* culture of various buds of mulberry trees (Table 1, Fig. 1). Winter buds from 1-year-old branches of mature trees in the non-dormant period, terminal buds in the growing season and axillary buds at various developmental growth stages have been aseptically isolated and cultured on Murashige and Skoogs' (MS) medium (13) containing growth substances.

As is common with other plants, mulberry buds have an absolute requirement for cytokinin for their growth. N6-benzyladenine (BA) was more effective than zeatin, while kinetin was unexpectedly of no use over a wide range of concentrations. The optimum concentration of BA was between 0.1 and 1.0 mg/l. The number of shoots produced from a bud explant varied in accordance with the level of BA. A single shoot developed at 0.1 mg/l, while at 1.0 mg/l the main apical meristem developed along with several axillary ones, giving rise to multiple shoot formation. Auxin combined with 1 mg/l BA had no effect on bud growth, but induced rooting of the explant when combined with 0.1 mg/l BA. Thus, a complete plantlet was produced in medium containing 0.1 mg/l BA and 0.1 mg/l indole-3-acetic acid (IAA).

Growth of these explants depended on the kind of sugars added to the medium (Table 2). The response to a sugar varied with the variety of mulberry tree tested. Some varieties developed shoots in the presence of sucrose as well as fructose, but others did so only in the presence of fructose, which was substituted for sucrose.

The concentration of agar (Table 2) was found to be a limiting factor in shoot tip culture (19). Higher concentrations (0.8%-1.0%) of agar gave poor growth, with only a few leaves opening and without further shoot development, whereas a concentration as low as 0.4% supported good shoot growth. This result was confirmed later with winter bud cultures (25). A promotive effect of agar at low concentration on tissue growth has also been reported with *Picea abies* (31), *Iris* (35), and *Malus* (34). This effect might be explained by (1) inhibiting substances in the agar; (2) availability of water and nutrients; (3) increased uptake of nutrients through leaves or tissue submerged in the medium.

Table 1. Mulberry species in organ culture

Species	Explant	Medium	References
Morus alba	Root	White	(16)
	Winter bud	MS	(22)
	Nodes	MS	(23)
	Leaf	MS	(27)
	Hypocotyl	MS	(20)
M. bombycis	Winter bud	MS	(22)
	Shoot tip	MS	(19)
M. multicaulis	Winter bud	MS	(22)
M. kagayamae	Shoot tip	MS	(17)

Table 2. Explant sources and conditions for mulberry bud cultures

Explant source	Medium[a]	Hormone	Sugar	Agar
Winter bud	MS	1 mg/l BA	3% Fructose	0.6%[b]
	MS	1 mg/l BA	3% Fructose	0.4%
Lateral bud				
without node	MS	1 mg/l BA	3% Fructose	0.4%
with node	MS	0.2 mg/l NAA	3% Sucrose	0.8%
Shoot tip	MS	1 mg/l BA	3% Fructose	0.6%
	MS	1 mg/l BA	3% Fructose	0.4%
Nodal and shoot explant grown *in vitro*				
for subculture	MS	1 mg/l BA	2-3% Sucrose	0.8%
for rooting	MS	None	2-3% Sucrose	0.8%

[a]pH 5.6 was adjusted before autoclaving.
[b]When fructose is used, 0.4% agar gives too soft a medium.

FIGURE 1-1. Shoot development from a winter bud explant of mulberry (*Morus alba*) on MS medium containing 3% fructose in the presence of 1 mg/l BA.
FIGURE 1-2. Stock culture of multiple shoots grown on MS medium with 1 mg/l BA.
FIGURE 1-3. Rooting of a shoot explant on MS basal medium without hormones after 30 days.
FIGURE 1-4. A mulberry plant transferred to a pot after 3 weeks at 28°C with 12-h photoperiods provided by cool white fluorescent light.

Shoot meristems from bud cultures could be subcultured repeatedly. Subcultured explants grew on a medium that did not contain growth substances, provided that explants had already developed roots. Rapid growth of new shoots was observed soon after the development of roots. The activity of roots plans an important role in the elongation of isolated buds.

2.2. Excised root culture

Although isolated root tips of many plants have been grown successfully (4), there are only a few examples of success with trees. Ohyama (16) studied factors affecting the growth of excised mulberry roots. Excised root tips (5 mm in length) grew well in White's (36) medium at 28°C in darkness. In mulberry, it was found that the isolated roots can be grown in a relatively simple medium containing inorganic salts, carbohydrates, vitamins and growth hormones. In the early period of culture, thiamine and pyridoxine were not needed for satisfactory growth of root tips. However, when the subculture of the main axis of root tips was attempted, the meristematic activity in the individual root tips was prolonged by adding thiamine and pyridoxine to the medium. Moreover, the response of isolated mulberry roots to IAA was similar to that of pine roots (4). Namely, in the case of an IAA concentration above 10^{-5} M, the elongation of roots and initiation of lateral roots were inhibited and dichotomy was initiated. In contrast, application of IAA at low concentration stimulated the development of lateral roots. By use of the above method, it may be possible to clarify what kinds of substances produced in the shoots whould be most effective for the growth of roots in mulberry trees.

3. ORGANOGENESIS

Organogenesis and plantlet regeneration in culture are common with herbaceous plant species. Despite intensive work, however, shoot bud formation and embryogenesis from mulberry callus have not been achieved (5, 32, 39). Oka and Ohyama (26, 27) demonstrated that adventitious buds were induced directly without intermediate callus formation in leaf explants of young mulberry plants. The location and frequency of bud formation were markedly influenced by the concentration of BA. Histological observations showed that bud initiation from the excised leaves was peripheral in origin.

Recently, Kim Ho-Rak *et al.* (7) also reported on organogenesis in mulberry explants. Cotyledon, leaf, hypocotyl and shoot tip explants of mulberry, derived from elongating embryos cultured with BA, were induced to form adventitious shoots when cultured in media containing high BA with or without a low concentration of naphthalenacetic acid (NAA). The adventitious shoots formed at the basal cut ends of the cotyledons and leaves. Histological analysis of the explants showed a wound cambium at the base of the bud primordia.

Cytokinins are generally required by most tissues for induction of buds. In some cases the kind of cytokinin has a significant effect on the differentiation and growth of the organ. Okamoto *et al.* (29) reported that N'-(2-chloro-4-pyridyl)-N'-phenyl-urea (4PU), which is a cytokinin-active urea, remarkably stimulated shoot formation in tobacco callus cultures. Ohyama and Oka (20) cultured hypocotyl segments of germinating embryos of mulberry on MS (13) medium supplemented with 4PU. In a few days both cut ends of all explant began to swell. However, multiple buds were produced only at the apical end. When transferred to 0.1 mg/l NAA, these buds rooted and whole plants were obtained. A histological study of the process of the multiple shoot formation is now in progress.

4. CALLUS AND CELL SUSPENSION CULTURE

Mulberry callus can be readily induced from various explants (Table 3). MS (13) and Linsmaier and Skoog (LS) (8) media have been used primarily and have proved better than White (36), Heller (6) or Wolter and Skoog (37) media (21, 38, 39). For inducing callus, auxins were required, whereas cytokinins were not necessarily important (32, 33, 39). Unlike organogenesis in organ explants, subculture of mulberry callus did require both an auxin and cytokinin (24).

Suspension cultures have been prepared from hypocotyl-derived (16) and embryo-derived (39) callus. Yamada and Okamoto (38) were the first to succeed in the large-scale liquid culture of mulberry callus and obtained a three to four times volume increase during a

Table 3. Callus and cell suspension cultures of mulberry

Species	Explant	Medium	References
Morus alba	Root, Hypocotyl	MS	(16)
	Embryo	LS	(39)
	Cambium	MS	(32)
	-	MS	(5)
	Young stem	MS	(21, 24)
	Vein, Petiole stem, Root	MS	(38)
M. bombycis	Young stem	MS	(9)
M. multicaulis	Embryo, Young stem Cotyledon	MS	(10)
	Endosperm	MS	(33)

2-week culture period in a jar-fermenter. They used leaf vein and young petioles of mulberry leaves as explants, and indolebutric acid (IBA) instead of 2,4-dichlorophenoxyacetic acid (2,4-D) or NAA. Cultured mulberry cells have replaced fresh leaves as the major ingredient in some artificial silkworm food. However, the cells needed to be cultured under light to produce a small amount of chlorophyll to make them palatable to the insect. Silkworms fed artificial food which included these cultured mulberry cells, spun good cocoons.

5. PROTOPLAST ISOLATION AND CULTURE

In 1971 Nagata and Takebe (15) reported that protoplasts isolated from leaves of tobacco plants could be induced to divide in culture and form callus tissue which could regenerate whole plants. Recently, many reports have suggested that protoplast isolation and regeneration of plants from protoplasts may become a powerful new tool for plant breeding. Protoplasts are useful for that purpose because they can absorb foreign DNA and form inter-specific hybrids by fusion.

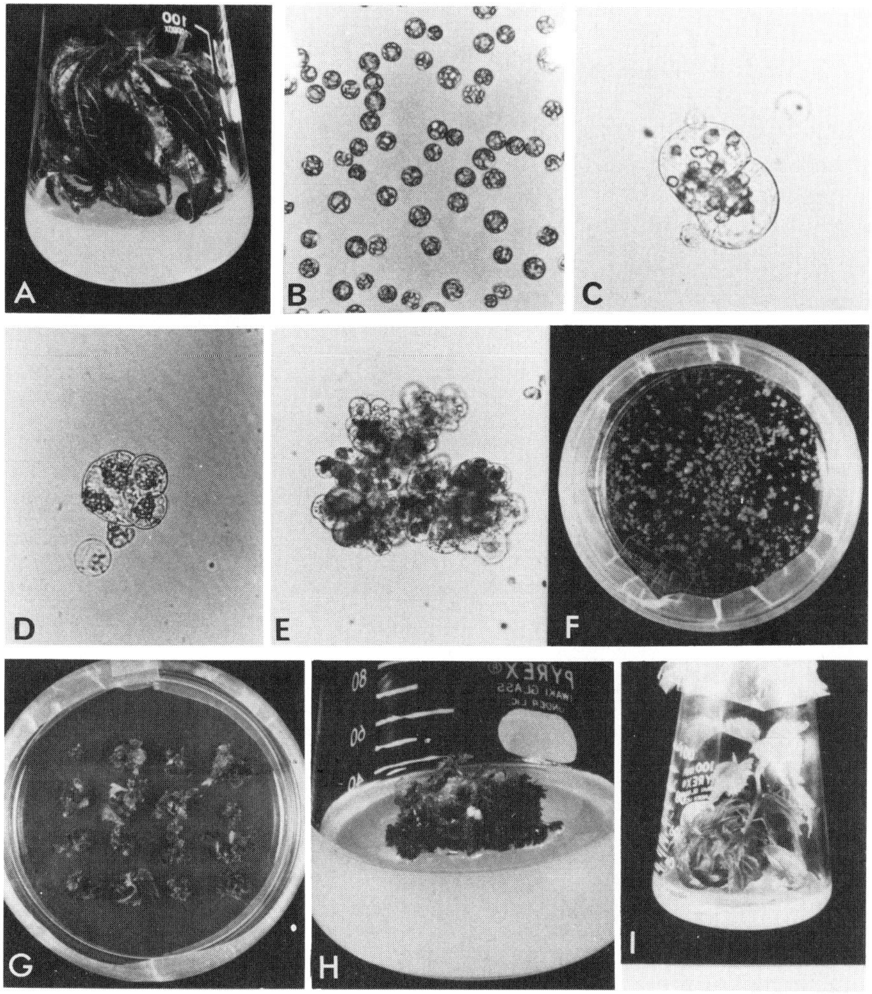

FIGURE 2. Plant regeneration from protoplasts of paper mulberry (Broussonetia kajinoki Sieb.). A. Shoot cultures, derived from shoot tip explants, used for protoplast isolation. B. Freshly isolated protoplasts. C-D. Various stages in the division of isolated protoplast cultured in a liquid nutrient medium. E. Protoplast-derived colony after 4 weeks in culture. F. Protoplast-derived callus macroscopically visible after 8 weeks in culture. G-H. Shoot bud formation from callus 4-5 weeks after transfer to regeneration medium. I. Young regenerated plant after 3 weeks.

Ohyama and Oka (18) studied cultures of cells and protoplasts, using both mulberry and other species. To date, protoplasts have been prepared from cotyledons, leaves and callus by using various enzyme solutions. Recently, Oka and Ohyama (28) reported successful production of whole plants from mesophyll protoplast cultures of the paper mulberry tree (Broussonetia kajinoki sieb.). (Fig. 2). The procedures were as follows: Leaf pieces from shoots in vitro were incubated in a mixture consisting of 0.05% pectolyase Y23, 1% cellulase onozuka R10, and 0.5 M mannitol at 25°C for 3 h. This yielded viable protoplasts, which after 1-2 days of culture in a modified MS medium, regenerated new cell walls and showed first divisions within 10-20 days. Small cell colonies were formed within 4 weeks. Reduction of the ammonium concentration and using glucose as a carbon source were essential for sustained cell divisions and colony formation. Shoot buds were produced from protoplast-derived callus on modified MS medium containing 0.1 mg/l NAA and 1 mg/l BA. These shoots rooted occasionally when transferred to MS medium supplemented with 0.05 mg/l NAA.

6. CONCLUSION AND PROSPECTS

Since the control of adventitious shoot development or somatic embryogenesis is difficult in mulberry tissues, meristem culture is a good alternative tool for micropropagation in vitro.

Following our standard methods, buds of various origin will elongate into shoots on MS medium supplemented with BA, with sugar and agar concentrations having a pronounced controlling effect. The shoots were separated into nodal and shoot sections and subcultured every month as stock culture. Between 3 and 5 shoots were reproduced from each explant and later rooted in the root-forming medium. In one experiment, we demonstrated that 150 shoots could be proliferated from one explant within 120 days through three subcultures.

Many problems remain before in vitro techniques can be used for large-scale clonal propagation. One of these is the establishment of simple and reliable methods for transferring plantlets to soil (1, 2). Not much information is available on transfer to soil, and we have not yet done much work in this area either.

The small scale micropropagation methods that are now available for mulberry can be used for the following:
(1) rapid cloning of plants selected for breeding programs,
(2) reproduction of plants from long-term stored bud explants,
(3) maintenance of virus-infected plants for host-pathogen studies.

Cell and tissue culture methods will eventually become an effective means, not only for micropropagation, but also for helping to overcome various problems of breeding. However, since a long time is generally required for woody plants to flower, one problem is low breeding efficiency. Therefore, it would be of great significance to the breeders if early selection in the test tube could become possible by applying tissue and cell culture methods.

7. REFERENCES

1. BIONDI S, TA THORPE 1981 Clonal propagation of forest tree species. In AN Rao, ed, Tissue Culture of Economically Important Plants. COSTED and ANBS, Singapore, pp 197-204
2. BONGA JM 1977 Application of tissue culture in forestry. In J Reinert, YPS Bajaj, eds, Application and Fundamental Aspects Plant Cell, Tissue and Organ Culture. Springer-Verlag, Berlin, pp 93-105
3. BONGA JM 1982 Vegetative propagation in relation to juvenility. In JM Bonga, DJ Durzan, eds, Tissue Culture in Forestry, Martinus Nijhoff/JUNK Publ, The Hague, pp 387-412
4. BUTCHER DN, HE STREET 1964 Excised root culture. Bot Rev 30: 513-586
5. GHUGALE DD, DD KULKARN, R NARASHIMHAM 1971 Effects of auxins and gibberellic acid on growth and differentiation of *Morus alba* and *Populus nigra* tissue *in vitro*. Indian J Exp Bot 9: 381-384
6. HELLER R 1953 Recherches sur la nutrition minerale des tissues végétaux cultives *in vitro*. Ann Sci Nat Biol Vég 14: 1-223
7. HO-RAK KIM, RP KAMLESH, TA THORPE 1985 Regeneration of mulberry plantlets through tissue culture. Bot Gaz (in press)
8. LINSMAIER EM, F SKOOG 1965 Organic growth factor requirements of tobacco tissue cultures. Physiol Plant 18: 100-127
9. MINAMIZAWA K, H HIRANO 1974 Effect of the concentrations of several sugars on differentiation in excised stem segments of mulberry trees cultured *in vitro*. J Seric Sci Jpn 43: 51-57
10. MINAMIZAWA K, H HIRANO 1974 Difference of differentiation among the kinds of explants of mulberry plants *in vitro*. J Seric Sci Jpn 43: 94-97
11. MINAMIZAWA K 1976 Science of Mulberry Cultivation - Fundamental and Practice. Meiho-sha Press, Tokyo
12. MOTT RL 1981 Trees. In BV Conger, ed, Principles and Practices of Cloning Agricultural Plants via *In Vitro* Techniques. CRC Press, Boca Raton, Florida, pp 217-254

13. MURASHIGE T, F SKOOG 1962 A revised medium for rapid growth and bio-assay with tobacco tissue cultures. Physiol Plant 15: 473-497

14. MURASHIGE T 1978 The impact of plant tissue culture on agriculture. In TA Thorpe, ed, Frontiers of Plant Tissue Culture 1978: Int Assoc Plant Tissue Cult, Univ Calgary, Alberta, Canada, pp 15-20

15. NAGATA T, I TAKEBE 1971 Planting of isolated tobacco mesophyll protoplasts on agar medium. Planta 99: 12-20

16. OHYAMA K 1970 Tissue culture in mulberry tree. Jpn Agric Quart 5: 30-34

17. OHYAMA K, H KAWAKITA 1975 In vitro culture of shoot tips infected with mulberry dwarf. J Seric Sci Jpn 44: 413-414

18. OHYAMA K, S OKA 1975 Isolation of protoplasts from mesophyll cells of Morus alba L. Proc Crop Sci Soc Jpn 44: 121-122

19. OHYAMA K, S OKA 1976 Regeneration of whole plants from isolated shoot tips of mulberry tree. J Seric Sci Soc Jpn 45: 115-120

20. OHYAMA K, S OKA 1982 Multiple shoot formation from mulberry (Morus alba L.) hypocotyls by N'-(2-chloro-4-pyridyl)-N'-phenylurea. In A Fujiwara, ed, Plant Tissue Culture 1982, Tokyo, pp 149-150

21. OKA S, K OHYAMA 1973 Induction of callus and effects of constituents of media on callus formation in mulberry trees. J Seric Sci Jpn 42: 317-324

22. OKA S, K OHYAMA 1974 Studies on in vitro culture of excised bud in mulberry tree. I Effect of growth substances on the development of shoot and organ formation from winter bud. J Seric Sci Jpn 43: 230-235

23. OKA S, K OHYAMA 1975 Studies on in vitro culture of excised bud in mulberry tree. II Effects of growth substances on the development of shoots from axillary bud. J Seric Sci Jpn 44: 444-450

24. OKA S, K OHYAMA 1976 A study on the medium for the subculture of mulberry callus. J Seric Sci Jpn 45: 385-391

25. OKA S, K OHYAMA 1978 Studies on in vitro culture of excised bud in mulberry tree. III Effects of agar concentration, pH and sugars of medium on the development of shoots from winter buds. J Seric Sci Jpn 47: 15-20

26. OKA S, K OHYAMA 1981 In vitro initiation of adventitious buds and its modification by high concentration of benzyladenine in leaf tissues of mulberry (Morus alba). Can J Bot 59: 68-74

27. OKA S, K OHYAMA 1978 In vitro culture of excised mulberry bud and dormancy analysis with it. In T Akihama, N Nakajima, eds, Long Term Preservation of Favourable Germ Plasm in Arboreal Crops. Fruit Tree Res Stn MAF, Japan, pp 136-141

28. OKA S, K OHYAMA 1985 Plant regeneration from leaf mesophyll protoplasts of Broussonetia kazinoki Sieb (paper mulberry). J Plant Physiol 119: 455-460

29. OKAMOTO T, K SHUDO, S TAKAHASHI, T YATSUNAMI, Y ISOGAI, K YAMADA 1978 Shoot formation of tobacco callus by various cytokinin-active urea and pyrimidines. Chem Pharm Bull 26: 3250-3252

30. OMURA S 1973 Silkworm Rearing Techniques in the Tropics. Textbook Ser No 35, OTCA, Tokyo

31. ROMBERGER JA, CA TABOR 1971 The Picea abies shoot apical meristem in culture. I Agar and autoclaving effects. Am J Bot 58: 131-140
32. SEKI H, M TAKEDA, K TSUTSUMI, Y USHIKI 1971 Studies on the callus culture of the mulberry tree. I The effect of concentration of auxin and kinetin on the callus culture of the mulberry stem. J Seric Sci Jpn 40: 81-85
33. SEKI H, M TAKEDA, T YAMAGUCHI 1974 Studies on the callus culture of the mulberry tree. II The callus culture of the embryo and endosperm of the mulberry seed. J Seric Sci Jpn 43: 487-491
34. SINGHA S 1982 Influence of agar concentration on in vitro shoot proliferation of Malus sp Almey and Pyrus communis seckel. J Am Soc Hortic Sci 107:657-660
35. STOLITZ LP 1971 Agar restriction on the growth of excised Iris embryos. J Am Soc Hortic Sci 96: 681-684
36. WHITE PR 1943 A Handbook of Plant Tissue Culture. Jacques Catlell Press, Lancaster, Pennsylvania
37. WOLTER K, F SKOOG 1966 Nutritional requirements of Fraxinus callus culture. Am J Bot 53: 263-269
38. YAMADA Y, A OKAMOTO 1977 A basic study of mulberry cell culture for silkworm feeding. Agric Biol Chem 41: 2261-2267
39. YAMAMOTO A, M KAMATA 1971 Studies on the tissue culture of mulberry tree. Part 1. Dedifferentiation of tissue of mulberry tree and culture of callus. Bull Fac Text Sci, Kyoto Univ Ind Arts Tex Fibers 6: 108-121

21. PLANTLET REGENERATION IN CALLUS CULTURES OF *LEUCAENA*

R. NAGMANI and S. VENKETESWARAN

1. INTRODUCTION

Research reports published by the National Academy of Sciences (USA) (1) advocate increased use of underexploited crops of promising economic value. One of these, the tree legume *Leucaena* is a promising forage and wood crop for the tropics (5,6). It provides good fire wood, with heating values of 8,300 BTU/lb dry weight (19.3 mj/kg), that is used as a fuel source in several developing countries. Brewbaker and Hutton (5) report a biomass yield of up to 10-15 dry tons per acre.

The genus *Leucaena* of the Leguminosae family consists of 10 species and is native in tropical America. They are mostly small trees and shrubs with a growth habit and foliage similar to that of *Acacia*. They have been planted on a large scale in Guatemala, Peru, Mexico, India, Philippines, and the Pacific Islands. In Java, species of *Leucaena* are interplanted with teak to provide cover for soil protection (9).

The ability of *Leucaena* species to grow on marginal soils holds promise for reforesting eroded wastelands. Additionally, it is estimated that *Leucaena* can add up to 500 kg nitrogen/ha per year to the soil (1), which makes it highly suitable as a cover plant in plantations. When used as soil cover in plantations it is grown as a catch crop or as an intercrop in pure stands (9).

According to a research report on *Leucaena* (4), most of the species are self-incompatible and produce very little inbred (true to type) seeds; with the probable exception of *L. leucocephala*. *In vitro* propagation may therefore be a useful method to clonally multiply these "true to type" inbred seeds. Since the value of this tree legume remained unrecognized until recently, there have not yet been many attempts to clonally propagate this genus on a large scale by tissue culture methods. The first attempt to propagate *Leucaena* sp. *in vitro* was by Peasely and Collins (18). They established callus cultures using hypocotyl segments of the seedlings. They did not report the differentiation of these callus cultures into plantlets. Recently, Goyal *et al*. (13) obtained multiple shoot differentiation in tissue cultures of *L. leucocephala*, by using single lateral bud explants from 2-3 m tall, 1-year-old plants. They reported the production of 22 shoots per bud.

We have attempted to initiate callus cultures and to induce organogenesis in callus cultures of two species of *Leucaena* (16,17). The first species was *L. leucocephala* (Lam) de Witt; c.v. Cunningham (K 500 & K 8). It is familiarly called the Ipil-ipil tree and has many varieties (2). One of them is the Hawaiian type (giant Ipil-ipil) that grows and thrives well in the State of Texas. Field trials indicate the *L. leucocephala* cultivar K 67 is promising in Texas for forage and woody biomass production (21,22). The tree is single trunked and fast growing, reaching a height of 20 m in 4-5 years. It grows on marginal lands and hillsides. The second species used was *L. retusa* Benth; Pl. Wright - a frost resistant variety.

2. MATERIAL AND METHODS

Seeds of *L. leucocephala* and *L. retusa* were pretreated with concentrated sulfuric acid for 30 min. The sulfuric acid was removed by repeated washing with sterile distilled water. Subsequently, the seeds were surface sterilised for 15 min. with 15% commercial clorox. After rinsing the seeds again with sterile distilled water, they were germinated on full strength Murashige-Skoog (15) (MS) or Gamborg et al. (B5) (10) media with sucrose at 2% (w/v) but without growth hormones. Sections of hypocotyl and cotyledons 5-10 mm long were excised from 7- to 10-day-old seedlings and placed on nutrient medium. The media used for culture were MS or B5 at full strength, with sucrose 2% (w/v), 2,4-dichlorophenoxyacetic acid (2,4-D) 0.5, 1, 2, 5, and 10 mg/l, naphthalene acetic acid (NAA) 0, 0.5, 1, and 2 mg/l, dimethylallylaminopurine (2iP) 0, 1, 2, and 5 mg/l, N-6-benzylaminopurine (BA) at 0, 1, 2, and 5 mg/l, and Difco Bacto agar 0.8% (w/v). The pH was adjusted with dilute HCl or NaOH to 5.8 before autoclaving at 120°C for 15 min. The autoclaved medium was poured into 100 x 15 mm diameter disposable petri dishes.

For each experiment 200 hypocotyl and 200 cotyledon discs were used (8-10 per dish). The petri dishes were then sealed with tape. The cultures were incubated at 27 $\pm$ 1°C in an incubator illuminated with fluorescent tubes (approximately 7000 lux) 12 h/day.

After the callus was initiated, it was removed and transferred to MS or B5 supplemented with BA at 0.5, 1, or 2 mg/l or 2iP at 0.5, 1, or 2 mg/l.

For root initiation, leafy shoots were transferred to MS or B5 supplemented with indolebutyric acid (IBA) at 0.5 and 1 mg/l.

In vitro-produced plantlets were transferred to sterile garden soil mixed with vermiculite and placed in a growth chamber with fluorescent light 12 h/day.

3. RESULTS

3.1. *L. leucocephala*

In both varieties of *L. leucocephala*, 5% of the cotyledonary segments produced very small leafy shoots from their cut ends on MS but not on B5 medium after 2 weeks in culture. Of the hypocotyl segments, 34% produced some callus from their cut ends on MS medium after 3 weeks in culture. Addition of 2,4-D (0.5-5 mg/l) to the MS medium did not bring about any significant increase in callus formation, although at 10 mg/l, a small amount of friable callus was formed in *L. leucocephala*.

Addition of NAA (0.5-10 mg/l) did not result in callus formation. However, a combination of NAA (0.5 mg/l) and BA (0.5-5 mg/l) induced callus proliferation in 60% of both the hypocotyl and cotyledonary explants (Fig. 1). This callus was dark green and friable and continued to proliferate through 3-5 subcultures.

After transfer of segments of these calli to MS medium containing BA at 2 mg/l small green leafy shoots developed in 25% of the cotyledonary callus cultures in about 2 weeks (Fig. 2). These leafy shoots elongated further after transfer to fresh medium with the same nutrient composition (Figs. 3,4).

The basal ends of 2% of the elongated shoots formed small adventitious roots on MS medium supplemented with IBA at 0.5 and 1 mg/l (Fig. 4). These rooted plants, after transfer to a sterile garden soil mixture, continued to grow for a period of 4-6 weeks, producing one or two more leaves (Fig. 5).

3.2. L. retusa

About 50% of the cotyledonary sections of L. retusa produced a compact, dark green callus on MS medium supplemented with NAA (0.5 mg/l) and BA (2 mg/l) after 3 weeks in culture (Fig. 6). The callus could be subcultured through 2-6 transfers. When portions of these calli were transferred to MS medium with 2iP at 1 mg/l, 20% of the calli formed small green leafy shoots 3 weeks after transfer. These leafy shoots failed to root on any of the auxin containing media.

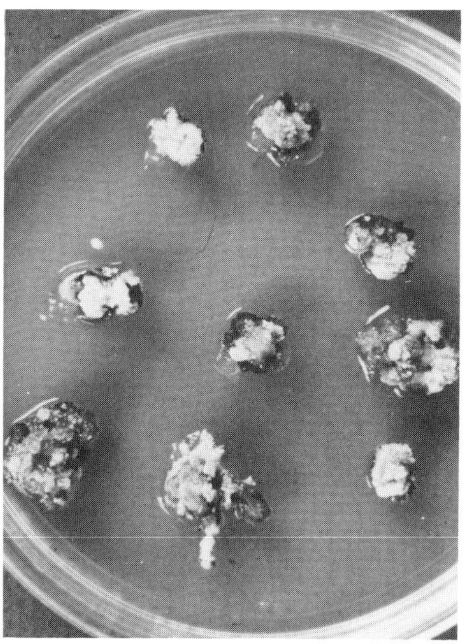

FIGURE 1. Callus formation from explants of L. leucocephala.
FIGURE 2. Initiation of leafy shoots from cotyledonary callus of L. leucocephala.

FIGURE 3. Fully differentiated leafy shoot of *L. leucocephala*, from callus of cotyledonary origin.
FIGURE 4. Elongated leafy shoot of *L. leucocephala* with an adventitious root at the base.

4. DISCUSSION

Differentiation of leafy shoots in callus cultures of two species of *Leucaena* (16,17) followed a sequence of events similar to that observed in tissue cultures of other tree legumes such as *Acacia* (20), *Alhagi* (3), *Albizzia* (11), *Dalbergia* (8), *Prosopis* (12) and *Sesbania* (14). Since the publications on tissue culture of *Leucaena* deal mostly with only one species, i.e., *L. leucocephala* (13,18,19), tissue culture of the genus should be extended to other species and hybrids.

At present, the advantages of propagation of *Leucaena* sp. by tissue culture methods over other methods of vegetative propagation remain to be determined.

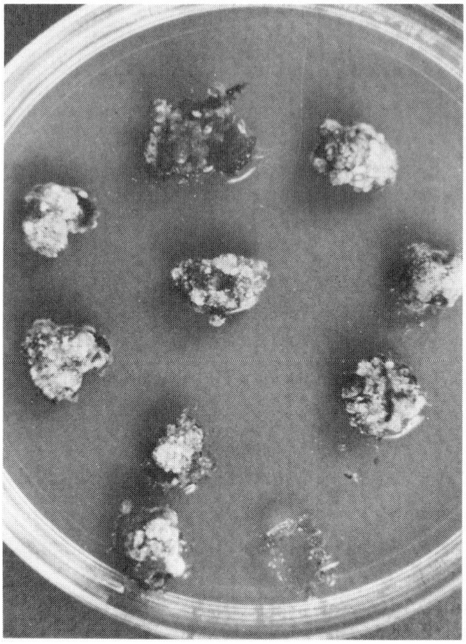

FIGURE 5. Two-month-old plantlet of L. leucocephala.
FIGURE 6. Callus formation from cotyledon explants of L. retusa.

5. REFERENCES

1. AD HOC PANEL 1979 Tropical Legumes: Resources for the Future. Board Sci Technol Int Dev Nat Acad Sci. Washington DC
2. ANON 1978 International Consultation on Ipil-Ipil Research Papers and Proceedings. pp 1-172
3. BHARAL S, A RASHID 1981 Tissue culture of Alhagi camelorum - a legume of high regenerative ability. Physiol Plant 53: 497-500
4. BREWBAKER JL 1982 Systematics, self-incompatibility, breeding systems and genetic improvement of Leucaena species. In Leucaena Research in the Asian-Pacific Region. Proc Workshop, Singapore 1982, Int Dev Res Center, Ottawa Publishers, pp 17-22
5. BREWBAKER JL, EM HUTTON 1979 Leucaena - Versatile tropical tree legume. In GA Ritchie, ed, New Agricultural Crops. Am Assoc Adv Sci, Westview Press, Colorado
6. BUCHANAN RA 1977 Leucaena - Promising Forage and Tree Crop for the Tropics. Nat Acad Sci Rep, Washington, DC
7. BUCHANAN RA 1979 Tropical Legumes: Resources for the Future. Nat Acad Sci Rep, Washington, DC
8. DATTA SK, K DATTTA, T PRAMANIK 1983 In vitro clonal multiplication of mature trees of Dalbergia sisso Rosb. Plant Cell Tissue Organ Cult 2: 15-20

9. FAO 1959 Choice of Tree Species. FAO For Dev Pap
10. GAMBORG OL, RA MILLER, K OJIMA 1968 Nutrient requirements of suspension cultures of soybean root cells. Exp Cell Res 50: 151-158
11. GHARYAL PK, SC MAHESHWARI 1983 _In vitro_ differentiation of plantlets from tissue cultures of _Albizzia lebbeck_ L. Plant Cell Tissue Organ Cult 2: 49-53
12. GOYAL Y, HC ARYA 1981 Differentiation in cultures of _Prosopis cinearia_ Linn. Curr Sci 50: 468-469
13. GOYAL Y, RL BINGHAM, P FELKER 1985 Propagation of the tropical tree, _Leucaena leucocephala_ K67, by _in vitro_ bud cultures. Plant Cell Tissue Organ Cult 4: 3-10
14. KHATTAR S, HY MOHANRAM 1982 Organogenesis in the cultured tissues of _Sesbania sesban_, a leguminous shrub. Indian J Exp Biol 20: 216-219
15. MURASHIGE T, F SKOOG 1962 A revised medium for rapid growth and bioassays with tobacco tissue cultures. Physiol Plant 15: 473-497
16. NAGMANI R, S VENKETESWARAN 1983 _In vitro_ culture of hypocotyl and cotyledon segments of _Leucaena_. Leucaena Res Rep 4: 88-89
17. NAGMANI R, S VENKETESWARAN 1983 Morphogenetic responses of cultured hypocotyl and cotyledonary segments of _Leucaena_. In Vitro 19: 265
18. PEASLEY EL, GB COLLINS 1980 Development of _in vitro_ culture system for _Leucaena_. Leucaena News Lett 1: 51
19. RAVISHANKAR GA, A WALLI, S GREWAL 1983 Plantlet formation through tissue cultures of _Leucaena leucocephala_. Leucaena Res Rep 4: 37
20. SKOLMEN RG, MO MAPES 1976 _Acacia koa_ Gray. Plantlets from somatic callus tissue. J Hered 67: 114-115
21. VENETESWARAN S, V GANDHI 1982 Mass propagation and genetic improvement of forest trees by biomass production by tissue culture. Biomass 2: 5-15
22. VIETMEYER N, B COTTON, FR RUSKIN 1977 _Leucaena_-Promising Forage and Tree Crop for the Tropics. Nat Acad Sci Rep, Washington, DC

22. TISSUE CULTURE OF LIQUIDAMBAR

HARRY E. SOMMER AND HAZEL Y. WETZSTEIN

1. INTRODUCTION

1.1. Species and taxonomy

The genus *Liquidambar* belongs to the Hamamelidaceae, a dicotyledonous family comprised of 23 genera of deciduous or evergreen trees and shrubs. *Liquidambar* is in the subfamily Liquidambaroideae (5, 7). The genus has 3 species (1) [twenty extinct species are known according to Peattie (16)]: *L. formosana* Hance, Formosan gum native to China and Taiwan; *L. orientalis* Mill., oriental sweetgum native to Turkey; and *L. styraciflua* L., American sweetgum or red gum, of the eastern United States with disjunct populations in Mexico and Central America (26). Interspecific hybrids of the three species have been produced (17). The species are monoecious, deciduous trees with simple, alternate, palmately 3-7 lobed leaves, and deciduous stipules. Flowers are apetalous in globular heads. Staminate flowers are in terminal racemes or panicles; pistillate flowers are pedunculate with numerous ovules. Fruits are coherent capsules, dehiscent at apex; seed is winged (5, 10, 12,).

1.2. Economic importance

Liquidambar is a valuable timber tree, and is one of the more important commercial hardwoods of the United States, particularly in the manufacture of utility and decorative plywood panels (9). *L. styraciflua* produces a heavy close-grained heartwood for furniture making; the lumber is at times referred to as "satin-walnut" (16).

An aromatic balsam, called styrax or storax, is obtained from *L. orientalis* and *L. styraciflua*. The derivation of Liquidambar is from the Latin liquidus, liquid and the Arabic ambar, amber in reference to the resinous exudate which exudes from the injured bark of the Asiatic and American species (9). Storax is used in perfumery, and in medicine as a topical protectant and expectorant.

The Liquidambars are valued as excellent ornamental shade trees, with brillant fall color. The rapid growth, pyramidal shape, and strong horizontal branches make Liquidambar a good choice for street or yard plantings. Also of horticultural interest is the development of corky outgrowth on the twigs. Selections of trees for various characteristics have been made, such as for specific fall coloration, variegation, fruitlessness, and corkiness. Twenty cultivars of L. styraciflua, and one for L. formosana (18) are recognized.

Due to its rapid growth rate and ability for coppice regrowth, L. styraciflua has been recognized as a species with potential for use in short rotation forestry, especially in the southeastern United States. The selection and propagation of superior clones suited for coppice rotation is an integral part of the short rotation program. The potential use of Liquidambar in biomass production has been one of the reasons for our interest in clonal multiplication and in vitro culture of this species.

1.3. Vegetative propagation

In spite of the economic importance of sweetgum, the literature on its vegetative propagation is very scanty, probably because of the many failures (2). Airlayering (3) and field grafting (15) have been used to propagate L. styraciflua, but are very restricted in use. Rooting of cuttings is difficult (2, 14). American sweetgum is difficult to root even with physiologically rejuvenated material taken from the main bole (11). However, it has been propagated using segments of lateral roots (4) and with softwood cuttings from naturally occurring root suckers or from suckers cultured on excised roots (2, 6). Kormanik and Brown (11) found that succulent stem cuttings maintained in outdoor mist either failed to root or underwent decline. However, between 0 to 70% rooting of epicormic branches was obtained if cuttings were placed in a modified growth chamber. Although American sweetgum has been rooted from juvenile material, attempts to root stem cuttings from mature ortets has been largely unsuccessful (8). Hare (8) obtained 90% rooting of Formosan sweetgum cuttings after 1 month. In contrast, only 5% of American sweetgum cuttings rooted within the same period; only 35% rooted after 2 months.

2. TISSUE CULTURE FROM APICES

Several workers have reported on the tissue culture of shoot tips of sweetgum. The most sophisticated work is that of A. J. Abbott and D. R. Costantine who cultured the apical meristem of sweetgum trees growing in Bristol. Their work is as yet unpublished, but was presented at the 8th Long Ashton Symposium on the Improvement of Vegetatively Propagated Plants in Bristol, England.

Two varieties, Morane and a variegated variety from the North Carolina Arboretum, have been placed in culture by Lineberger and Brand (personal communication). Morane is a cold-hardy patented variety produced by budding. The cultures were established from the shoot tips of grafted adult shoots. Multiplication rates of 10-15 times were obtained in 8 weeks. The buds and shoots tended to remain rosetted with sporadic elongation. Those that did elongate rooted with an 80% success rate in nonsterile perlite-peat and high humidity. Shoots that remained rosetted did not root on agar or under nonsterile conditions. The rooted shoots would make 1 flush of growth, developing good caliper, then stop and form huge terminal buds. In contrast, shoots from cultures of the variegated variety did not rosette but elongated and grew continuously. Multiple shoots from the shoot tips were obtained on woody plant medium (WPM) (13) with 1 ppm N^6-benzyladenine (BA). The multiplication rate was dependent on the length of the subculture periods.

Due to its desirable horticultural characteristics, sweetgum is a widely planted street tree; however, its roots can be responsible for damage to sidewalks and curbs. Wager and Barker (23) found in the San Francisco Bay area, that sweetgum caused the most damage to sidewalks of 10 species surveyed. Some of the sweetgum trees were found to do less damage than the others. They proposed selecting and propagating these trees by tissue culture.

Sutter and Barker (22) cultured explants from mature and seedling sweetgum buds at 25°C with a 16 hour photoperiod and a light intensity of 60 $\mu Em^{-2}sec^{-1}$. The method for the culture of 1-year-old seedlings called for the excission of lateral buds as explants. Growth and multiple shoots were obtained on WPM (unspecified as to composition) with 0.2 mg/l or 1.0 mg/l BA, however higher concentrations of BA caused rosetting. When the WPM was used at 1/2 and 3X normal strength, it was found that if $CaCl_2$ was reduced to 1/3 concentration more vigorous growth resulted. For the culture of buds from mature trees, shoots were removed from the tree prior to bud break in the spring, and forced to break. Explants consisting of actively growing buds were taken from the upper 3-4 inches of the shoots. The culture medium was WPM with 1/3 $CaCl_2$ concentration and 0.2 mg/l BA. It was found that the buds had to be transferred every 3-4 days for the 2 weeks for proper growth. Buds from one source remained stunted, while those from three others grew and differentiated multiple shoots. Differences in growth were attributed to genotype. Rooting of the shoots on WPM with 0.5 ppm indole-butyric acid (IBA) was dependent on the source and varied from 20-100%. The stunted genotype gave the poorest rooting.

Since their establishment in culture, plantlets have been transferred to containers for further growth (Barker, personal communication). About 40 one-inch-tall plantlets of 2 genotypes and 80 of a third had been produced by November 1984. These plants overwintered in a cold greenhouse, without dropping their leaves. In March the plantlets were placed in 1 x 1 x 8 inch deep Tinus containers and forced. About 18 inches of growth occurred in that year.

Sommer and Wetzstein (19, 20, 21) have cultured shoot tips of 1-3 month old seedlings of shoot tips of _L. styraciflua_ on a modified Blaydes (20) medium with either 1 ppm BA plus 1 ppm kinetin or 1 ppm BA and 1 ppm N^6-isopentenyl adenine (2iP). Multiple shoots were usually obtained from 70-80% of the explants. However with seedlings from some half-sib seed lots, the procedure was successful in only 0-33%. This procedure was also successful with shoot tips of _L. formosiana_. Multiplication rates of 2-4 times were obtained. Sommer has also attempted to obtain multiplication of shoot tips from 5-30 year old sweetgum trees. Occassionally multiple shoots have been obtained, however, not in sufficient numbers to recommend a specific medium.

3. TISSUE CULTURE VIA ADVENTITIOUS BUD DIFFERENTIATION

The only reports found in the literature on the tissue culture of sweetgum resulting in adventitious bud production are those of Sommer and Wetzstein (19, 20, 21). Their procedure is relatively simple. Hypocotyl sections about 3 mm long are removed from sterile seedlings. These are placed on a modified Risser and White's (20) medium modified with 6 times the amount of KNO_3 plus 5 mM NH_4Cl, 1 ppm indole-3-acetic acid (IAA), and 5 ppm 2iP. On this medium a small amount of callus is produced and 3-5 adventitious shoots differentiate within 8 weeks (Fig. 1). These cultures can then be transferred to a liquid version of modified Blaydes medium with 0.01 ppm naphthaleneacetic acid (NAA), 0.5 ppm BA, and 1% sucrose. In 1-2 months the callus proliferates and numerous shoots are produced in 10-20% of the cultures. In 125 ml flasks with 40 ml of medium, typically 25-50 shoots may be harvested monthly provided the medium is likewise renewed monthly. Vitrification was not a problem during the first 3-6 months on liquid medium. The excised shoots loose excessive succulence within a few days upon transfer to Risser and White's solidified basal medium. On this medium, rooting starts in one month. This culture system has been used extensively to study changes in morphology and anatomy

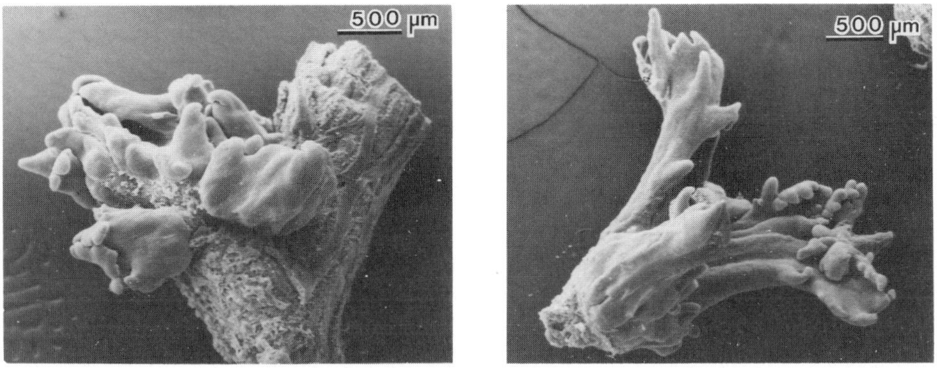

FIGURE 1. a) Adventitious buds differentiated on a hypocotyl section. b) More developed adventitious bud showing expanding leaves.

of new leaves produced during acclimatization (24, 25). Under the conditions found in culture the stomata appear raised, and apparently nonfunctional. A pallisade area is not differentiated and chloroplast internal membranes are not organized in granal stacks. Leaves produced after acclimatization show a more normal morphology.

Plantlets have been successfully transplanted into nursery beds and into the field. Growth of plantlets in nursery beds has been equivalent to that of seedlings. Plantlets are currently under observation for evidence of somaclonal variation.

4. FINAL COMMENTS

The tissue culture of sweetgum has been successful at least on a small scale. Clones in sufficient number for research purposes have been produced. However, the ready availability of large quantities of seed makes the propagation of Liquidambar by tissue culture economically unviable unless advantages of superior selections can be justified. The production of sweetgum by tissue culture awaits further improvement.

Once methods for the production of superior individuals by shoot culture have been developed, a reevaluation of the propagation of trees with superior volume production and survival ability can be made.

5. REFERENCES

1. BAILEY HORTORIUM 1976 Hortus Third. MacMillan, New York
2. BILAN MV 1974 Rooting of Liquidambar styraciflua cuttings. N Z J For Sci 4:177-180
3. BONNER FT 1963 Some southern hardwoods can be air-layered. J For 12:923
4. BROWN CL and RG MCALPINE 1964 Propagation of sweetgum from root cuttings. Georgia For Res Pap 24
5. ERNST WR 1963 The genera of Hamamelidaceae and Plantanaceae in the southeastern United States. J Arnold Arbor 44:193-210
6. FARMER JR RE 1965 Propagation of sweetgum by softwood stem cuttings. In Proc 8th South For Tree Improv Conf pp 123-124
7. GOLDBLATT P and PK ENDRESS 1977 Cytology and evolution in Hamamelidaceae. J Arnold Arbor Harv Univ 58:67-71
8. HARE RC 1976 Rooting of American and Formosan sweetgum cuttings taken from girdled and nongirdled cuttings. Tree Planters' Notes 17:6-7 and 33
9. HARLOW WM and ES HARRAR 1968 Textbook of Dendrology, McGraw-Hill, New York
10. HUTCHINSON J 1967 The Genera of Flowering Plants (Angiospermae) Dicotyledons. Vol II Oxford Univ Press, Oxford
11. KORMANIK PP and CL BROWN 1973 Modified growth chamber enhances vegetative propagation of selected sweetgum and yellow-poplar. In Proc 12th South For Tree Improv Conf pp 259-266
12. LEE S-C 1935 Forest Botany of China. Commercial Press, Shanghai, China
13. LLOYD G and B MCCOWN 1980 Commercially feasible micropropagation of mountain laurel, Kalmia latifolia, by use of shoot-tip culture. Int Plant Propagators Soc Comb Proc 30:421-427
14. MAISENHELDER LC 1957 Propagation of some delta hardwoods by rooting. In Proc 4th South For Tree Improv Conf pp 55-58
15. MCALPINE RG 1965 Vegetative propagation methods for hardwoods. In Proc 8th South Conf For Tree Improv pp 14-20
16. PEATTIE DC 1950 A Natural History of Trees. Houghton Mifflin, Boston
17. SANTAMOUR FS 1972 Interspecific hybridization in Liquidambar. For Sci 18:23-26

18. SANTAMOUR FS JR and AJ MCARDLE 1984 Cultivar checklist for *Liquidambar* and *Liriodendron*. J Arboric 10:309-312
19. SOMMER, HE 1981 Propagation of sweetgum by tissue culture. In Proc 16th South For Tree Improv Conf pp 184-188.
20. SOMMER HE 1983 Organogenesis in woody angiosperms: applications to vegetative propagation. Bull Soc Bot Fr 130, Actual Bot 983:79-85
21. SOMMER, HE, HY WETZSTEIN and N LEE 1985 Tissue culture of sweetgum (*Liquidambar styraciflua* L.). *In* Proc 18th South For Tree Improv Conf pp 42-50.
22. SUTTER E AND P BARKER 1984 Tissue culture propagation of selected mature clones of *Liquidambar styraciflua*. Int Plant Propagators Soc Comb Proc 33:113-117
23. WAGAR JA and PA BARKER 1983 Tree root damage to sidewalks and curbs. J Arboric 9:177-181
24. WETZSTEIN HY and HE SOMMER 1982 Leaf anatomy of tissue-cultured *Liquidambar styraciflua* (Hamamelidaceae). Am J Bot 69:1579-1586
25. WETZSTEIN H Y and HE SOMMER 1983 Scanning electromicroscopy of in vitro cultured *Liquidambar styraciflua* plantlets during acclimatization. J Am Soc Hortic Sci 108: 475-480
26. WISNIEWSKI M and AL BOGLE 1982 The ontogeny of the inflorescence and flower of *Liquidambar styraciflua* L. (Hamamelidaceae). Am J Bot 69:1612-1624

23. TEAK

AF MASCARENHAS, SV KENDURKAR, PK GUPTA, SS KHUSPE, DC AGRAWAL

1. INTRODUCTION

Some of the important physical characteristics of wood are determined by the thickness of the cell walls and the relative size and proportions of the vessels, tracheids and fibres. Gymnosperms are known as 'softwoods' because their wood is composed entirely of tracheids. Angiosperms containing both vessels and fibres are 'hardwoods.' This classification, however, does not mean that all angiosperms have 'harder' wood than gymnosperms. Hardness of wood is a reflection of the sturdiness of the individual cell walls and other factors such as the amount of lignin they contain.

It has been emphasized that tree planting in the tropics is no longer almost wholly for industrial purposes. Although this remains the dominant reason, planting for firewood to reduce soil erosion, to control of water run-off, to combat desertification, and to provide shelter and shade, are all becoming increasingly important (8). All these uses clearly point out the urgent need of either expanding the current area under forestation or to improve the quality of forests by planting uniform, genetically superior varieties with higher yields.

Conventional methods of propagation, for example by cuttings, grafts etc. have their limitations especially when large quantities of propagules are required. Cuttings from various heavy hardwood trees have not rooted in spite of repeated experiments under various growth conditions using a wide variety of root induction hormones. The tissue culture method offers an alternate, novel possibility and the technique has been successfully applied to many of the tree species, especially to certain gymnosperms and a few angiosperms of temperate regions (5).

The work done with tropical trees is very limited when compared to that with temperate ones (16, 20).

In this chapter we will be dealing with teak. Another tropical hardwood tree, tamarind, will be discussed in the chapter following this one.

1.1. Distribution

Teak (*Tectona grandis*) (Fam. Verbenaceae) is an important forest tree, highly valued for its wood. It is a tree with a rounded crown very variable in size, depending on its habitat.

It is indigenous to Burma, Central and South India and Thailand. Trees were introduced with varying degrees of success in Malaysia, Indonesia, Africa and the Philippines. Teak is one of the exotic species for large scale plantations in the tropics (9) and does well even in areas much beyond its natural limits.

1.2. Economic importance

1.2.1. Timber. Teak has a worldwide reputation as a quality timber on account of its remarkable physical and mechanical properties, particularly shape retention, durability, resistance to fungi and chemicals (33). The most important property is its natural resistance to termites due to the presence of sesquiterpenes (21). Several uses of the timber are given in Table 1.

1.2.2. Leaves. Leaves are used to produce a dye (yellow or red) for wool and silk.

1.3. General botanical characteristics

1.3.1. Morphology. The tree attains a large size, with a tall clean and cylindrical bole, carrying its girth well up the stem when grown on good sites. The bark is fibrous, light brown to grey and 4-18 mm thick. Its leaves are broadly elliptical or obovate. The flowers are small, white, sweet and scented. The globular teak fruit is a drupe with a tetralocular, hard, bony endocarp. Each fruit contains 1 to 3 or rarely 4 marble white, ovate, 4-8 mm long seeds. The tree starts bearing fruit at the age of 20 years (33).

1.3.2. Climate and rainfall. Teak thrives best and reaches its largest dimensions in a fairly moist, warm, tropical climate (33). It can withstand drought very well, but is sensitive to frost

Table 1. Uses of teak wood.

Marine constructions	: Ship building particularly for decking, piles in harbours
Railways	: Carriages, wagons, coaches, etc.
Bridge constructions	
Furniture and cabinets	
Chemical factories	: Bench tops, cases and vats for shipping, corrosive liquids
Foundries	: Patterns
Musical instruments	
Construction timber	: Poles, beams, trusses, columns, roofs, doors, windows, frames, flooring, planking, panelling, staircases, general carpentry
Plywood	: General purposes plywood, decorative plywood, tea-chest plywood, faces and cross bands of marine plywood
Wood wastes	: Particle, plastic and fibre boards, pulp for writing and printing paper, fuel

(21). The areas with a normal annual rainfall from 125-250 cm and a marked dry season of 3 to 5 months are favourable for its development (33).

1.3.3. Temperature, soil and light conditions. Teak can withstand an absolute maximum shade temperature of 48°C and an absolute minimum shade temperature of 2°C. The mean annual temperature which suits it best is between 23°C to 28°C (35). Teak generally prefers rich soil with good drainage and a dry sub soil. It thrives well in the fertile lower slopes of the hills where the soil is deep. It is intolerant of shallow soils.

1.4. Diseases and pests

The tree is affected by numerous diseases such as root rot caused by *Peniophora rhizomorpho - sulphurea*, leaf rust caused by *Olivea tectonae* and *Uncinula tectonae* (33). In dry teak forests,

50% of the trees exhibit hollowness due to decay of heartwood by *Polyporus zonalis* and *Fomes lividus* and also *Nectria haematococoa*. More than fifty insect pests attacking the living plants have been recorded. Havoc is often caused by the teak skeletonizer, *Hapalia machaerates* and the teak defoliator *Hyblaea purae* (33). Some other pests such as *Pyrausta macheralis* and *Pagyda salvalis* damage the inflorescences very badly and thus cause poor fruit formation (3).

1.5. Conventional methods of propagation

Teak is generally propagated through seeds. Various methods of vegetative propagation have been used, such as budding and grafting (17), rooting of cuttings (15) and rooting of buds cut from stock stumps raised in polypots (18). For large scale plantation of teak the method of planting of stumps from 30-40 cm long seedlings about one year old is often used. The stump is 1-2 cm in diameter and includes about 2-4 cm of the stem portion above the first pair of buds; its tap root is cut to a length of 20-25 cm. The stumps are bundled, packed and transported to the plantation sites, where they are planted in the soil leaving only a 1-2 cm portion of them above the ground (23).

2. NEED FOR TISSUE CULTURE STUDIES IN TEAK

The recent trend in genetic improvement of forest trees emphasizes the need for the identification and multiplication of superior genotypes. These trees have certain desirable traits and are called 'elite' or 'plus' trees. As with other tree species 'elite' teak trees have been identified in several forests. They were selected on the basis of total height, clean bole height, crown height, crown diameter, cylindrical unfluted trunk, and self pruning habit.

The conventional methods of vegetative propagation have limitations. They are slow and time consuming. Teak being a cross pollinated species, seed raised progenies show wide variations. Teak trees also have an irregular seed bearing habit and the production of seeds is much lower than the actual requirements (16). Seed viability is poor and is affected by the season when collected and the storage conditions (32). Teak seeds being enclosed in hard coats, show very poor germination rates in spite of several pretreatments (4, 6, 13).

The number of seedlings required annually by the forest corporations of India alone amounts to over a few billions. They are produced mainly through seeds from trees collected in areas with superior trees. For instance, in Maharashtra State, in the West of India, the plantation target is 20,000 hectares per year, which requires about 200-400 tonnes of fruits to fulfill the requirement of 40 million plants per year (3). In order to achieve the above aim, there is a need to increase production of viable seeds from 'plus' trees and to get higher germination rates. In addition, methods can be devised whereby a large population of genetically superior seedlings is clonally produced for large-scale plantations. This is where tissue and organ culture can play an important role. Recently several reports and reviews have become available on tissue culture of trees (1, 2, 10, 14, 24, 27, 28, 29, 30, 31, 34, 37).

3. TISSUE CULTURE STUDIES: PROTOCOLS FOR SEEDLINGS AND MATURE TREES

The method developed in this laboratory for micropropagation of teak seedlings and 100-year-old 'elite' trees, and the results of a statistically designed field trial, are presented below.

3.1. Culture media. White's medium (W) (36) and Murashige and Skoog's medium (MS) (22) modified as reported earlier (12) were used. These media were supplemented with the following hormones: Kinetin (K); Benzyladenine (BA); Indole-3-acetic acid (IAA); Indole-3-butyric acid (IBA); Indole-3-propionic acid (IPA).

The media were prepared according to the composition given below, with the growth hormone concentrations (mg/l) shown in brackets.

MS-1 : MS + K (0.15) + BA (0.15) semisolid
W-1 : W (No hormones) liquid
W-2 : W + IAA (1.0) + IBA (1.0) + IPA (1.0) liquid
W-3 : W + IAA (2.0) + IBA (2.0) + IPA (2.0) liquid
W-4 : W (No hormones) semisolid

The steps followed are given below:

1. Adjust pH of all media to 5.8 with 1N NaOH or 1N HCl.

2. Solidify the MS-1 and W-4 media with 0.8% bactoagar (Difco).

3. Pour 20 ml media/tube (25 mm x 15 mm, Corning brand). Use filter paper supports for liquid media.

4. Sterilize media at 15 psi for 20 min and steam for 30 min at atmospheric pressure on the subsequent day.

5. Filter sterilize (millipore filter, membrane size 0.22 μ m) IAA, IBA and IPA and add aseptically to W-2 and W-3 media.

3.2. Collection and surface sterilization of seeds

1. Select fresh stock of teak fruits and wash under tap water.

2. Treat with 1N KOH for 12 h.

3. Wash thoroughly under running tap water to remove KOH.

4. Break the fruits using a nut cracker and collect healthy seeds in distilled water.

5. Wash seeds with detergent for 2-3 min followed by a wash in 70% ethanol for 2-3 min. Finally sterilize seeds with 0.07% mercuric chloride for 2 min.

6. Wash off sterilant with sterile distilled water under aseptic conditions.

3.3. Collection and surface sterilization of buds from mature trees

1. Collect fresh, green, healthy, terminal buds (from March to May) about 1-2 cm long, from the upper branches.

2. Put buds in a non-sterile conical flask, wash with detergent for 2-3 min and thoroughly wash with distilled water.

3. Transfer to a sterile conical flask. Surface sterilize with 0.15% mercuric chloride for 10 min.

4. Under aseptic conditions wash thoroughly with sterile distilled water and inoculate immediately.

3.4. Inoculations

3.4.1. Seedlings

1. Transfer the seeds to a sterile petri plate and allow to dry for 2-3 min.

2. Inoculate 1 seed/tube on W-4 medium.

3. Incubate in the dark at 27°C. Seeds germinate within 2-3 weeks (the percent germination is 70 to 80%).

3.4.2. Mature trees

1. Remove some of the leaves enclosing the shoot tip with the help of sterile forceps and scalpel and excise out the apex (3-5 mm).

2. Inoculate on MS-1 medium.

3. After 48 h blackening is observed in the medium just below the bud. Shift the bud to another position in the same tube.

4. Incubate at 28 ± 2°C in light (16 h photoperiod, 1000 lux).

5. Buds sprout and turn green in 10-15 days.

6. At this stage shift the buds to fresh tubes on MS-1 medium. On this medium buds elongate and develop new leaves with 1 or 2 internodes within 2 weeks (Fig. 1).

3.5. Multiplication in subculture

3.5.1. Seedlings

1. Cut the seedling shoot into small segments so that each segment includes a nodal region with a pair of leaves.

2. Put segments in a sterile 100 ml flask containing 0.75% of polyvinylpyrollidone (PVP) (insoluble, M.W. 36,000, crosslinked, Sigma Chemical Company) with 2% sucrose.

3. Keep the flask on a rotary shaker (at 120 rpm) for 30 min.

4. Wash off PVP with sterile distilled water and put the segments in sterile petri plates.

5. Inoculate on MS-1 medium taking care that the axillary buds are in contact with the medium.

6. Incubate the cultures at 28°C in the light (16 h photoperiod 1000 lux).

7. Within 20-25 days about 3 to 4 shoots develop from each segment, except the ones with apical buds. These shoots can be either subcultured for further multiplication or kept for rooting.

3.5.2. Mature trees

1. Take elongated shoots with 1 to 2 internodes.

2. Follow steps 1 to 6 as in the protocols for seedings.

3. After about 2-3 weeks each segment gives rise to 1 to 2 shoots, which can be further subcultured (Fig. 2).

4. Shoots which have undergone 2 to 3 passages can be taken for rooting since the rooting ability increases with subculture.

3.6. Rooting

3.6.1. Seedling cultures

1. Separate the multiple shoots by excision at the base.

2. Keep each shoot on a filter paper support placed in medium W-2 so that the 2-3 mm portion at the base of the shoot dips into the medium.

3. Incubate under the same conditions as were used for the shoots for 48 h.

4. Transfer the shoots to W-1 medium.

5. After incubation for about a week, roots start emerging. It takes about 15-20 days to get plantlets (50-60% rooting).

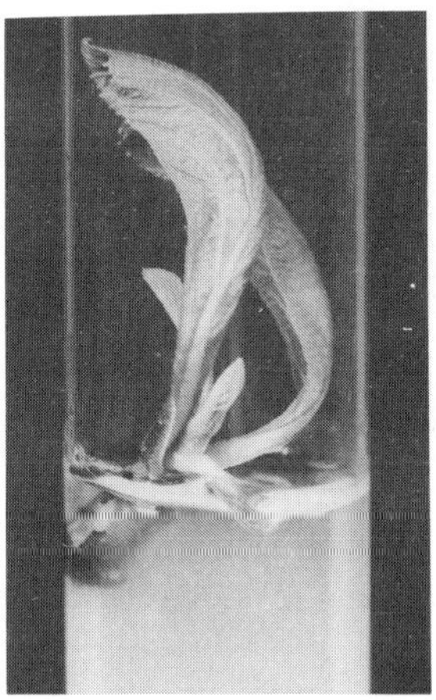

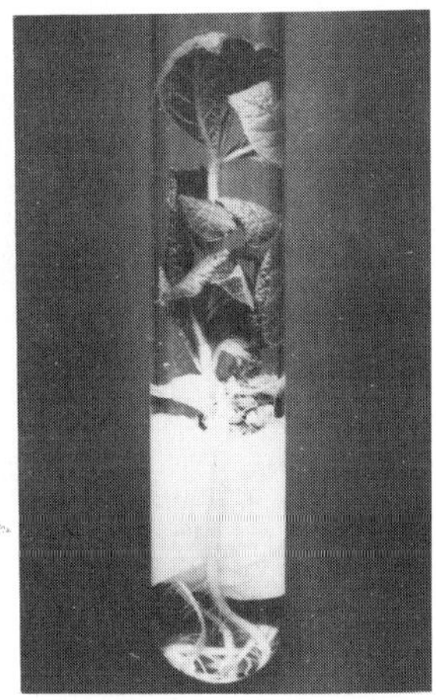

FIGURE 1. Elongated teak bud (25 days in culture)
FIGURE 2. Teak bud forming multiple shoots on subculture
FIGURE 3. Rooted plantlet (one month old)

3.6.2. Mature tree cultures

1. Carefully observe the shoots in culture. Some shoots in the clump elongate more than others. Subculture the smaller ones as in the procedure described earlier. Shoots 4-5 cm long must be used for rooting.

2. Follow steps 2 to 4 as above for seedlings except that the W-3 medium is used in step 2.

3. Root emergence is seen within the first week. The plantlets are ready for transfer to pots or flats within 32-35 days (40-50% rooting) (Fig. 3).

3.7. Transfer to pots

1. Transfer the rooted plantlets to flats or small plastic pots (60 mm diameter) containing sterile soil:sand:compost (3:3:1, v:v:v) (Fig. 4).

FIGURE 4. Teak plant in pot

2. Incubate at 30°C in light (16 h photoperiod 1000 lux).

3. After 15-20 days (when plantlets are acclimatized, transplant to polybags (size 15 x 25 cm) containing soil:sand:compost (3:3:1).

3.8. Transfer to field

1. After about 40-60 days, when plantlets attain a height of about 30-40 cm, they are ready for transfer to the field.

2. Fill pits (30 cm cubes), dug at 3 m intervals with 3.5 m between rows, with a mixture of good soil (river silt) and farmyard manure.

3. Before transplanting, drench the pits with 0.1% copper fungicide to avoid fungal attack.

4. After transplanting, water plants at 3-day-intervals initially. Later they can be irrigated once a fortnight in summer season (March to June) and once a month in winter (October to February).

5. Urea 100 gm/tree can be applied at 2-month intervals.

4. COMPARATIVE FIELD TRIALS

4.1. Growth data

The field trial (Fig. 5) was conducted with plants obtained from three sources; T1 - control seedlings obtained from germinated seeds of 14 'plus' trees; T2 - tissue culture plants raised from seedlings of 'plus' trees; and T3 - tissue culture plants raised from 14 'elite' trees. A random block design (RBD) was followed with 7 replicates. After planting the height measurements and the diameter at breast height (DBH) of all the experimental plants were recorded at monthly intervals. The DBH measurements were recorded only after the second year. All data obtained were statistically analyzed (see Table 2).

4.1.1. Height. Maximum height increase was attained by tissue culture raised plants from 'elites' (T3) up to the third year, the height increment becoming more or less uniform by the fourth year. This was followed by height increase in plants raised by tissue culture from seedling explants (T2). The lowest height increments were from seeds (T1-control). In the first year (1980-81), the increase in height was 18% higher in tissue culture raised plants from 'elites' over seedling tissue culture plants and 40% higher over the controls respectively. This gain in height in the tissue culture plants from 'elite' trees gradually reduced in the following years, becoming very marginal in 1984.

An interesting observation was that even tissue culture plants raised from seedling explants (T2) showed a higher growth increment than those raised from explants from seedlings raised from seed (T1-control).

Table 2. Comparative growth data of tissue culture and control teak (*Tectona grandis* L.) plants.

Statistical design : RBD
No. of replicates : 7
Total no. of plants : 42
Date of planting : June 1980

Treatment	Height increment (cm)				DBH increment (cm)	
	1981	1982	1983	1984	1983	1984
T1	83.03	166.9	90.1	115.9	5.30	1.93
T2	114.50	198.3	105.4	139.3	5.80	2.01
T3	140.00	216.3	130.9	140.9	6.97	2.15
S E ±	10.20	7.39	4.33	9.75	0.84	0.078
C D 0.5%	31.18	22.77	13.32	N S	2.58	N S

T1 = Control
T2 = Tissue culture (seedling)
T3 = Tissue culture (100-year-old 'plus' trees)

4.1.2. Diameter at breast height (DBH) increment. Similarly there was a significant increase in diameter (DBH) in tissue culture T3 plants raised from 'elite' trees followed by tissue culture plants raised from seedlings (T2) and the control (T1). The increase was 30% higher in T3 plants over the controls and 19% over the tissue culture plants raised from seedlings (T2). In the fourth year the DBH increments were marginal and insignificant.

4.2. Flowering and fruiting

In natural populations the earliest flowering occurs after about 20 years. However, the earliest flowering in experimental teak plants was observed within 2 years of planting in the field, when 1 plant out of 14 each from T3 and T2 flowered. Till 1984 out of 14 plants under each treatment, 9 tissue culture raised plants from 'elite' (T3), 10 plants from seedling explant raised by tissue culture (T2), and 3 plants from seed raised controls (Tl) flowered.

Flowering commenced in June and lasted till August-September every year. The inflorescences bore normal fruits with viable seeds.

FIGURE 5. Teak field

5. FUTURE PROSPECTS

By the process developed in this laboratory it is possible to produce over 500 plants from a single bud of a mature plant or 3000 plants from a seedling in a year. Over 1000 plants produced by tissue culture of seedlings and 'elite, trees were supplied to the Forest Development Corporation of Maharashta, India, and planted in the teak growing areas. Stumps of standard size, to be used for conventional clonal propagation as described earlier (Section 1.5.), were also prepared from tissue culture raised plantlets and successfully transported. These are now undergoing field trials at different locations in the teak growing areas in the same regions from where the hardwood material was collected. This is important since the high survival rate of stumps in transport reduces the problems of transport over long distances. This can cut down the transport cost of plants, particularly when tissue culture laboratories are situated a long distance away from the teak growing areas.

Only a few reports are available regarding the field performance of tissue culture plants from mature woody trees e.g. teak (19), Eucalyptus citriodora (20).

The growth data obtained from the field trials on teak clearly indicate that tissue culture raised plants grow faster, and flower very early as compared to trees in natural populations. This observation suggests the immediate application of tissue culture in tree improvement programs in teak for raising seed orchards. However, field trials for longer periods are still necessary to determine whether the early high rates will affect the quality of the physicochemical properties of the wood.

The cost per plant, which works out to about 30 cents, is still high particularly if one considers application of this method for mass propagation. Improvements and refinements in the technique will be necessary. Development of somatic embryos and encapsulation with enriched polymers and production of synthetic seeds could be a solution to this problem of economics of the plants. Durzan (7) has made some, interesting calculations with cell suspension cultures, i.e., if 80% of the cells in 100 liters of culture medium could produce embryos and plants this would be sufficient to plant 100,000 acres at a 12 x 12 foot spacing. To date, various reports are available on somatic embryogenesis of woody tree species such as sandalwood (26). Paulownia (25), Albizzia lebbeck (11), sweetgum (27, 29). Very recently plantlets have also been regenerated through somatic embryogenesis from protoplasts of sandalwood trees (26) (cf Tulecke volume 2).

New developments in genetic engineering procedures, including the recent somaclonal, gametoclonal and embryo rescue technologies, show promise for their inclusion and applications in tree breeding programs for the genetic improvement of forestry. These technologies which are described in Volume 1, when combined with the methods of micropropagation for raising seed orchards, could greatly reduce the time normally taken in conventional tree breeding programs which is in the range of 20-30 years.

Several 'elites' of teak have been identified for different traits in Burma, India and also in various other countries. Germ plasm of this valuable material could be preserved in tissue culture banks for future use in breeding when necessary.

Judging from the recent progress one can visualize within the next decade, the production of a whole new range of improved forest trees, having superior characteristics for uniform growth rates, wood quality and resistance to disease, herbicides and extreme climatic conditions.

7. REFERENCES

1. BIONDI S, TA THORPE 1981 Propagation of forest tree species. In AN Rao, ed, Tissue Culture of Economically Important Plants. COSTED, Singapore, pp 197-204
2. BONGA JM 1977 Application of tissue culture in forestry. In J Reinert, YPS Bajaj, eds, Applied and Fundamental Aspects of Plant Cell, Tissue and Organ Culture. Springer-Verlag, Berlin, pp 93-108
3. DABRAL SL, PW AMIN 1975 Poor fruit formation in teak in Chanda Forests of Maharashtra. Indian For 101: 616-619
4. DABRAL SL 1976 Extraction of teak seeds from fruits, their storage and germination. Indian For 102: 650-658
5. DAVID A 1982 *In vitro* propagation of gymnosperms. In JM Bonga, DJ Durzan, eds, Tissue Culture in Forestry. Martinus Nijhoff, The Hague, pp 72-108
6. DAVIDSON J, J FAIRLAMB 1976 Germination of teak seeds - preliminary evidence of a chemical germination inhibitor. In Seed Problems, Proc 2nd Int IUFRO Symp Physiol Seed Germination, Japan
7. DURZAN DJ 1976 Biochemical changes during gymnosperm development. Symp on juvenility in Woody perennials. Acta Hortic 56: 183-194
8. EVANS J 1979 The effects of leaf position and leaf age in foliar analysis of *Gmelina arborea*. Plant Soil 52: 547-552
9. EVANS J 1982 Plantation Forestry in the Tropics. Oxford Univ Press
10. GEORGE EF, PD SHERRINGTON 1984 Plant Propagation by Tissue Culture. Handbook and Directory of Commercial Laboratories. Eastern Press Reading, Berks, Great Britain, pp 432-434
11. GHARYAL BK, SC MAHESHWARI 1981 Somatic embryogenesis in tissue cultures of *Albizzia lebbeck*. Naturwissenschaften 68: 379-380
12. GUPTA PK, AL NADGIR, AF MASCARENHAS, V JAGANNATHAN 1980 Tissue culture of forest trees - clonal multiplication of *Tectona grandis* (Teak) by tissue culture. Plant Sci Lett 17: 259-268
13. JOSHI MD, SP KELKAR 1971 Germination of seed in dry teak (*Tectona grandis* L.) (1) Preliminary studies in fruit development and seed dormancy. Indian For 97: 210-212
14. KARNOSKY DP 1981 Potential of forest tree improvement via tissue culture. Bioscience 31: 114-118

15. LAHIRI AK 1974 Preliminary study on rooting of green wood cutting of teak. Indian For 100: 559-560
16. LEE SK, AN RAO 1981 In vitro plantlet development in tropical trees - Calophyllum inophyllum and Eugenia grandis. In AN Rao, ed, Tissue Culture of Economically Important Plants. COSTED, Singapore, pp 185-190
17. MAHMOOD HUSAIN AM, TR SOMASUNDARAM 1975 Scions on stumps - an innovation in the vegetative propagation of teak (Tectona grandis L.f.). South For Rangers Coll Mag 51: 1-8
18. MAHMOOD HUSAIN, TR SOMASUNDARAM, KN SUBRAMANIAN 1976 A recent advance in teak culture. Indian For 102: 531-532
19. MASCARENHAS AF, PK GUPTA, VM KULKARNI, U MEHTA, RS IYER, SS KHUSPE, V JAGANNATHAN 1981 In AN Rao, ed, Tissue Culture of Economically Important Plants. COSTED, Singapore, pp 175-179
20. MASCARENHAS AF, S HAZRA, U POTDAR, DK KULKARNI, PK GUPTA 1982 Rapid clonal multiplication of mature forest trees through tissue culture. In A Fujiwara, ed, Proc 5th Int Cong Plant Tissue Cell Culture, Tokyo, Japan, pp 119-121
21. MASLEKAR AR 1981 Forester's Companion (Indian Forestry Handbook). Jugal Kishore and Company, India
22. MURASHIGE T, F SKOOG 1962 A revised medium for rapid growth and bioassay with tobacco tissue culture. Physiol Plant 15: 473-493
23. OKA AF 1976 Teak Nursery Practices in Maharashtra State. Research and Education Circle, Nagpur, India
24. POWLEDGE L 1984 Biotechnology touches the forest. Biotechnology 2: 763-772
25. RADOJEVIC L 1970 Somatic embryos and plantlets from callus cultures of Paulownia tomentosa. Z Pflanzenphysiol 91: 57-62
26. RAO PS, P OZIAS-AKUS 1985 Plant regeneration through somatic embryogenesis in protoplast cultures of sandalwood. Protoplasma 124: 80-86
27. SOMMER HE 1983 Organogenesis in woody Angiosperms: Application to vegetative propagation. Bull Soc Bot Fr 130, Actual Bot 983: 79-85
28. SOMMER HE, CL BROWN 1979 Application of tissue culture to forest tree improvement. In WR Sharp, PO Larsen, EF Paddock, V Raghavan, eds, Plant Cell and Tissue Culture, Principle and Applications. Ohio State Univ Press, Columbus, pp 461-491
29. SOMMER HE, CL BROWN 1980 Embryogenesis in tissue culture of sweetgum. For Sci 26: 258-267
30. SOMMER HE, CL BROWN 1982 Vegetative propagation of dicotyledonous trees. In JM Bonga, DJ Durzan, eds, Tissue Culture in Forestry, Martinus Nijhoff, The Hague, pp 107-149
31. SOMMER HE, LS CALDAS 1980 In vitro methods applied to forest trees. In TA Thorpe, ed, Plant Tissue Culture Applications to Agriculture, Academic Press, New York, pp 349-358
32. Storage of Teak Seeds 1962 Extr Rep Dep For Papua N Guinea For Abst 23: 1879
33. The Wealth of India 1976 Vol X, Publ Inf Directorate, CSIR, India
34. THORPE TA 1984 Biotechnological applications of tissue culture to forest tree improvement. In MM Young, ed, Biotechnological Advances, Pergamon Press, Vol 1, pp 1-16

35. WATT G 1972 The Dictionary of Economic Products of India Vol VI, Part IV. Periodical Experts, Delhi, India
36. WHITE PR 1962 The Cultivation of Animal and Plant Cells, The Ronald Press Co, New York
37. WINTON LL 1978 Morphogenesis in clonal propagation of woody plants. In TA Thorpe, ed, Frontiers of Plant Tissue Culture, 1978, Univ Calgary, Alberta, Canada, pp 419-426

8. ACKNOWLEDGEMENT

The authors wish to acknowledge Miss S Rao for her contribution during the course of this work.

24. TAMARIND

A.F. MASCARENHAS, SRILATA NAIR, V.M. KULKARNI, D.C. AGRAWAL, S.S. KHUSPE, U.J. MEHTA

1. INTRODUCTION

In the INTRODUCTION of the previous chapter we have outlined the importance of tissue culture of tropical hardwoods. In this chapter we will be dealing with one species, tamarind, which we have extensively propagated *in vitro*.

1.1. Distribution

Tamarindus indica (Fam. Leguminosae) is a common and important tree of India. It is also grown in the West Indies and Southern Florida (2), Costa Rica, Puerto Rico and some states of Mexico (8).

1.2. Economic importance

Almost every part of the tamarind tree finds some use. The wood yields a valuable timber, but the volume of heartwood is small. The sapwood is yellowish, often with brownish red streaks. The heartwood is dark brown, usually streaked and mottled with black, particularly in old trees. The wood is close-grained, strong, very hard and heavy (weight: 913-1282 kg/m^3). It is moderately refractory, shows cracking and splitting during air-seasoning and is unresponsive to preservative treatment. It is durable under cover and is resistant to insect attack. It bends well and takes on good polish. The wood is used mostly for agricultural implements, tool-handles, wheels, mallets, rice pounders, oil mills and turnery. It is suitable for printing blocks and tent plugs. It yields a decorative plank for panelling and furniture. In South India it is employed for construction work and has been recommended as a substitute for teak and sal (*Shorea robusta*) for beams, rafters, purlins and trusses. The wood is also used for fuel and is preferred for making gunpowder charcoal.

The fruit contains a sweetish, acidic pulp, the tamarind of commerce, which is widely used for souring curries, sauces, chutneys and certain beverages, and also as a local medicine. The kernel powder, commercially known as tamarind kernel powder (TKP), finds extensive use as a sizing material in the textile industry. When the powder is boiled a gruel is obtained which on drying in thin layers produces strong, smooth, continuous and elastic films, which is useful for sizing. The fatty oil from the kernel resembles peanut oil and is reported to be useful in the preparation of paints and varnishes and for burning in lamps. The average yield of tamarind fruit from a mature tree is 30-50 kg annually, although some trees have been identified which yield 180-225 kg per year. India is perhaps the chief producer of this fruit in the world with an essential production of about 250,000 tonnes per annum. The entire demand for the pulp in India is met by indigenous production.

1.3. General botanical characteristics

Tamarind is a moderately sized to large, evergreen tree growing up to 24 m in height and 7 m in girth. It is cultivated or found naturalized, throughout the plains and sub-Himalayan tracts of India, particularly in the south. The tree has a brownish or dark grey bark, longitudinally and horizontally fissured. Leaves are paripinnate and the flowers are small and yellowish in colour with pink stripes. The tree grows with a spreading crown. Pods are more or less constricted between the seeds, slightly curved and brownish-ash coloured. The tree generally prefers a warm climate and can withstand drought, but is sensitive to frost (4).

1.4. Diseases and pests

The tree is affected by a number of rots such as saprot (caused by *Xylaria euglona*), brownish saprot (caused by *Polyporus calcuttensis*) and white rot (caused by *Trametes floccosa*) (7); powdery mildew (caused by *Oidium* sp.) (6); hypocotyl rot (caused by *Sclerotium rolfsii*) (3). Several insects attack the fruits and seeds of tamarind, both on the tree and when stored for marketing. These include larvae which feed on the sap of the fruits and the sap of the young twigs and branches.

1.5. Conventional methods of propagation

The main method of propagating tamarind is by seeds, though it can also be reproduced from cuttings and by budding. Propagation by means of "gootee" is also followed. "Gootee" is another term used for air-layering. In this method roots form on the arterial part of the plant where the stem has been girdled or slit at an upward angle. The injured portion is enclosed at the point of injury with a rooting medium which is maintained continuously moist. The tree is long lived and attains a large size, but the rate of growth after the seedling stage is slow. The tree begins to bear fruit at the age of 13-14 years and continues to yield abundant crops for more than 60 years (7).

2. NEED FOR TISSUE CULTURE STUDIES IN TAMARIND

Although tamarind has been in commercial demand for centuries there has been no research to improve it as a crop plant. Throughout much of the tropics, the plants grow untended along roadsides, in backyards and on wasteland. These untended, unimproved tamarind trees are nonetheless commercially exploited and a number of countries, Costa Rica and Puerto Rico for example, produce canned tamarind pulp and juice. The Mexican government has recently planted large areas with tamarind in the state of Guerro, but to date the only mature plantations are those in India (8).

Tamarind has an attractive commercial future for producing drinks, jams and confections on an industrial scale. The tree is so well adapted to both dry savanna regions and monsoon areas with well drained soils, that it deserves greater research attention and extensive organized plantings.

Conventional methods of propagation of tamarind are faced with various limitations:

1. Tamarind is suitable for growing only in regions that have extended spells of dry weather. In the humid tropics, where rain falls evenly the year round, the tree refuses to bear and fails to grow unless the soil is well drained.

2. The tree is sensitive to frost and requires protection when small.

3. It is slow growing and late to reach bearing age. Commonly it gains about 0.5 to 0.8 m in height per year and takes 10-12 years to mature and yield fruit.

4. Seed-borne progeny undoubtedly show considerable variation in the size and quality of their fruit. This accounts for the different varieties which have been noted by many workers.

5. None of the vegetative methods of propagation (gootee, grafting, etc.) have been very successful.

6. There is no collection of germ plasm from the various regions that stretch from Senegal across sub-Sahelian Africa to the Sudan. Collections of Indian tamarind types are not available. There are interesting genetic types in Thailand and elsewhere in Southern Asia which should be preserved.

7. Lack of efficient harvesting techniques.

By using the method of tissue and organ culture at least some of the above problems can be circumvented and quicker means of propagation can be visualized.

Given below are the experiments and results of various studies carried out with seedling tissues of tamarind, using the *in vitro* technique.

3. TISSUE CULTURE STUDIES - METODS AND RESULTS

Different types of explants were used to initiate cultures of tamarind. Different types of responses were obtained depending upon the media composition and other physical parameters. A brief outline of the various experiments are given below:

3.1. Protocols

3.1.1. *Culture media*. The two basal media used were White's (W) (9) and Murashige and Skoog's (MS) (5), both with minor modifications as reported earlier (1). The following growth regulators and vitamins were added: Benzylamino purine (BAP); Biotin (Bio); Calcium pantothenate (CalP); Indole-3-acetic acid (IAA); Indole-3-butyric acid (IBA); Indole-3-propionic acid (IPA); Kinetin (Kn); α-Naphtalene acetic acid (NAA). The various culture media used are given below; the concentrations of hormones (in parentheses) are in mg/l.

MS-1: MS + Kn(0.2) + BAP(0.5) + NAA(0.5) + CalP(0.1) + Bio(0.1).
MS-2: MS + Kn(0.2) + BAP(0.5) + CalP(0.1) + Bio(0.1).
W-1: W (no hormones) liquid.
W-2: W + IAA(0.1) + IBA(1.0) + IPA(1.0) + NAA(1.0) liquid.

Observe the following steps during media preparation:

1. Adjust pH of all media to 5.8 with 1 N NaOH or 1 N HCl before autoclaving.

2. Solidify using 0.8% (w/v) Bacto agar (Difco) for MS-1, MS-2 and W media.

3. Pour 20 ml media/tube (25 mm x 150 mm, Corning brand). Use filter paper supports for liquid media.

4. Sterilize media at 15 psi for 20 min, followed by steaming at atmospheric pressure for 30 min on the subsequent day.

5. Filter sterilize (millipore filter, membrane size 0.22 μm) hormones used for root initiation like IAA, IBA, IPA and NAA, and add aseptically to W-1 medium.

3.1.2. Germination of seeds

1. Wash healthy tamarind seeds with a detergent (Tween-20, 10%, v/v), followed by a wash in 70% ethanol for 1 min and finally in 0.1% $HgCl_2$ for 20 min.

2. Remove all traces of the sterilant by repeated washings with distilled water. Inoculate the seeds on W basal medium.

3. The seeds germinate within 10 days of culture.

3.1.3. Callus initiation

1. Excise segments (10-15 mm) from different regions of *in vitro* grown seedlings (viz. stem, leaf, root and cotyledon) and inoculate onto MS-1 medium.

2. Callus formation is seen from all explants after 15-20 days of culture. Callus obtained from stem, leaf and root is whitish in colour and soft in nature while the callus from cotyledons is nodular and light brown in colour. Attempts to differentiate the various calli into roots, shoots, etc., proved unsuccessful.

3.1.4. Regeneration of shoots from seedling explants

1. Sterilize seeds and germinate them as described earlier.

2. Use only seedlings of approximately 50-60 mm height for these studies. Inoculate 8-10 mm long explants of hypocotyl, shoot tips, nodal segments, cotyledons and roots on MS-2 medium.

3. Take all observations after a 30 day culture period. Hypocotyl explants develop 4-5 shoots per tube as compared to nodal and shoot tip explants which produce 1 and 2 shoots/tube respectively. Cotyledon and root explants form only callus without any shoot regeneration.

4. Shoots regenerated _in vitro_ attain a height of about 20-30 mm in 30 days. Subculture the shoots once in 6 weeks onto the same MS-2 medium; 3-4 shoots regenerate per tube at each subculture.

3.1.5. Rooting of in vitro grown shoots and transfer to soil

1. Choose elongated shoots (40-50 mm long) for rooting.

2. An auxin mixture comprising IAA, IBA, IPA, and NAA in water at a concentration of 1 mg/l each is used.

3. Add the auxin mixture aseptically to W-1 medium contained in tubes.

4. Excise the shoots and keep for treatment in W-2 medium in tubes so that the basal cut end (5 mm) of the shoot dips in the liquid medium and is supported firmly by the filter paper bridge.

5. Incubate cultures for 72 h in the dark.

6. After auxin treatment shift the shoots to W-1 medium. Incubate these cultures now in light (18 h photoperiod, 1000 lux intensity).

7. Root initiation is observed after 2 weeks of culture. Rooting efficiency is 60-70%.

3.1.6. Transfer of plantlets to field

1. Remove plantlets which have attained a height of about 50-60 mm from tubes.

2. Wash the root system thoroughly in running tap water to remove all traces of medium adhering to the plantlet.

3. Transfer the individual plantlets into pots containing a mixture of sterile sand, soil and compost (3:3:1).

4. Incubate the potted plants at 28°C under a 16 h photoperiod.

5. Transfer the plants to the field after they reach a height of 300-350 mm in pots.

6. Prepare pits in the field measuring 30 cm^3, refill with soil and plant the seedlings in the centre. Of the plantlets, 50-60% survive in the field.

4. FIELD DATA OF TAMARIND TREES

The tissue cultured plants of tamarind which were transferred to the field were given regular irrigation and a nitrogen fertilizer (urea). Weeding was performed at regular intervals. Initially the plants were watered twice a week but after one year of growth irrigation once a week in summer (March to June) and once in two weeks in winter (October to February) was found sufficient. Plants were observed for their growth performance and other morphological characteristics such as flowering, branchiness, spread etc. The results obtained from tissue culture plants were compared with those of the controls (Table 1).

Table 1. Comparative field data of tissue culture plants derived from seedling explants of tamarind. Field planting, October 1980. Last observation, May 1985. Number of tissue culture plants = 14. Number of seed raised plants (control) = 14. Statistical design, paired plot.

Observation	Control	Tissue culture
1. Average height (m)	2.47	3.73[a]
SE (±)	0.02	0.02
CD (5%)	0.06	0.06
2. Branching habit	Moderate	Profuse
3. Average spread (m)	5.18	7.89
4. Date of first flowering	Not yet in flower	May 1984[b]
		April 1985[c]

[a] Statistically significant.
[b] 4 plants flowered.
[c] 11 plants flowered.

Although the tissue culture plants were raised initially from seedling explants they exhibited characteristics of full grown trees in that they flowered early (Fig. 1). The first flowers appeared in 4 of the 14 plants in May 1984, that is in three and a half years after planting. In the following year, i.e., in April 1985, 7 more plants flowered and set fruit. In contrast, seed raised trees take about 13-14 years to reach the flowering state (7). This is a very important field observation. It reveals the potential use of tissue culture technique for clonal multiplication of elite tamarinds, i.e., of mature trees which bear high yields (180-225 kg of fruit/plant) and with exceptionally large pods well filled with pulp of a lower acid content than others and referred to as "sweet". It is also useful for propagation of seedlings from seeds collected from mature selected trees.

FIGURE 1. Tamarind trees in the field ($4\frac{1}{2}$-year-old). a) Tissue culture raised plant showing early flowering (see inset). b) Seed raised plant (control).

It has also been observed that the average spread (Table 1) of the tissue culture plants is better than that of their controls. The branching was profuse in tissue culture raised plants as compared to the moderate branching of their seed raised counterparts. At an average height of only 3.7 m, flowering was observed among the tissue culture raised progeny. On the other hand the height attained by trees raised from seed often reaches up to an average of 8 m before flowering occurs.

5. FUTURE PROSPECTS

The application of the *in vitro* technique for propagation of tamarind has definite advantages over the conventional methods from seed. This is evident even though the plants regenerated were from juvenile seedling tissues. The immediate advantage will be in the early flowering of the tissue culture raised plants and the height and spread of the branches. Tamarind trees grow to an average height of about 10-12 m, with a wide spreading habit. The fruits develop, and generally grow all around the tree, attached to the extreme ends of the branches. Fruits are ripened on the tree before harvest. Pedicels that hold the fruit to the tree are very tough. They cannot be broken off by hand without damaging the fruit and must be clipped off. However, in India the general practice followed for harvesting the fruits is a dangerous, time consuming and labour-intensive operation. A climber goes up the tree and strikes down the fruits from the periphery of the branches using long bamboo poles.

From the present results of the field grown tissue culture raised plants it appeasr as if the harvesting operations can be simplified and carried out by safer procedures or even by mechanical methods. The tissue culture trees grow as small bushes, barely 3-4 m in height as compared to the large trees raised from seed that may reach a height of 25 m. The short stature of the trees thereby permits planting of a much higher population per given area (about 156 plants with a 9 x 9 m spacing per hectare). This could substitute for the present planting practice where trees are planted at a 12.5 x 12.5 m spacing with 65 plants per hectare.

The method described permits a rapid rate of plant production with a tree species where conventional vegetative methods of propagation are not yet well developed. No attempts have been made to do a cost analysis of the plants produced, but based on the results of the field trials, it is not expected that the costs would be excessive. Even if the cost is higher than with seed raised plants, the use of tissue culture plants may well be justified by their subsequent performance with early flowering and fruiting cycles.

The potential application of plant tissue culture to crop and tree improvement is now astonishing. The tissue culture technique described for tamarind could be integrated into breeding programs. This could result in the development of unique germ plasm of this important tropical tree.

6. REFERENCES

1. GUPTA PK, AL NADGIR, AF MASCARENHAS, V JAGANNATHAN 1980 Tissue culture of forest trees: Clonal multiplication of *Tectona grandis* L. (teak) by tissue culture. Plant Sci Lett 17: 259-268
2. HILL AF 1979 Economic Botany. McGraw Hill Pub Co
3. LINGARAJU S, AL SIDDARMAIAH, RK HEDGE 1979 Pre and post emergence rot of tamarind seedling. Indian For 105: 526
4. MASLEKAR AR 1981 Forester's Companion (Indian Foresrty Handbook). Jugal Kishore Co, India
5. MURASHIGE T, F SKOOG 1962 A revised medium for rapid growth and bioassays with tobacco tissue cultures. Physiol Plant 15: 473-493
6. SIDDARAMAIAH, AL, S KULKARNI 1982 Control of powdery mildew disease of Tamarind seedlings. Indian For 108: 361-364
7. The Wealth of India 1976 Vol X. Publ Directorate, CSIR, India
8. Tropical Legumes - Resources for the Future 1979 Nat Acad Sci, Washington
9. WHITE PR 1963 The Cultivation of Animal and Plant Cells. The Ronald Press Co, New York

25. HEVEA TISSUE CULTURE

K. PARANJOTHY

1. INTRODUCTION

The genus *Hevea* is a member of the Euphorbiaceae and is endemic to the Amazon basin and its neighbouring parts including Matto Grosso, Upper Orinoco and the Guianas. The taxonomy of the genus has been intensively studied (29). Nine species are now recognized in the genus and of these only *H. brasiliensis* is important economically. Virtually all the natural rubber produced to-day is derived from this species.

Hevea is cultivated both on a plantation basis and in small-holdings, mostly in high rainfall areas within 15°N and 10°S of the equator. Rubber growing countries in Asia include Malaysia, Indonesia, Thailand and Sri Lanka. Rubber is also grown in southern parts of India and in the Peoples Republic of China in Hainan island. It is also cultivated in Africa, but to a lesser extent, principally in Liberia and Nigeria. Cultivation in the South American tropics has not been successful, attempts generally being severely limited by leaf blight caused by the fungus *Microcyclus ulei*.

Total natural rubber production in 1984 amounted to 4.2 M tonnes (13), production from various countries being estimated as follows: Malaysia (1.51 M), Indonesia (1.14 M), Thailand (0.60 M), India (0.19 M), China (0.18 M), Sri Lanka (0.15 M) and other countries (0.43 M).

Chemically, the rubber obtained from *H. brasiliensis* is cis-1-4-polyisoprene $(C_5H_8)_n$ and is present in the form of spherical, ovoid or pear-shaped particles (ranging in size from 5-6 µm in diameter or length) in laticifers distributed throughout the tree, but only rarely in the wood. The bark is the only tissue that is commercially exploited, the bark being tapped at periodic intervals by removal of a thin shaving. Rubber particles comprise about 90% of the dry weight of the latex exudate.

Breeding progress has been remarkable, with modern clones yielding in excess of 1637.5 kg/hectare/year compared to a yield of about 512.5 kg/hectare/year in unselected seedlings (28). Chromosome counts of somatic cells of all members of the genus studied to-date indicate 2n to be 36. It has been suggested (12) that Hevea is an allotetraploid, with x=9.

Propagation for commercial planting is largely by bud-grafting. Under plantation conditions, the trees may reach heights of 25 metres after about 25 to 30 years. The phenomenon of juvenility (see for example, ref. 7) is clearly reflected in many aspects of the growth and development of Hevea. Growth during the first 3 to 5 years is purely vegetative and thereafter (in Peninsular Malaysia) the trees usually flower twice a year. There evidently are rapid physiological changes even during the vegetative phase. Thus cuttings from young plants root readily, whilst this ability is reduced as the plants age. Wiersum (31) found that leaf cuttings, carrying one leaf and one bud on a piece of stem approximately 1-2 cm long rooted only when taken from seedlings less than 6 months old. He also noted that cuttings taken from young seedlings produced pseudotaproots while Yoon and Leong (34) found that cuttings taken from source bushes generally lacked taproots. Buds taken progressively lower down the bole of seedlings produce buddings with correspondingly greater seedling-like stem conicity (16). Buds taken from source bushes (as for commercial propagation) produce buddings with boles that are uniformly cylindrical in diameter, unlike the concial stem of seedlings.

2. POSSIBLE USES OF IN VITRO METHODS IN HEVEA

One possible use of tissue culture, albeit a limited one, is in propagation. Though Hevea is easily propagated by bud-grafting, the routine use of clonal rootstocks is limited by the absence of taproots on cuttings from mature plants (or source bushes) and by the difficulty of rooting some clones. Thus the production of clonal rootstocks with tap roots might enable exploitation of useful stock/scion interactions.

Another anticipated use of tissue culture methods is in the creation of somaclonal variants. The question of whether it will be possible to produce useful variants is an important one. It is likely that the production of variants will be required on a large scale as most will be undesirable. Success will depend to a large extent on the type of change desired. For example, it should be easy to spot morphological deviants such as dwarfs. Features such as disease resistance would be more difficult to identify as this will require screening and selection. It is conceivable that cells *in vitro* could be screened against the pathogen (or toxin) and that plants could then be developed from resistant cells. However, it is likely that factors conferring disease resistance in *in vitro* cultures may often not be operative in mature plants in the field.

The production of haploids from pollen is another use in genetic improvement of *Hevea*. This may be viewed as particularly useful since *Hevea* is not only a perennial but also an outbreeder. Haploids are relatively rare in outbreeding species, except in amphidiploids (10). The successful production of haploids by anther culture methods (5) lends credence to the view that *Hevea* is probably a tetraploid. Chen (4) points out that it is virtually impossible to obtain pure lines by successive inbreeding in *Hevea* and suggests that anther culture now provides a means of obtaining homozygous diploid lines, which after assessment for combining ability can be used for production of superior uniform hybrid lines.

Novel methods of crop improvement are of interest especially in crops where progress through conventional breeding is slow as in many perennial crops. In this respect protoplasts may be relevant. Nevertheless, in *Hevea* it is difficult to identify clearly their use in crop improvement at present, though the potential use of protoplasts in genetic engineering is well recognized.

Aseptic culture methods have been recommended by quarantine experts as a means of avoiding transfer of undesirable microbes together with *Hevea* germplasm. Thus methods for aseptic culture of seeds or embryos may have useful applications. Aseptic culture of immature embryos may also have applications in overcoming barriers in interspecific crosses or difficult crosses (less than 5% success rate) between clones within _H. brasiliensis_ itself. Finally _Hevea_ seeds are short-lived and cyropreservation of excised embryos might be potentially useful in germplasm storage.

3. ORGAN CULTURES

3.1. Seed embryos

Muzik (15) found that mature embryos developed rapidly in La Rue's (11) medium but immature embryos cultured at various stages of their development failed to develop further. Toruan and Suryatmana (30) reported that seedlings were obtained from decotyledonized embryos in Murashige Skoog (MS) medium and subsequently established in soil. Others (15, 19) however, found that mature embryonic axes usually fail to grow unless a portion of the cotyledon is attached, though improved growth was observed in liquid media. Genetic differences in embryos used could partly account for these different observations.

3.2. Seed culture

Investigations on aseptic methods for use in international transfer of _Hevea_ germplasm have been carried out in the Rubber Research Institute of Malaysia (RRIM) (21, 25). Dehusked seeds from pods or freshly fallen seeds collected from the ground germinated normally when cultured in nutrient media containing 3% sucrose - the nutrients and sucrose being utilized for detection of microbial growth. Surface-sterilants such as sodium hypochlorite and mercuric chloride were useful in reducing contamination in cultures only when fresh seeds were used. Dehusked seeds from older seed batches gave rise to high contamination levels. Bromine, used at 2% for 5 minutes, was found to be superior to sodium hypochlorite and mercuric chloride for surface-sterilization of dehusked seeds from older seed batches, where contamination was presumably deep seated. Seed germination was not adversely affected by treatments with bromine and seedling growth appeared normal. Attempts to reduce sterility by prolonged treatment with sodium hypochlorite or mercuric chloride inevitably led to adverse effects on seed germination. The toxicity of the latter sterilants in relation to bromine is probably related to the relative difficulty in the removal of the sterilants after surface-sterilization. Seedlings were maintained for about 3 weeks in test-tubes, at which time plumule height averaged 9.3 cm. The seedlings were then established in soil.

3.3. Shoot apices

Paranjothy and Ghandimathi (18, 19) attempted culture of 2-3 cm long shoot apices derived from 2 to 3 week old aseptically grown seedlings. Growth of the apices was poor in solid media. However, the apices grew into rooted plantlets within 4 weeks of transfer to liquid MS media (23). Mascarenhas *et al.* (14) reported development of about 3 shoots/explant when terminal buds from 10-20 year old trees were cultured in solid MS media supplemented with 0.5 ppm kinetin, 2.0 ppm N6-benzyladenine (BA), 200 ppm casein hydrolysate, 0.1 ppm calcium pantothenate and 0.1 ppm biotin. Excised shoots were induced to root by treatment in White's medium containing 10 ppm each of indole-3-acetic acid (IAA), indole-3-butyuric acid (IBA), indole-3-pyruvic acid (IPA) and naphthalene acetic acid (NAA). A rooting rate of 60% was achieved.

3.4. Stem nodes

Carron and Enjarlic (3) reported that nodes from young greenhouse plants sprouted in basal media (devoid of growth substances) supplemented with 0.5% activated charcoal and 60 g/l of sucrose. To obtain sprouting of axillary buds the explants had to be pretreated in a solution containing 5 ppm IBA and 10 ppm BA for 2 hours. Rhizogenesis in the sprouts was induced by dip treatment in a solution containing IBA and NAA (5 ppm each) for 5 days. Roots were developed subsequently on nutrient media devoid of growth substances.

4. SOMATIC CALLUS CULTURES

Though callus cultures of *Hevea* can usually be initiated without much difficulty from a variety of explants, cultures are not always amenable to subculture or are slow growing. Explants derived from young *Hevea* seedlings tend to reflect very strongly the capacity for root formation seen in juvenile cuttings (6, 18, 19, 33). Thus on media containing IAA, NAA or β-naphthoxyacetic acid (NOA) hypocotyl explants will form roots but little callus (18,

19). In media containing 2,4-dichlorophenoxy acid (2,4-D) callus formation is usually seen in these explants; root formation is evident when sucrose concentration is raised to 10%. The callus obtained from hypocotyl explants often produces roots on subculture. The callus from hypocotyl explants is usually hard and compact and some samples exude latex when cut for subculture. Cotyledon explants also proliferate roots, even in media without growth substances (18, 19). Root formation is supressed in these explants in media containing 2,4-D or p-chlorophenoxyacetic acid (p-CPA), callus proliferation usually being evident. This callus, however, is usually not amenable to subculture. The site of root formation is clearly constrained by polarity as rooting always occurs at the petiolar end. Triiodobenzoic acid (TIBA), which is well known to inhibit polar movement of auxin, is strikingly inhibitory to root production in these explants, even at low concentrations in the region of 0.1 ppm.

Carron and Enjarlic (3) reported that callus could be obtained from leaf explants and "maternal tissues from seed" on basal media of high salt concentration supplemented with auxins (2,4-D and IAA at 0.3 to 2 ppm and 1 to 5 ppm respectively) and cytokinins (kinetin and BA at 1 to 5 ppm each). When callus was allowed to age for 5 months without subculture, on media containing 2 ppm NAA and 2 ppm BA, numerous embryoids appeared.

Callus formation in stem explants has been observed by several investigators (2, 6, 33). Wilson and Street (33) found that newly initiated callus cultures from stem explants spontaneously initiated roots. On serial subculture this property was lost, and the cultures became heterogenous, i.e., they formed light proliferating segments and darker compact non-growing segments. After two passages in liquid medium, large cell aggregates, when transferred back to solid medium yielded a highly friable light coloured fast-growing homogenous callus which retained its character on subculture. This callus, when transferred back to agitated liquid medium, yielded a fine rapidly growing cell suspension culture. When the suspensions were maintained for several months without subculture the larger cell aggregates which developed gave rise to embryo-like structures. Attempts to promote further development of these embryo-like structures into plantlets were unsuccessful.

Callus from the somatic tissue of anthers has been found to be readily established and also to be amenable to routine subculture (5, 18, 26, 27). The nutritional requirements of anther wall derived callus have been studied in detail (19). An auxin and a cytokinin are required for both initiation of the callus and its sustained growth. Coconut milk was inhibitory to growth of the callus. Amongst a large number of carbohydrates tested sucrose appeared to be the best source of carbon for growth of the callus, with fructose and raffinose supporting growth to a lesser extent. Surprisingly, glucose was relatively ineffective in supporting growth of the callus. Thiamine promoted growth of the callus markedly, the optimal concentration being 5 ppm. Amongst the trace elements tested, Fe, B, Mn, Zn, I, Mo and Cu were essential for callus initiation and growth. Cobalt did not appear to be required for callus initiation or growth.

Differentiation of embryoids from the anther derived callus was achieved by transfer of fresh callus, initiated on MS basal medium containing 0.1-10 ppm 2,4-D and 1.5 ppm kinetin, to media containing 0-0.5 ppm 2,4-D and 0.1-5 ppm kinetin. This procedure generally involved transfer of callus to media containing a reduced concentration of auxin. In addition to the sequential transfer, it was found that embryoids appeared only when high levels of sucrose (10%) were incorporated in both the callus initiation and differentiation media (17). Embryoid initiation is evidently a clonal feature. Embryoid induction also may depend on the physiological state of the donor plant. An analysis of experiments carried out over three years in the RRIM indicated that embryoids were most frequently obtained in anthers cultured in August (Paranjothy, Unpublished).

The embryoids that develop from the anther wall derived callus are usually nodular or spindle shaped with clear evidence of bipolarity. They are white and shiny. The presence of an epidermis and the development of procambial strands is clearly evident in anatomical sections. The embryoids generally turn green when kept under light and some develop a pair of cotyledon-like structures. Shoot development appeared to depend on the development of large cotyledon-like structures (20, 24).

5. SUSPENSION CULTURES

Supsension cultures have been initiated from anther-wall derived callus (19) and from callus initiated from stem explants (33). Suspension cultures initiated from anther-wall tissue have been used for isolation of protoplasts (22) and suspensions developed from stem segment callus have been used in studying metabolism of ethephon (1).

Suspension cultures initiated from anther-wall derived callus were reported to increase a hundred fold in fresh weight within six weeks in media supplemented with 1 ppm 2,4-D, 1 ppm BA and 3% sucrose (19). This growth was comparable to that seen in callus cultured on solid media. The suspension cultures turned green when the MS medium was modified by doubling the concentration of KH_2PO_4 and reducing the levels of NH_4NO_3 and KNO_3 to half the original concentration.

Callus initiated freshly from stem segments did not yield cell suspension cultures (33). However, large aggregates, recovered after two passages in liquid medium, yielded a highly friable light coloured fast growing homogenous callus when again grown on solid medium. This callus retained its distinctive character on subculture. When transferred back to agitated liquid medium this callus yielded a fine rapidly growing cell suspension culture which could be serially propagated in the same medium as that used for callus culture. When the suspensions were maintained for several months without subculture the larger aggregates which developed gave rise to embryo-like structures.

It was suggested (32) that the changes resulting from subculture in liquid medium are related to those undergone by callus which becomes habituated. Estimation with Salkowski's reagent and by bioassay showed enhanced auxin production in the rapidly growing callus.

6. PROTOPLASTS

Rohani and Paranjothy (22) attempted isolation of protoplasts from a variety of *Hevea* tissues and found that pith from young shoots and suspension cultures of anther-derived callus were the most promising sources. Fine suspension cultures were necessary for isolation of viable protoplasts. The fine cell suspensions were obtained by repeated serial subculture of the anther-wall derived callus in liquid media. The isolated protoplasts regenerated cell walls within seven days but cell divisions were not observed.

7. HAPLOIDY

7.1. Anther culture

Satchuthananthavale (26) and Paranjothy and Ghandimathi (8, 19) reported isolated cases of multicellular pollen in cultured anthers. The first production of pollen plantlets in *Hevea*, however, was reported by Chen *et al*. (5) in 1979. Numerous papers on haplogenesis in rubber by investigators in China have since appeared, many of the papers in Chinese with English summaries. Indeed several hundred pollen plantlets have been established in soil. Recent reviews (4, and Chen, Volume 2 of this book) summarize most of these investigations, and provide detailed protocols. The following account is based on the latter review and attempts to highlight the important aspects of *Hevea* anther culture.

Anthers used for inoculation should have well developed microspores and, therefore, male-sterile clones should be avoided. Inviable microspores may also be encountered during those periods of the year when climatic conditions may be unfavourable. The uninucleate stage is optimal for inoculation of *Hevea* anthers. The majority of flower buds at the uninucleate state are 3-3.5 mm in length. When the corollas have turned yellow the microspores in the anthers are at the binucleate stage.

Proliferation of both haploid and somatic cells in cultured anthers is common and an important strategy in obtaining haploid plantlets eventually is suppression of somatic callus proliferation. This is achieved partly by incorporating coconut water and partly by use of sucrose at a level of 7-8% in the medium. Lower levels of sucrose (3%) favour somatic callus proliferation. Both the N composition and concentration is evidently important for the dedifferentiation of microspores in preference to that of somatic cells. A decrease of NO_3 ion concentration in MS medium favours the development of pollen embryos, but a higher concentration of total nitrogen is required for induction of pollen embryos. The concentration of KH_2PO_4 is evidently also important, the highest frequency of callus and embryo induction being obtained at the highest concentration tested, i.e., triple the concentration in MS medium. Kinetin and 2,4-D is sufficient to induce a high frequency of callusing anthers, but the addition of three growth substances is required for dedifferentiation of microspores (1 ppm K, 1 ppm 2,4-D and 1 ppm NAA).

The time of transfer of callusing anthers to fresh media for differentiation is also critical. After 20-25 days in culture, most of the dividing cells are diploid, but after about 50 days the latter start senescing and are replaced by callus of microspore origin. Thus, transfer of callusing anthers after 50 days to the differentiation medium is recommended. Three growth substances (K, NAA and GA_3) are required in the differentiation medium at concentrations of 1, 0.2 and 0.07 ppm respectively. Sucrose at a concentration of 7-8% is again optimal for differentiation.

Visible embryos emerge about one month after callus has been transferred to the differentiation medium. Embryo development is complete after about 2-3 months when they are transferred to fresh media for plantlet development. For the latter phase of growth the optimal concentration of sucrose is 4-6%. Incorporation of GA_3, IAA and 5-bromouracil is required for plantlet development.

7.2. Ovule culture

Guo *et al*. (9) reported obtaining two plantlets from cultures of unpollinated ovaries. The protocol used was similar to that for anther culture described above. The chromosome number of these plantlets was not reported.

8. CONCLUSION

Several aspects of *Hevea* tissue culture are in a well advanced experimental stage. Notable developments include the production of plantlets from both somatic and gametophytic cells through embryogenesis and production of multiple shoots by culture of excised stem apices. Application of these findings for improvement of *Hevea* is being attempted at several centres of rubber research.

Ongoing research is aimed at increasing the frequency of plantlet regeneration from pollen, somatic callus and from shoot apices. The regeneration of plantlets from cell suspension cultures and protoplasts of Hevea is yet to be achieved but potential applications through genetic engineering now provide research incentives in this area.

9. REFERENCES

1. AUDLEY BG, HT WILSON 1978 Metabolism of 2-chloroethylphosphonic acid (Ethephon) in suspension cultures of Hevea brasiliensis . J Expt Bot 29: 1329-1336
2. BOUCHOU JG 1953 La culture in vitro des tissues d'Hevea. Arch Rubb Cult 30: 50-53
3. CARRON MP, F ENJARLIC 1982 Studies on vegetative micropropagation of Hevea brasiliensis by somatic embryogenesis and in vitro microcutting. In A Fujiwara, ed, Plant Tissue Culture 1982. Proc 5th Int Cong Plant Tissue Cell Cult, Tokyo, pp 751-752
4. CHEN Z 1984 Rubber. In DA Evans, WR Sharp, PV Ammirato, Y Yamada, eds, Handbook of Plant Cell Culture, Vol 3, Macmillan, New York, pp 546-571
5. CHEN Z, F CHEN, C CHIEN, C WANG, S CHANG, H HSU, H OU, Y HO, T LU 1979 A process of obtaining pollen plants of Hevea brasiliensis Muell-Arg. Sci Sin 22: 81-90
6. CHUA ES 1966 Studies on tissue culture of Hevea brasilensis. J Rubb Res Inst Malaysia 19: 272-276
7. DOORENBOS J 1965 Juvenile and adult phases in woody plants. In W Ruhland, ed, Encyclopedia of Plant Physiology, Vol 15, Springer Verlag, Berlin, pp 1222-1235
8. GHANDIMATHI, H, K PARANJOTHY 1975 Anther culture - attempts at induction of haploidy in Hevea and other plants. Proc Nat Plant Tissue Cult Symp, Kuala Lumpur 1975, pp 32-35
9. GUO G, X JIA, L CHEN 1982 Induction of plantlet from ovule in vitro of Hevea brasiliensis. Hereditas 4: 27-28
10. KIMBER G, R RILEY 1963 Haploid angiosperms. Bot Rev 29: 480-531
11. LA RUE CD 1936 The growth of plant embryos in culture. Bul Torr Bot Club 64: 97-102
12. MAJUMDER SK 1964 Chromosome studies of some species of Hevea. J Rubb Res Inst Malaya 18: 269-275
13. MALAYAN RUBBER REVIEW 1985 MRRDB, Kuala Lumpur, Malaysia 1: 1-35
14. MASCARENHAS AF, S HAZARA, U POTDAR, DK KULKARNI, PK GUPTA 1982 Rapid clonal multiplication of mature forest trees through tissue culture. In A Fujiwara, ed, Plant Tissue Culture 1982. Proc 5th Int Cong Plant Tissue Cell Cult, Tokyo, pp 719-720
15. MUZIK TJ 1956 Studies on the development of the embryo and seed of Hevea brasiliensis in culture. Lloydia 19: 86-91

16. PAARDEKOOPER EC 1956 The occurrence, properties and possible application of juvenile type buddings. Arch Rubb Cult 33:141-157
17. PARANJOTHY K 1974 Induced root and embryoid differentiation in *Hevea* tissue cultures. 3rd Int Cong Plant Tissue Cell Cult, *Univ* Leicester. Abstr No 67
18. PARANJOTHY K, H GHADIMATHI 1975 Morphogenesis in callus cultures of *Hevea brasiliensis* Muell Arg. Proc Nat Plant Tissue Cult Symp, Kuala Lumpur 1975, pp 19-25
19. PARANJOTHY K, H GHANDIMATHI 1976 Tissue and organ culture of *Hevea*. Proc Int Rubb Conf, Kuala Lumpur 1975, II: 59-84
20. PARANJOTHY K, O ROHANI 1978 Embryoid and plantlet development from cell culture of *Hevea*. 4th Int Cong Plant Tissue Cell Cult, Univ Calgary, Canada, Abstr No 134
21. PARANJOTHY K, TM LIM, O ROHANI, H ISMAIL, SAK AZIZ, LS KWAI 1979 Transfer of *Hevea* seed germplasm. IRRDB Meet Techn Experts, Kuala Lumpur 9-11 Oct 1979
22. ROHANI O, K PARANJOTHY 1980 Isolation of *Hevea* protoplasts. J Rubb Res Inst Malaysia 28: 61-66
23. RUBBER RESEARCH INSTITUTE OF MALAYSIA 1974 Annu Rep p 58
24. RUBBER RESEARCH INSTITUTE OF MALAYSIA 1977 Annu Rep p 67
25. RUBBER RESEARCH INSTITUTE OF MALAYSIA 1979 Annu Rep p 70
26. SATCHUTHANANTHAVALE R 1973 *Hevea* tissue culture. Q J Rubb Res Inst Ceylon 50: 91-97
27. SATCHUTHANANTHAVALE R, ZE IRUGALBANDRA 1972 Propagation of callus from *Hevea* anthers. Q J Rubb Res Inst Ceylon 49: 65-68
28. SIMMONDS NW 1979 Principles of crop improvement. Longman, London
29. SCHULTES RE 1970 The history of taxonomic studies in *Hevea*. Bot Rev 36: 197-276
30. TORUAN NL, N SURYATMANA 1977 Kultur jaringan *Hevea brasiliensis* Muell-Arg. Menara Perkebunan 45: 17-21
31. WIERSUM LK 1955 Observations on the rooting of *Hevea* cuttings. Arch Rubb Cult 32: 213-243
32. WILSON HM, MA EISA, SWB IRWIN 1976 The effects of agitated liquid medium on *in vitro* cultures of *Hevea brasiliensis*. Physiol Plant 36: 399-402
33. WILSON HM, HE STREET 1975 The growth, anatomy and morphogenetic potential of callus and cell suspension cultures of *Hevea brasiliensis*. Ann Bot 39: 671-682
34. YOON PK, SK LEONG 1976 Induction of pseudotaproots of cuttings and production of clonal rootstocks in *Hevea*. Proc Int Rubb Conf Kuala Lumpur 1975, II: 85-108

26. PALMS

B. TISSERAT

1. INTRODUCTION

Previous reviews have already expounded on the uses of tissue culture to aid in the more effective utilization of Arecaceae (i.e., palm family) (9, 18, 20, 51, 58, 84, 98, 104, 116). The palm family consists of over 200 genera and about 2500 species (9, 84, 104). Palms are woody tree-like monocots ubiquitous in the tropics. However, several important species also are native to mediterranean and desert climates. A great deal of variation in palm structure, morphology, and size occurs within the palm family although two monocotyledenous characteristics are common; i.e., the trunk consists of numerous scattered cortical layers interdispersed with vascular bundles and fibers, and fruits are either drupes or berries occurring in discreet compact inflorescence packages terminating at the end of inflorescence buds.

1.1. Economics

Detailed reviews concerned with the economic importance of palms in world agriculture have recently been published (84, 104). In subsistence agriculture, prior to the twentieth century, palms provided food, shelter, fuel, fiber, clothing, and material for furniture and other implements. Changing agricultural conditions, such as the advent of large plantation culture systems, have eliminated the practice of full palm utilization. Specialized palm crop products include: palm hearts, palm oils (from coconut (*Cocos nucifera* L.) and oil palms (*Elaeis quineensis* Jacq.), dates (*Phoenix dactylifera* L.) and coconut fruits, ornamental trees, timber, and waxes (104). Palm oil has become the most important vegetable oil on the world food scene today (104). Due to its agricultural requirements and high crop yield oil palm has become the most efficient oil producing crop in the world (104).

1.2. Husbandry and breeding problems in this family

The genetics, morphogenesis, morphology, and physiology of palms have been less systematically studied than that of other tree crops because their agricultural uses were not fully appreciated by earlier agriculturalists. Increased world food demand has resulted in the recognition of the inherent value of indigenous plants as potential food producers. Consequently, within the last decades several palm research centers and agricultural industries have developed.

Most palms are long-lived, requiring several years of juvenile development to reach the adult flowering stage. Thus, for dioecious species one must wait several years to determine sex and then evaluate the fruiting qualities. Trees are highly heterozygotic, and, therefore, do not breed true. Consequently, breeding disease resistant or high crop yielding trees is a long-term, expensive project. To compound crop improvement problems, vegetative propagation is difficult, if not impossible (62, 80, 81). Many palms (e.g. coconut and oil palm) can not be clonally propagated since they lack vegetative axillary buds. Grafting is not possible because monocots do not have a vascular cambium. Without an adequate method to clonally propagate palms, uniform yields are not possible in plantations.

1.3. Potential benefits of tissue culture

Serious interest in palm tissue culture as a means to mass produce high-yielding or disease resistant clones in large numbers for plantation cultivation began in the 1960's (3, 27, 66, 67, 69, 79, 113). Numerous attempts to determine the parameters necessary to culture palms *in vitro* were conducted (1, 4, 7, 10, 11, 24, 32, 33, 34, 35, 37, 39, 53, 55, 61, 80, 81, 82, 88, 90, 92, 93, 94, 115). By the late 1970's, a number of *in vitro* methodologies had been presented for potentially cloning palms (5, 21, 23, 29, 47, 48, 65, 70, 77, 83, 85, 86, 91, 96, 97, 108). In the 1980's these reports were reconfirmed and expanded upon in numerous studies (6, 13, 14, 15, 22, 31, 43, 44, 46, 52, 59, 63, 64, 100, 101, 121). Initially, coconut, oil and date palms, being the most economically important palms were target plants for early tissue culture studies. However, the methodologies employed to tissue culture these palms may be used to culture other palms as well (41, 46, 85, 122).

Most palm tissue culture studies may be divided into three categories each with distinct objectives: 1) clonal propagation, 2) embryo culture, and 3) physiological studies of growth and development. Clonal propagation of palms which is pursued to produce numerous genetically uniform trees (19), clearly is currently drawing the most interest and research efforts (2, 9, 102). Potentially, clonal propagation can be accomplished through either asexual embryogenesis, i.e., plantlet formation by the initiation and germination of a somatic embryo (104), or organogenesis, i.e., the sequential formation of roots from shoots (86) or vise versa (91). Embryo culture, i.e., the excision and germination of an isolated embryo, is employed to produce rare incompatible crosses, or increase seedling populations through improved germination (8, 25, 38, 45, 68, 71, 72, 74, 75, 89). Tissue culture techniques may be employed to study various aspects of palm growth, morphogenesis, and physiology (16, 17, 26, 54, 73, 74, 78, 107). Germplasm preservation through aseptic culture or cryogenic storage coupled with tissue culture is possible (36, 58, 60, 110, 111, 112).

2. MATERIALS AND METHODS

Although the palm family is large and diverse, a general similarity appears to exist regarding the response of palm tissues to *in vitro* conditions (41, 122). Since it is not possible to address each palm study individually in this paper, I will only discuss procedures that were primarily developed for the date palm. However, these procedures should not be taken as absolute but as a preliminary workable guide which will undoubtedly be modified through later research efforts for other palms.

2.1. Embryo culture

2.1.1. *Explant selection and sterilization techniques*. Both immature and mature seeds can be successfully germinated *in vitro* and produce seedlings through stages of development that correspond to that found in the seed (57, 98). Two longitudinal incisions will usually suffice in opening the fruit to obtain the seed. Mature seeds usually have to be scrub brushed to remove fruit debris which, when left on seeds, will increase contamination problems. A harsh disinfectant, such as 2.6% (w/v) sodium hypochlorite (NaOCl) (122), or 0.2% aqueous mercuric chloride ($HgCl_2$) (60), must be employed to surface sterilize palm seeds. To obtain excised embryos immersion of seeds in a 0.5% HgI_2, or $HgCl_2$, or 3% I_2, in 95% ethanol for 30 minutes is recommended (56).

2.1.2. Excision and medium composition. Mature seeds are soaked for 24 to 48 hours in tap water to hydrate the embryo and endosperm and facilitate subsequent opening of the seed. Some palm seeds such as oil palm have such a hard seed coat that soaking does not soften the endosperm. In such cases, seeds may be opened by wrapping in a towel or bag and cracking slowly in a vise. The seed (minus seed coat) is positioned between thumb and index finger and using an anvil hand cutter (presoaked in 95% ethanol and flame treated) a 2 cm longitudinal incision is administered at the opposite ends of the furrow. The halved seed will fracture apart allowing the embryo to become exposed without damage. Using a surgeon's scalpel fitted with a #11 surgical blade the exposed embryo is removed from the halved seed either by 1) piercing the haustorium end and the embryo (i.e. the end most embedded within the seed), or 2) lifting the embryo out of the endosperm-embryo cavity by applying pressure, using the blunt side of the scalpel blade (this procedure will avoid embryo damage). As with seed sterilization, a harsh disinfectant is required to surface sterilize the surface of palm explants compared to those treatments employed for other tissue cultured plants (32, 82, 96, 121). It is suggested that 2.6% NaOCl be employed for 15 min without a sterile water rinse. Following aseptic preparation of the tissue an additional NaOCl soak is employed prior to explant introduction to the medium (96). The excised embryo is placed longitudinally on the surface of the agar medium, making sure that embryo damage or immersion of the embryo into the agar medium is avoided. The embryos are incubated on Murashige and Skoog inorganic salts, 0.4 mg l^{-1} thiamine HCl, 100 mg l^{-1} meso-inositol, 3% sucrose, and 0.8% Phytagar (basal medium, BM), under 2.25 Wm^{-2} light intensity provided by Gro-lux fluorescent lights at 28°C in an environmental chamber. Activated charcoal may be added at the rate of 0.3% to enhance germination and reduce browning of medium and explant (114, 118). Addition of hormones was usually found to be inhibitory to normal embryo germination (8, 122) although this is not always the case (12, 66). Embryos enlarge and germinate within 1 to 2 weeks after planting. Following 8 weeks in culture, plantlets will be 2-6 cm in length with a primary root system and a first foliar leaf. It may be necessary to reposition the embryo during this first culture transfer so that the root portion is embedded in the agar medium and the leaves grow upwards. The embryos are recultured every 8 weeks. Isolated coconut embryos were found to germinate poorly *in vitro* but if recultured in medium containing high sucrose (8%) embryo and shoot growth was much enhanced (28).

2.2. Asexual embryos via callus

2.2.1. Explant selection and contamination problems. A variety of explants have been employed to successfully establish embryogenic callus. Highly meristematic tissues such as meristems, shoot tips and axillary buds should be employed as the explants of choice since they are more likely to behave in a highly morphogenetic fashion and should be more likely to exhibit clonal qualities (121). Excised zygotic embryos generally produce morphogenetic callus in about 20% of the cultures with the rest showing only partial morphogenetic ability (122). In no case is a 100% embryogenic callus obtained from every cultured explant. Immature leaves and roots have served as the source of embryogenic callus in oil palms with usually less than 5% producing such callus (18). Immature inflorescence strands obtained from unopened spathes have served as a source of embryogenic callus in date palm and in *Chamaedorea costaricana* (85). Tissues of more mature inflorescences of date palm were less morphogenetically plastic (96). Contamination in palms may occur days, weeks or even months after establishment *in vitro*. It is suspected that there are either internally lodged contaminants or spore borne surface types which resist sterilization techniques. Frequently, when cultures are under stress (e.g. not frequently recultured) contamination is most likely to occur.

2.2.2. Medium composition and cultural techniques. Explants are cultured on BM supplemented with 30-100 mg l^{-1} 2,4-dichlorophenoxyacetic acid (2,4-D), 0-3 mg l^{-1} N^6-(2-isopentenyl) adenine (2iP) and 0.3% activated charcoal. Production of callus from palm explants was achieved through the inclusion of an auxin source in the nutrient medium (14, 15, 85, 100), and was further enhanced by smultaneous inclusion of a cytokinin (such as 2iP) (85). Callus growth usually is slow and may require several weeks or months prior to becoming evident on the explants (106). Explants are recultured at 8 week intervals. When white-yellow, friable nodular callus becomes prominent, 1 cm^3 pieces are subcultured on BM without growth regulators. Asexual embryos and green plantlets usually will become apparent within 2-4 weeks in culture. Once this friable embryogenic callus is obtained it may be proliferated for prolonged periods of time through reculture on medium devoid of growth regulators under light. Apparently, light strongly promotes further callus and plantlet development from subcultures (103).

2.3. Axillary budding in vitro

2.3.1. Explant selection and treatment. Currently, axillary budding is being seriously explored as a type of micropropagation with date palms only (41, 65, 105, 120). Axillary budding *in vitro* is infrequent in coconut palms (38), and *P. canariensis* L. (41), presumably because these species do not produce vegetative axillary buds in nature. In date palm, axillary budding can be induced from offshoot axillary buds, plantlets from somatic callus, shoot tips, or germinated embryos *in vitro* (107). Shoot production *via* inflorescence tissue through reversal of the generative meristem to a vegetative form has been reported in coconut (34) and date palm (84). The same protocol to surface sterilize the explants that produce embryogenic callus is employed to obtain sterile explants of buds and tips to produce axillary bud proliferations.

2.3.2. Medium composition and cultural techniques. Tips and bud explants are established on BM with 0.1 mg l^{-1} 1-naphthaleneacetic acid (NAA) and 5-15 mg l^{-1} 2iP, kinetin or 6-benzylaminopurine

(BA). Incubate cultures under 2.25 Wm^{-2} illumination using a 16 hour photoperiod at 28°C in an environmental chamber. The cultures will initiate leaves and enlarge considerably in size within the next 4-6 weeks in culture. The explant is recultured on fresh media every 8 weeks. Initially axillary budding from cultured tip and bud explants is not as prolific as in some herbaceous species. But through culture habituation the number of buds produced rises per reculture. Generally 1-3 axillary buds can be produced from a single tip every 8 weeks.

2.4. Establishment of free-living plantlets

2.4.1. Plantlet handling in vitro. Once plantlets have been obtained *in vitro* either from germinating zygotic embryos or through somatic embryogenesis or axillary tip development, the mode of their further development follows a similar pattern. Therefore, all plantlets can be treated alike for the purpose of producing free-living palms. Plantlets should be isolated from the tissue of origin and established on basal medium containing 0.1 mg l^{-1} NAA. Under 4.5 Wm^{-2} illumination rooting is obtained from somatic plantlets derived from callus if the primary root is retained (87, 103). Oil palm plantlets derived from callus were initially found to root poorly; this deficiency was overcome through light treatments (19, 21). The plantlets are recultured two or three times every 8 weeks until they reach a length of 10 cm, with 2 to 3 leaves, and has a well-developed adventitious root system. Alterations of the BM composition concentrations were not found to be beneficial to plantlet development in date palm (99).

2.4.2. Plantlet handling in vivo. Establishing palm plantlets in soil requires initial continuation of high humidity levels found in the culture environment followed by a gradual lowering of these levels. This is accomplished in high humidity chambers or with fogging systems. Plantlets should be carefully removed from agar medium without damage to the root systems and soaked in distilled water for 15 minutes to avoid dehydration and to remove excess adhering media. It is desirable for plantlets to be rinsed 3 times with distilled water, and to be sprayed with 0.5% (w/v) benlate (DuPont, Wilmington, Delaware) fungicide solution before transfer to soil medium. The soil medium consists of peat moss and vermiculite in a 1:1 v/v ratio, although a variety of other soil mixtures have been employed without any notable difference in growth responses (99). Plantlets were planted in either 3" diameter plastic pots or jiffy peat pots and enclosed within a transparent tent composed of two interlocking clear polystyrene tumblers. The plants are initially incubated in an environmentally controlled chamber under 500-800 foot-candle light intensity, 16 hour photoperiod at 28°C for 2 weeks, and then transferred to a shaded greenhouse. The plantlets are gradually acclimatized to the greenhouse humidity conditions by punching holes in the plastic cover. After 2 months, the covers may be removed and the plants may be treated as palm seedlings thereafter. When a fogging system is employed plantlets should be kept in 95% relative humidity (RH) for 1-2 weeks in a shaded environment, followed by transfer to 90%, then 80% RH at one week intervals while increasing light intensities. The RH is decreased to 65% using a periodic misting system for 1 to 2 weeks; then the plantlets are transferred to a shaded greenhouse at 40-60% RH for 2-8 weeks; followed by transfer to the outdoors under a shade cloth at 30-40% RH for 2-8 weeks prior to exposure to full sunlight. Once a week 0.5% benlate is administered to the foliage to minimize fungal growth. The pots are watered once a week with one-fourth strength Hoagland's solution during the first two months of development.

2.5. Microbreeding systems: anther, flower and protoplast culture

2.5.1. Explant selection and treatment. Production of embryos from anthers in culture has been reported for dates (15) and coconut (95). Anthers were obtained from flowers prior to their opening. The procedures employed were similar to those used in other anther culture experiments (95). *In vitro* flowering of plantlets derived from callus, germinating zygotic embryos, and cultured shoot tips has been reported (107). In all cases, flowers were produced from axillary buds produced in response to the tissue culture environment. Culture of preformed inflorescence structures, whether as isolated flower buds or as whole immature unopened inflorescence buds has been accomplished in date palm (30, 96, 107). Surface sterilization procedures were the same as employed in the other protocols, except that inflorescence buds could be planted directly without further treatment following surface sterilization of the unopened spathe. In date palms, protoplast production invariably is more successful when using a friable callus than when using leaves and shoot tips (40). Prior soaking of date shoot tips for 24 hrs in BM with 0.3% charcoal, 100 mg l^{-1} NAA, and 3 mg l^{-1} 2iP improved protoplast yields in comparison to no presoaking (40). Rachillae from immature coconut inflorescences also produced protoplasts (42, 51).

2.5.2. Medium composition and cultural techniques. Anthers obtained from unopened coconut spadices were incubated on a modified Blade's or Keller medium supplemented with 6 or 9% sucrose, 15% (v/v) Coconut milk, 0.5% activated charcoal and 2 mg l^{-1} NAA. Androgenic embryos were observed in anther squashes after 6-9 weeks of culture (95). In both date and coconut palms, these embryos failed to develop beyond the early stages of embryogenesis (15, 95). Cultured individual flowers and whole inflorescences buds were short lived structures although they essentially recapitulated the major morphogenetic events common in their *in vivo* counterparts (107). Isolated flowers derived from male inflorescences produced latent female morphological traits (i.e., production of carpels) (30). Protoplast production was accomplished using common conventional techniques and media (8, 40, 51). Callus was derived from coconut, *Erythea edulis* L., and date palm protoplasts.

3. RESULTS AND DISCUSSION

3.1. Survey of palm species cultured in vitro

3.1.1. Growth responses obtained from various species and explants. Approximately 62 species of 38 genera have been tissue cultured to date (104). Of this number 19 species have produced free-living plantlets from germinated zygotic embryos, 4 from asexual plantlets, and 5 from cultured shoot tips (40, 104, 122).

Various abnormalities in seedling and plantlet development have been noted to occur *in vitro*. Plantlets may become disoriented in development resulting in the shoot and root systems becoming intertwined. Other aberrant growth patterns observed are retarded germination, abnormal leaf curling, and leaf swelling. It is not known if these abnormalities are genetic or environmentally-induced in origin.

3.1.2. Similarities and differences of morphogenetic responses obtained among palm species. Recultured palm callus, regardless of the species, was composed of two distinct co-existing types of tissues, friable and nodular components (15, 84). The nodular callus bodies are the source of asexual embryos or are themselves asexual embryos (18, 106). The friable callus appears to be more supportive and nutritional in function and less morphogenetic (106). Root formation from callus often occurs from non-embryogenic callus lines (121). A great deal of genetic variation in the callus population apparently exists (18). Fast growing callus lines in oil palm were usually not embryogenic although exceptions have been noted (18, 76). In date palm, embryogenic callus develops from initially slow growing callus. After 4-6 months in culture this callus is isolated and then rapidly multiplied as callus/embryo mass subcultures. Generally, each palm explant formed a clone *in vitro* with unique growth characteristics (101). Similarly, explants obtained from shoots of clones often exhibit this individualism. Inherent explant character rather than medium composition is the determining factor in obtaining growth or organogenesis from palm tissue cultures (120).

3.2. Practical uses of current tissue culture techniques

3.2.1. Clonal multiplication. Two viable methods of clonal propagation of palms exist: asexual embryogenesis *via* callus, and axillary budding from shoot tips. Both methods can be used to propagate date palm while the other palms are still restricted to the embryogenesis method (41). The latter is much faster than the axillary budding method. Genetic instability of callus derived palm plantlets has yet to be reported (19, 49, 50, 98, 117).

3.2.2. Studies of physiology and morphogenesis. Date palms may reproduce several life-cycle events within a few months time-span *in vitro* while in nature these same events require years to occur. For example in date palms, flowering and suckering occurs in plantlets *in vitro* in 6-12 weeks (107), while in nature, they usually do not produce suckers till their late juvenile-stage at about 3-7 years of age. Flowering and the adult-life cycle stage in nature may occur at 5-7 years. Production of lateral bud outgrowths from palm plantlets derived from zygotic embryos, asexual embryos and shoot tips cultured *in vitro* has been reported (107). These axillary branches may be either vegetative or reproductive in nature.

3.2.3. *Microbreeding*. A number of potential avenues exist in which to induce variation in palms through tissue culture techniques. These include: chemical mutation of cultures, somaclonal variation in plantlets derived from callus, protoplast fusion, flowering/ pollination *in vitro* and anther culture. However, none of these methods are currently being exploited for production of new cultivars.

3.3. Remarks on future emphasis of tissue culture studies to improve techniques to culture palms in vitro

Plant tissue culture has been explored as a technique which could be employed to mass produce desirable palms. Date and oil palm tissue culture techniques are more developed than those for other palms because: 1) they have been the focus in several intense research programs and 2) meristematic tissues of both date and oil palms appear to be more totipotent than those of other palms (15). Coconut palm, although continually studied over the last few decades, has only relatively recently been cloned through tissue culture (15). A number of other palm species were cultured with the *in vitro* techniques developed for date palm using either zygotic embryo or other explants (41, 122). Generally, date palm techniques can be used for other palms to obtain plantlets from germinated excised embryos, shoot tips, or callus. Those palms which grew poorly using currently available techniques, are candidates for more intensive study.

Understanding the mechanism of lateral bud initiation *in vitro* may help to understand morphogenesis in palms grown in the field and greenhouse. Potentially, breeding and physiological studies can be performed year-round with *in vitro* systems (107).

To date, tissue culturists have been preoccupied with micropropagation and have ignored the fact that crop yield and not plant numbers is the objective in agriculture. Suitable product source is immaterial; cost of production is the overriding consideration.

Interest should be focussed on applying morphogenetic knowledge to develop agricultural products *in vitro*. Essential *in vitro* product-design objectives should include: 1) determination of the parameters necessary to produce fruit or other desired products *in vitro*, and 2) develop technology appropriate for efficient economical mass products growth. In order to minimize time, cost, and labor requirements for maximized plantlet and culture development for long-term micropropagation or agri-morphogenetic studies new tissue culture technology should be developed. Existing culture techniques are not suitable to produce large numbers or large-sized sterile palms or palm products efficiently. Use of an automated tissue culture system coupled to enlarged culture vessels, whereby plantlets and cultures grow optimally for months without interference, should be employed (109).

4. REFERENCES

1. ANONYMOUS 1971 Tissue culture of the oil palm (*Elaeis guineensis* Jacq.). CELOS Bull 13: 10-11
2. ANONYMOUS 1981 Yield increase expected from cloned palms. Ceres 14: 10-11
3. ABRAHAM A, KJ THOMAS 1962 A note on the *in vitro* culture of excised coconut embryos. Indian Coconut J 15: 84-88
4. AL-MEHDI AA, L HOGAN 1979 *In vitro* growth and development of papaya (*Carica papaya* L.) and date-palm (*Phoenix dactylifera* L.). Plant Physiol (Suppl) 63: 100
5. AMMAR S, A BENDADIS 1977 Multiplication végétative du palmier dattier (*Phoenix dactylifera* L.) par la culture de tissus de jeunes plantes issues de semis. CR Acad Sci Paris 284: 1789-1792
6. AMMAR S, N DRIRA 1980 Multiplication végétative du palmier dattier (*Phoenix dactylifera* L.) à l'aide des cultures de tissus. Physiol Vég 18: 188-189
7. APAVATJRUT P, J BLAKE 1977 Tissue culture of stem explants of coconut (*Cocos nucifera*). Oléagineux 32: 267-271
8. BALAGA HY, EV DE GUZMAN 1970 The growth and development of coconut "Makapuno" embryos *in vitro*. II. Increased root incidence and growth in response to media composition and to sequential culture from liquid to solid medium. Philipp Agric 53: 551-565
9. BLAKE J 1983 Tissue culture propagation of coconut, date and oil palm. *In* JH Dodds, ed, Tissue culture of Trees. AVI Publ Co, Westport, pp 29-50
10. BLAKE J, CJ EEUWENS 1981 Culture of coconut palm tissues with a view to vegetative propagation. *In* AN Roa, ed, Tissue Culture of Economically Important Plants. COSTED and ANBS, Singapore, pp 145-148

11. BLAKE J, CJ EEUWENS, WW SCHWABE, P APAVATJRUT, RM FULFORD, S JUSTIN, RW SMITH 1975 Progress in the vegetative propagation of coconut, using tissue culture techniques. In Fourth Session FAO Tech Working Party Coconut Prod, Prot and Proc, Kingston, 14-25 September 1975, Food Agric Organ United Nat, Rome, pp 1-6
12. BOUVINET J, H RABÉCHAULT 1965 Effets de l'acide gibberellique sur les embryons du palmier à huile (Elaeis quineensis Jacq. var. dura) en culture in vitro. CR Acad Sci Paris 260: 5336-5338
13. BRANTON RL, J BLAKE 1983 A lovely clone of cononuts. New Sci 98: 554-557
14. BRANTON RL, J BLAKE 1983 Development of organized structures in callus derived from explants of Cocos nucifera L. Ann Bot 52: 673-678
15. BROCHARD P 1981 Cultures de tissus de palmier dattier. Rapport de recherche 1975-1981. Stn Exp De Sidi Mahdi Rep. Inst Nat Rech Agron d'Algérie, pp 1-96
16. BUFFARD-MOREL J 1968 Recherches sur la culture in vitro des embryons de palmier à huile (Elaeis quineensis Jacq. var. dura). V. Effets du glucose, du lévose, du maltose et du saccharose. Oléagineux 23: 707- 711
17. BUFFARD-MOREL J, R VAN DEN DRIESSCHE, B VAZART 1963 Étude de l'influence du bore sur la germination in vitro du pollen de palmier à huile. Oléagineux 18: 35-43
18. CHOO WK, WC YEW, RHV CORLEY 1981 Tissue culture of palms — A review. In AN Roa, ed, Tissue Culture of Economically Important Plants. COSTED and ANBS, Singapore, pp 138-144
19. CORLEY RHV 1977 First clonal oil palms planted in the field. Planter, Kuala Lumpur 53: 331-332
20. CORLEY RHV 1981 Vegetative Propagation. In WR Stanton, M Flach, eds, Sago, The Equatorial Swamp as a Natural Research. Proc 2nd Int Sago Symp, Martinus Nijhoff Publ, The Hague, pp 98-109
21. CORLEY RHV, JN BARRETT, LH JONES 1976 Vegetative propagation of oil palm via tissue culture. Malays Int Agric Oil Palm Conf 1976: 1-7
22. CORLEY RHV, CY WONG, KC WOOI, LH JONES 1980 Early results from the first oil palm clone trials. Malays Int Agric Oil Palm Conf 1980: 1-19 (A-14)
23. CORLEY RHV, KC WOOI, CY WONG 1979 Progress with vegetative propagation of oil palm. Planter, Kuala Lumpur 55: 377-380
24. CUTTER VMJR, KS WILSON 1954 Effect of coconut endosperm and other growth stimulants upon the development in vitro of embryos of Cocos nucifera. Bot Gaz 115: 234-240
25. DE GUZMAN EV 1971 The growth and development of coconut Macapunco embryo in vitro. I. The induction of rooting. Philipp Agric 53: 65-78
26. DE GUZMAN EV, DA DEL ROSARIO 1964 The growth and development of Cocos nucifera L. Makapuno embryo in vitro. Philipp Agric 48: 82-94
27. DE GUZMAN EV, DA DEL ROSARIO 1972 Tissue culture — a tool in coconut propagation and varietal improvement. Kalikasan, Philipp J Biol 1: 241-242
28. DE GUZMAN EV, DA DEL ROSARIO, EC EUSEBIO 1971 The growth and development of coconut Makapuno embryo in vitro. III.

Resumption of root growth in high sugar media. Philipp Agric 53: 566-579

29. DE GUZMAN EV, AG DEL ROSARIO, EM UBALDE 1979 Proliferative growths and organogenesis in coconut embryo and tissue cultures. Philipp J Coconut Stud 7: 1-10
30. DE MASON DA, B TISSERAT 1980 The occurrence and structure of apparently bisexual flowers in the date palm, Phoenix dactylifera L. (Arecaceae). Bot J Lin Soc 81: 283-292
31. DRIRA N 1983 Multiplication végétative du palmier (Phoenix dactylifera L.) par la culture in vitro de bourgeons axillaires et de feuilles qui en dérivent. CR Acad Sci Paris 296: 1077-1082
32. EEUWENS DJ 1976 Mineral requirements for growth and callus initiation of tissue explants from mature coconut palms (Cocos nucifera) and cultured in vitro. Physiol Plant 36: 23-28
33. EEUWENS DJ 1978 Effects of organic nutrients and hormones on growth and development of tissue explants from coconut (Cocos nucifera) and date (Phoenix dactylifera) palms cultured in vitro. Physiol Plant 42: 173-178
34. EEUWENS DJ, J BLAKE 1977 Culture of coconut and date palm tissue with a view to revegetative propagation. Acta Hortic 78: 277-286
35. EL-HENNAWY HM, YA WALLY 1978 Date palm Phoenix dactylifera bud differentiation in vitro. Egypt J Hortic 5: 81-82
36. FINKLE B, JM ULRICH, B TISSERAT 1982 Responses of several lines of rice and date palm callus to freezing at -196°C. In PH Li, A Sakai, eds, Plant Cold Hardiness and Freezing Stress. Academic Press, New York, pp 643-660
37. FISHER JB, JH TSAI 1978 In vitro growth of embryos and callus of coconut palm. In Vitro 14: 307-311
38. FISHER JB, JH TSAI 1979 A branched coconut seedling in tissue culture. Principles 23: 128-131
39. FULFORD RM, SHFW JUSTIN, AJ PASSEY 1976 Vegetative propagation of coconuts. Rep East Malling Res Stn, 1975, Abstr 06009
40. GABR MF, B TISSERAT 1984 Parameters involved in the isolation, culture, cell wall regeneration and callus formation from palm and carrot protoplasts. Date Palm J 3: 359-365
41. GABR MF, B TISSERAT 1985 Propagating palms in vitro with special emphasis on the date palm (Phoenix dactylifera L.). Sci Hortic 25: 255-262
42. HAIBOU TK, A KOVOOR 1981 Regeneration of callus from coconut protoplasts. In AN Roa, ed, Tissue Culture of Economically Important Plants. COSTED and ANBS, Singapore, pp 149-151
43. HANOWER J, P HANOWER 1984 Inhibition et stimulation en culture in vitro, de l'embryogenèse des souches issues d'explants foliaires de palmier à huile. CR Acad Sci Paris 298: 45-48
44. HANOWER J, C PANNETIER 1983 In vitro vegetative propagation of the oil palm, Elaeis guineensis Jacq. In A Fujiwara, ed, Plant Tissue Culture, 1982. Proc 5th Int Cong Plant Tissue Cell Cult, Tokyo, pp 745-746
45. HODEL D 1971 Notes on embryo culture of palms. Principes 21: 103-108
46. HUETE FV, OM ARIAS 1981 Propagacion vegetativa del pejibaye. Asbana 10-13

47. JONES LH 1974 Plant cell culture and biochemistry: studies for improved vegetable oil production. In B. Spencer, ed, Industrial Aspects of Biochemistry. North-Holland Publ Co, Amsterdam-London 2 (part II), pp 813-833
48. JONES LH 1974 Propagation of clonal oil palms by tissue culture. Oil Palm News 17: 1-9
49. JONES LH 1984 Novel palm oils from cloned palms. JAOCS 61: 1717-1719
50. JONES LH, D BARFIELD, J BARRETT, A FLOOK, K POLLOCK, P ROBINSON 1982 Cytology of oil palm cultures and regenerant plants. In A Fujiwara, ed, Plant Tissue Culture 1982. Proc 5th Int Congr Plant Tissue Cell Cult, Tokyo, pp 727-728
51. KOVOOR A 1981 Palm tissue culture: State of the art its application to the coconut. In FAO Plant Prod Prot Pap, Food Agric Organ United Nat, Rome, pp 1-69
52. LIORET C 1980 Vegetative propagation of oil palm by somatic embryogenesis. Malays Int Agric Oil Palm Conf 1980: 1-10 (A-13)
53. MARTIN JP, S CAS, H RABÉCHAULT 1972 Cultures aseptiques de racines de palmier à huile (*Elaeis quineensis* Jacq. var. *dura* Becc. et son hybride *dura* x var. *pisifera* Becc.). CR Acad Sci Paris 274: 2171-2174
54. MARTIN JP, S CAS, H RABÉCHAULT 1972 Note préliminaire sur la culture *in vitro* de racines de palmier à huile (*Elaeis quineensis* Jacq. var. *dura* Becc. et son hybride *dura* x var. *pisifera* Becc.). Oléagineux 27: 303-305
50. MATER AA 1983 Plant regeneration from callus cultures of *Phoenix dactylifera* L. Date Palm J 2: 57-77
56. MAURICE V, CE VANDERCOOK, B TISSERAT 1985 Automated plant surface sterilization system. Physiol Vég 23: 127-133
57. MINIANO AP, EV DE GUZMAN 1978 Responses of non-makapuno embryos *in vitro* to chloride supplementation. Philipp J Coconut Stud 3: 27-44
58. NWANKWO BA, AD KRIKORIAN 1982 Water as a storage medium from *Elaeis quineensis* (*pisifera*) seeds under aseptic conditions. Ann Bot 50: 793-794
59. NWANKWO BA, AD KRIKORIAN 1983 Morphogenetic potential of embryo-and-seedling-derived callus of *Elaeis quineensis* Jacq. var. *pisifera* Becc. Ann Bot 51: 65-76
60. NWANKWO BA, AD KRIKORIAN 1983 Aseptic storage of *Elaeis quineensis* from *pisifera* seeds. Principes 27: 34-37
61. ONG HT 1977 Studies into tissue culture of oil palm. Malays Int Agric Oil Palm Conf 1977: 9-14
62. OPPENHEIMER C, O REUVENI 1972 Development of a method for quick propagation of new and superior date varieties. Agric Res Organ, The Volcani Cent, Bet Dagan, pp 1-92
63. PANNETIER C, J BUFFARD-MOREL 1982 Production of somatic embryos from leaf tissues of coconut, *Cocos nucifera* L. In A Fujiwara, ed, Plant Tissue Culture 1982. Proc 5th Int Cong Plant Tissue Cell Cult, Tokyo, pp 755-756
64. PARANJOTHY K, R OTHMAN 1982 *In vitro* propagation of oil palm. In A Fujiwara, ed, Plant Tissue Culture 1982. Proc 5th Int Congr Plant Tissue Cell Cult, Tokyo, pp 747-748
65. POULAIN C, A RHISS, G BEAUCHESNE 1979 Multiplication végétative en culture *in vitro* du palmier-dattier (*Phoenix dactylifera* L.). CR Acad Agric Fr 11: 1151-1154

66. RABÉCHAULT H 1962 Recherches sur la culture *in vitro* des embryons de palmier à huile (*Elaeis guineensis* Jacq.) I. Effets de l'acide indolyl-acétique. Oléagineux 17: 757-764
67. RABÉCHAULT H 1967 Relations entre le comportement des embryons de palmier à huile (*Elaeis guineensis* Jacq.) en culture *in vitro* et la teneur en eau des grains. CR Acad Sci Paris 264: 276-279
68. RABÉCHAULT H, J AHÉE 1966 Recherches sur la culture *in vitro* des embryons de palmier à huile (*Elaeis guineensis* Jacq.). III. Effets de la grosseur et de l'âge des graines. Oléagineux 21: 729-734
69. RABÉCHAULT H, J AHÉE, G GUÉNIN 1968 Recherches sur la culture *in vitro* des embryons de palmier à huile (*Elaeis guineensis* Jacq.). IV. Effets de la teneur en eau des noix et de la durée de leur stockage. Oléagineux 23: 233-237
70. RABÉCHAULT H, J AHÉE, G GUÉNIN 1970 Colonies cellulaires et formes embryoides obtenues *in vitro* à partir de cultures d'embryons de palmier à huile (*Elaeis guineensis* Jacq. var. *dura* Becc.). CR Acad Sci Paris 270: 233-237
71. RABÉCHAULT H, J AHÉE, G GUÉNIN 1972 Recherches sur la culture *in vitro* des embryons de palmier à huile (*Elaeis guineensis* Jacq. var. *dura* Becc.). VIII. Action du lait de coco autoclavé en présence ou non de gélose et de lumière et en raison de l'âge des graines. Oléagineux 27: 249-254
72. RABÉCHAULT H, J AHÉE, G GUÉNIN 1976 Recherches sur la culture *in vitro* des embryons de palmier à huile (*Elaeis guineensis* Jacq.). XII. Effets de substances de croissance à des doses supraoptimales. Relation avec le brunissement des tissus. Oléagineux 31: 159-160
73. RABÉCHAULT H, S CAS 1974 Recherches sur la culture *in vitro* des embryons de palmier à huile (*Elaeis guineensis* Jacq. var. *dura* Becc.). X. Culture de segments d'embryons. Oléagineux 29: 73-78
74. RABÉCHAULT H, G GUÉNIN, J AHÉE 1969 Recherches sur la culture *in vitro* des embryons de palmier à huile (*Elaeis guineensis* Jacq. var. *dura* Becc.). VI. Effets de la déshydratation naturelle et d'une réhydratation de noix dormantes et non dormantes. Oléagineux 24: 263-268
75. RABÉCHAULT H, G GUÉNIN, J AHÉE 1970 Recherches sur la culture *in vitro* des embryons de palmier à huile (*Elaeis guineensis* Jacq. var. *dura* Becc.). VII. Comparaison de divers milieux minéraux. Oléagineux 25: 519-524
76. RABÉCHAULT H, G GUÉNIN, J AHÉE 1973 Recherches sur la culture *in vitro* des embryons de palmier à huile (*Elaeis guineensis* Jacq. var. *dura* Becc.). IX. Activation de la sensibilité au lait de coco par une réhydratation des graines. Oléagineux 28: 333-336
77. RABÉCHAULT H, JP MARTIN 1976 Multiplication végétative du palmier à huile (*Elaeis guineensis* Jacq.) à l'aide de cultures de tissus foliaires. CR Acad Sci Paris 283: 1735-1737
78. RABÉCHAULT H, JP MARTIN, S CAS 1972 Recherches sur la culture des tissus de palmier à huile (*Elaeis guineensis* Jacq.). Oléagineux 27: 531-534
79. REUVENI O 1979 Embryogenesis and plantlets growth of date palm (*Phoenix dactylifera* L.) derived from callus tissues. Plant Physiol (Suppl) 63: 138

80. REUVENI O, Y ADATO, H LILIEN-KIPNIS 1972 A study of new and rapid methods for the vegetative propagation of date palms. Date Growers' Inst Rep 49: 17-24
81. REUVENI O, H LILIEN-KIPNIS 1969 Date palm. Div Subtrop Hortic Volcani Inst Agric Res 1960-1969. Bet Degan, pp 143-180
82. REUVENI O, H LILIEN-KIPNIS 1974 Studies of the in vitro culture of date palm (Phoenix dactylifera L.) tisssues and organs. Volcani Inst Agric Res, Pamphlet No 145, Bet Degan, pp 1-4
83. REYNOLDS JF 1979 Morphogenesis of palms in vitro. In Vitro 15: 210
84. REYNOLDS JF 1982 Vegetative propagation of palm trees. In JM Bonga, DJ Durzan, eds, Tissue Culture in Forestry. Martinus Nijhoff/Junk Publ, The Hague, pp 182-207
85. REYNOLDS JF, T MURASHIGE 1979 Asexual embryogenesis in callus cultures of palms. In Vitro 15: 383-387
86. RHISS A, C POULAIN, G BEAUCHESNE 1979 La culture in vitro appliquée à la multiplication végétative du palmier-dattier (Phoenix dactylifera L.). Fruits 34: 551-554
87. SAJISE JU, EV DE GUZMAN 1972 Formation of adventitious roots in coconut macapuno seedlings grown in medium supplemented with naphthaleneacetic acid. Kalikasan, Philipp J Biol 1: 197-206
88. SCHROEDER CA 1970 Tissue culture of date shoots and seedlings. Date Growers' Inst Rep 47: 25-47
89. SENTO T 1976 Studies on germination of palm seeds. Mem Coll Agric, Ehime Univ 21: 1-78
90. SHARMA DR, R KUMARI, JB CHOWDHURY 1980 In vitro culture of female date palm (Phoenix dactylifera L.) tissues. Euphytica 29: 169-174
91. SMITH SN 1975 Vegetative propagation of the date palm by root tip culture. Bull Agron Saharienne 1: 67
92. SMITH WK, LH JONES 1970 Plant propagation through cell culture. Chem Ind London 44: 1399-1401
93. SMITH WK, JA THOMAS 1973 The isolation and in vitro cultivation of cells of Elaeis guineensis. Oléagineux 28: 123-127
94. STARTISKY G 1970 Tissue culture of the oil palm (Elaeis guineensis Jacq.) as a tool for its vegetative propagation. Euphytica 19: 288-292
95. THANH-TUYEN NT, EV DE GUZMAN 1983 Formation of pollen embryos in cultured anthers of coconut (Cocos nucifera L.). Plant Sci Lett 29: 81-88
96. TISSERAT B 1979 Propagation of date palm (Phoenix dactylifera L.) in vitro. J Exp Bot 30: 1275-1283
97. TISSERAT B 1979 Tissue culture of the date palm. J Hered 70: 221-222
98. TISSERAT B 1981 Date palm tissue culture. Adv Agric Tech, West Reg, Ser No 17, USDA, ARS, pp 1-50
99. TISSERAT B 1981 Production of free-living date palms through tissue culture. Date Palm J 1: 43-54
100. TISSERAT B 1982 Development of new technology to aid in the cultivation and crop improvement of date palms. In YM Makki, ed, Proc 1st Symp Date Palms, Al-Hassa, 23-25 March, 1982, King Faisal Univ, Al-Hassa, pp 126-139
101. TISSERAT B 1982 Factors involved in the production of plantlets from date palm cultures. Euphytica 31: 201-214

102. TISSERAT B 1983 Tissue cultured palms – a new method to propagate an ancient crop – and a short discussion of the California date industry. Principes 27: 105-117
103. TISSERAT B 1984 Clonal propagation: palm. In IK Vasil, ed, Cell Culture and Somatic Cell Genetics of Plants. Academic Press, New York 1: 74-80
104. TISSERAT B 1984 Palm tissue culture. In DA Evans, WR Sharp, PV Ammirato, Y Yamada, eds, Applications of Plant Tissue Culture Methods for Crop Improvement. MacMillan Press, New York 2: 505-545
105. TISSERAT B 1984 Propagation of date palms by shoot tip cultures. HortScience 19: 230-231
106. TISSERAT B, DA DE MASON 1980 A histological study of the development of adventive embryos in organ culture of *Phoenix dactylifera* L. Ann Bot 46: 465-472
107. TISSERAT B, DA DE MASON 1985 Occurrence and histological structure of offshoots and inflorescences produced from *Phoenix dactylifera* L. plantlets *in vitro*. Bull Torrey Bot Club 112: 35-42
108. TISSERAT B, G FOSTER, DA DE MASON 1979 Plantlet production *in vitro* from *Phoenix dactylifera* L. Date Growers' Inst Rep 54: 19-23
109. TISSERAT B, PD GALLETTA, CE VANDERCOOK 1985 Automation of various plant tissue culture activities using micromachines and microcomputers. In BC Klein-Helmuth, JL Hufnagel, eds, Abstracts of Papers, 1985 AAAS Annu Meet, Los Angeles, 26-31 May 1985, AAAS, Washington DC, pp 129
110. TISSERAT B, MD NELSON, JM ULRICH, BJ FINKLE 1983 Cryostorage techniques to preserve date palm germplasm. In YM Makki, ed, Proc 1st Symp Date Palms, Al-Hassa, 23-25 March 1982, King Faisal Univ, Al-Hassa, pp 108-120
111. TISSERAT B, JM ULRICH, BJ FINKLE 1981 Cryogenic preservation and regeneration of date palm tissue. HortScience 16: 47-48
112. ULRICH JM, BJ FINKLE, BH TISSERAT 1982 Effects of cryogenic treatment of plantlet production from frozen and unfrozen date palm callus. Plant Physiol 69: 624-627
113. VENTURA PF, LC ZUNIGA, JE FIGUEROA, FD LAZO 1966 A progress report on the development of coconut embryo in artificial media. Philipp J Plant Ind 31: 81-87
114. WANG PJ, LC HUANG 1976 Beneficial effects of activated charcoal on plant tissue and organ cultures. In Vitro 12: 260-262
115. WALLY YA, SA HABIB, HM EL-HENNAWY 1980 Anatomical study on date palm *Phoenix dactylifera* shoot tip growth *in vitro*. Egypt J Hortic 7: 221-227
116. WOOI KC, WC YEW, RHV CORLEY 1982 Tissue culture of palms – a review. In AN Rao, ed, Tissue Culture of Economically Important Plants. COSTED and ANBS, Singapore, pp 138-144
117. WOOI KC, CY WONG, RHV CORLEY 1982 Genetic stability of oil palm callus cultures. In A Fujiwara, ed, Plant Tissue Culture 1982. Proc 5th Int Cong Plant Tissue Cell Cult, Tokyo, pp 749-750
118. ZAID A 1984 *In vitro* browning of tissues and media with special emphasis to date palm cultures – A review. Date Palm J 3: 269-275

119. ZAID A, M DJERBI 1984 Up to date bibliography on tissue culture and vascular wilt diseases of date palm and some other palm species. Inst Nat Rech Agron, Organ Nations Unies Alimentation Agric, Food Agric Organ United Nations, Rome, pp 1-51
120. ZAID A, B TISSERAT 1983 _In vitro_ shoot tip differentiation in _Phoenix dactylifera_ L. Date Palm J 2: 163-182
121. ZAID A, B TISSERAT 1983 Morphogenetic responses obtained from a variety of somatic explant tissues of date palms. Bot Mag 96: 67-73
122. ZAID A, B TISSERAT 1984 Survey of the morphogenetic potential of excised palm embryos _in vitro_. Crop Res 24: 1-9

27. ASEPTIC MULTIPLICATION OF CHINESE TALLOW (*Sapium sebiferum* (L.) Roxb.)

A. D. KRIKORIAN and R. P. KANN

1. INTRODUCTION

Most countries are faced with the long term need for independence with regard to energy supply and strategic materials. One important source of such materials is an expanded forestry/agriculture base. This expansion could be achieved, in part, from increased efficiency and overall productivity. However, more importantly, it could be attained by use of lands not currently under cultivation and which are often unsuitable for conventional crops. The key to the use of such areas will be the utilization of plants which are adapted to marginal soils and the development of appropriate cropping systems for them (4, 15, 19).

The Chinese Tallow tree *Sapium sebiferum* (L.) Roxb. (Euphorbiaceae) has been utilized to a greater or lesser extent, especially in its center of origin, China. The tree is now naturalized in many warmer parts of the world. At one time it enjoyed fairly large scale cultivation for its tallow and oil but interest in these products has generally diminished in more recent years. It is currently being re-evaluated in the western hemisphere for its potential as an oilseed and firewood crop and even as a petroleum substitute crop (1, 10, 14). The seeds of this species are of particular interest. Mature seeds borne in capsules are covered with a grayish-white, aril-like, triglyceride-rich coat resembling tallow. Moreover, the seed or kernel contains a liquified drying oil. This is referred to, especially in commerce, as stillingia oil. It is rich in conventional triglycerides as well as a rarer tetraester triglyceride or estolide fraction (15, 16). An especially attractive feature of *Sapium* is its adaptability to saline, poorly drained or otherwise disturbed soils unsuited to ordinary agriculture. In the U.S.A., the Chinese Tallow Tree is frequently grown as an ornamental especially in the southern coastal areas since it is one of the very few deciduous species that shows brilliant red autumn coloration in hot climates (1, 16).

One of a number of problems that will ultimately need to be addressed before the potential for development of Chinese Tallow as a modern corp is fully discernible involves its high variability. Since the tree is outbreeding and the offspring show tremendous variation, the identification and multiplication of appropriate genotypes with specific growth characteristics or qualities is critical. The plant does not, however, readily root from cuttings and when it does, roots are frequently few and not very vigorous. Grafting onto seedling rootstock can be achieved and is one way of achieving more or less uniform plantings but it is a time-consuming procedure. Although grafting is routinely practiced in China (8, 18), aseptic culture methods should have the potential to provide an easier means of producing strictly clonal material in large quantities for evaluation and for further multiplication of material selected in field trials. Aseptic culture techniques applied experimentally to mature trees would provide useful information on the problems that need to be addressed before fully functional industrial propagation systems can be devised. Such experimental work would also have the added advantage of yielding clues as to how to deal with other members of the Euphorbiaceae. This family is of additional interest to tissue culturists because of its widespread capacity to produce latex. Latex production *in vitro* in itself presents a challenge in micropropagation attempts, because it is well-known that latex and its components can be, and often are, antagonistic to vigorous growth of explanted tissues and cells in aseptic culture.

Some studies have been published of the behavior of seedling material of Chinese Tallow in culture (5,21). Seedlings rarely serve as models for older specimens (2, 3, 11). Our experience with Chinese Tallow tree embryos and seedlings (6) has convinced us that from the perspective of clonal multiplication, they provide little information that could be used for the propagation of mature, elite trees. Therefore, future efforts are best directed to elite specimens.

While there are several potentially useful approaches to aseptic propagation systems for mature woody plants, our initial plan has been to assess the formation of multiple shoots via stimulation of precocious axillary branching (7). Also, attempts have been made to assess the role of adventitious bud development as a prospective multiplicative system. A crucial feature of either strategy is that a suitable growth equilibrium needs to be achieved in culture which results in the continued formation of fully differentiated but minimally expanded shoots from the primary explant. When these shoots are separated from the proliferating mass by cutting, they should further develop upon adjusting the medium. Moreover, new proliferations should grow to replace the ones removed. Ideally, the extension growth and rooting of the "micro-cutting" should be achieved in a single step. Also it would be desirable if the shoot could be further reduced to nodal segment explants which, in turn, could re-initiate the entire process (cf. 6).

In this chapter we provide an overview of the potential for micropropagation of Chinese Tallow from mature trees. Special attention is drawn to the problem areas that still need to be investigated.

2. MATERIALS

Chinese Tallow trees do not tolerate the winter conditions of Long Island, New York and, in our hands, have at best repeatedly died back in the field to a perennial rootstock. We have relied on working with plants generated from dormant hardwood cuttings of mature trees from Texas shipped by air to Stony Brook. These have been rooted with varying degrees of success in equal parts of sphagnum, perlite and sand, using commercially available rooting preparation (Hormodin #2 - Merck & Co., Rahway, N.J.-containing 0.3 percent w/w/ IBA, i.e. 300 mg g^{-1} of powder, in an inert talcum carrier). On rooting, these are potted into soil and grown in environmental growth chambers with a 15 h light:9 h dark cycle (c. $21.x10^3$ lu/m^2; 28^oC day and 24^oC night). Vigorous flushing of new growth occurs readily under these conditions and inocula of stem tip and lateral bud explants can be made with virtually no contamination by sterilizing 4 to 6 cm stem lengths as follows. The stem sections are submerged for ten min in a diluted solution of commercial Clorox (1 part bleach:19 parts water or .26 percent sodium hypochlorite) containing a few drops of a wetting agent such as Tween 80. After rinsing the sections four times with sterile glass distilled water, the stem tips are excised and the remaining stems are reduced with a scalpel to small segments, each with one axillary bud. All are placed upright in a medium made semi-solid with 0.7 percent Difco Noble agar.

3. EFFECTS OF DIFFERENT BASAL MEDIA

We have tested a number of basal media: the salts of Murashige and Skoog (13) at full or half-strength (BMS or BMS 1/2), the "Woody Plant Medium" salts of Lloyd and McCown (9)--BWP, and the salts of Schenk and Hildebrandt (17)-- BSH. The BWP has also been tested when supplemented with 170 or 340 mg l^{-1} of $NaH_2PO_4 \cdot H_2O$. Two or three percent sucrose, inositol 100 mg l^{-1}, and full strength vitamins or thiamine.HCl alone have been used throughout our studies; chelated iron is also used (20). pH was adjusted to 5.6 or 5.7 in the case of BSH and BMS and BMS 1/2 respectively; pH 5.2 has been used for BWP prior to autoclaving for 20 min at 120 C.

Our early attempts to work with BMS or BMS 1/2, led to occasional necrosis or even a tendency to form some callus. While the formation of callus in itself is not necessarily deleterious, there is much literature to suggest that a substantial effort would need to be expended to assure that any plantlets generated via callus are clonally stable (12). BSH has given similar results but the plants become chlorotic if they are not frequently transferred. BWP, in general, gives a healthy, more vigorous growth and for that reason, the vast majority of our later experiments have been made using BWP.

4. MULTIPLICATION PHASE

The most useful medium to date for establishing cultures along the lines desired has been BWP supplemented with 44.4 to 88.8 μM benzylaminopurine (BAP) (10 to 20 mg l^{-1}) and 0.2 percent w/v activated charcoal. After several days in this medium, the buds will sprout and elongate. If the cytokinin levels are low in relation to the activated charcoal levels, the buds will merely expand and give rise to individual unbranched shoot systems. However, our objective was to foster a tight, compact branching habit with short internodes, because this would facilitate manipulation and render transfers and subculturing more manageable. More importantly, multiple shoots tend to be succulent and their growth rapid. To reach these objectives higher cytokinin levels are needed. Table 1 shows an evaluation of the "titration" of cytokinin against activated charcoal with notations on the growth mode. There is a fairly wide range of options available to an investigator seeking to stimulate a compact, succulent growth. However, the most persistent problem encountered remains the difficulty with which the desired mode may be sustained for long periods. The biggest problem, one encountered by many investigators, is that there is a substantial tendency for compact cultures to revert to a more normal mode of growth after a certain number of transfers. During this reversion the internodes elongate, petiole extension is prominent and the laminae tend to be broader and more expanded. Even so, multiplication can be achieved here by continued explantation of stem segments with a lateral bud. It is, however, slower. Rooting of over half of the shoots, obtained by forcing of branching, will occur within two weeks if cut and exposed to fresh basal medium supplemented with .05 to 5.4 μM (.01 mg l^{-1} or .1 mg l^{-1}) naphthaleneacetic acid (NAA) without activated charcoal. (Many more will root within three to four weeks). Kinetin or BAP at 8.8 μM or 8.8 to 22.2 μM (2 mg l^{-1} or 2-5 mg l^{-1})

Table 1. Interaction of activated charcoal and BAP on the growth mode of *Sapium* Key: N=normal; FC=fleshy compact growth; D=necrosis followed by death

	activated charcoal (percent w/v)					
BAP (μM)	0	.005	.01	.02	.05	.2
0	N	N	N	N	N	N
.88	FC	N	N	N	N	N
8.8	D	FC→N	N	N	N	N
44.4	D	D	FC	FC→N	N	N
88.8	D	D	D	FC	FC→N	N
222.0	D	D	D	D	FC	-

respectively, does not interfere with root formation and root growth, as long as .05 percent w/v activated charcoal is provided. However, the frequency of root formation and thickness are slightly increased if exogenous cytokinins are absent in the root-inducing medium. Moreover, the speed of root formation is dramatically accelerated when cytokinins are eliminated from the medium. Once rooting has occurred, plantlets are removed from the culture containers, and their roots are freed of agar by thorough washing. Then the plantlets are placed in a high humidity environment away from direct sunlight.

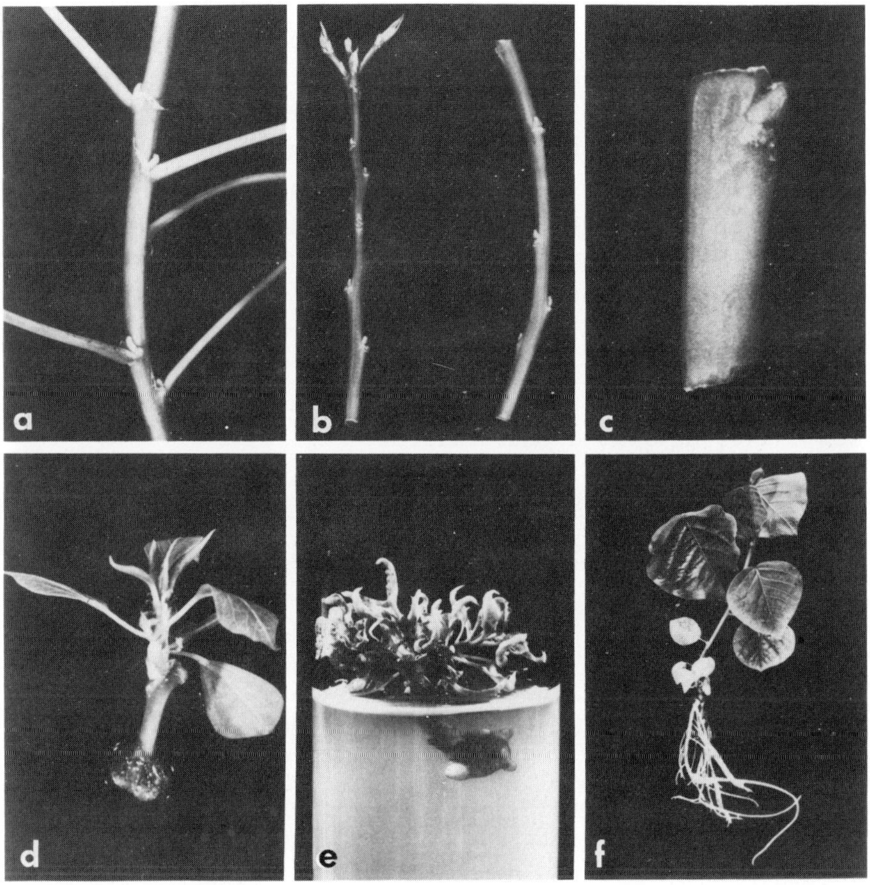

FIGURE 1. Sequence of procedures followed in the multiplication of Sapium from explanted lateral buds. (a) Young branch showing size and stage of development of buds used for primary explantation, X 1.8; (b) the same branch after removal of petioles and cutting to size, ready for surface sterilization, X 1.3; (c) typical primary explant ready to be placed on nutrient medium, X 6.3; (d) typical explant after 25 days growth, X 1.3; (e) another after 57 days growth and a single subculture, X 1.1; (f) rooted plantlet, X 50.

Many investigations have shown that excised aseptically cultured stem tips or axillary buds of woody species are capable of yielding an entire plant. In some cases, only one plant is obtained per stem tip or lateral bud explant and the development of roots from such excised buds or apices is sometimes slow (2, 11). But work on several hardwoods has demonstrated that roots can form promptly enough to render the excised lateral bud or stem tip technique quite useful (cf. 11 and refs. there cited). More significant however is the finding that judicious use of exogenous cytokinins can induce multiple shoots by precocious axillary branching. The multiple shoots which grow from such explants can be separated by cutting, and a profusion of functional roots can be induced to form on them in a few weeks, using exogenous auxin and/or activated charcoal (cf. Fig. 1).

Explanted lateral buds can grow well on BWP with exogenously added cytokinin. BAP ranging from 44.4 to 88.8 μM (10-20 mg l^{-1}) and .02 percent charcoal has been very useful. Frequently, however, a substantial darkening of the material occurs and a high loss is encountered when cultures are subcultured onto a medium with the same composition. This phenomenon was initially observed when cultures were grown on BWP and .88 μM BAP (.2 mg l^{-1}) in the absence of activated charcoal. When the cultures were transferred to the same medium with activated charcoal, darkening or browning of cultures was substantially reduced but the type of growth reverted to the non-branching mode - i.e. the internodes elongated, the laminae expanded etc. This mode is not compatible with multiplication and the desired system of open-ended growth is not achievable if this kind of habit is entrained.

Subculturing on a medium with high enough levels of BAP to maintain the compact, precocious branching habit has for the most part resulted in darkening of cultures but a number do survive the higher cytokinin concentration and even thrive. If transferred to a medium with a lower ratio of BAP to activated charcoal, the browning reactions again seem to be minimized, and the growth obtained has a healthy green color, but here also the desired fleshy and compact type of growth reverts to the non-branching mode. The larger shoots take the lead, and if not subcultured, the smaller shoots cease growth even though they may remain green for well over three months. Even on BWP with 88.8 μM BAP (20 mg l^{-1}) and .03 percent activated charcoal, roots are obtainable on the larger, more normal appearing shoots after two months. This emphasizes that increasing the level of cytokinin either by itself or by adjustment of the level of activated charcoal levels on the downward side in the medium does not invariably lead to a compact multiplicative mode. Attempts to stimulate coppicing along the lines sometimes achievable under field conditions, by cutting stems growing in aseptic culture have not been promising; generally only the uppermost of the axillary buds break their dormancy and grow.

5. ROOTING

BWP as a basal salt medium can support Sapium shoot growth if provided with supplements of sucrose and vitamins. After a few months' growth roots will form occasionally if the shoots had prior exposure to cytokinins. If NAA is added at around .054 μM (.01 mg l^{-1}) (without activated charcoal) about 50 percent of the shoots will root within 2 weeks. The same is true of BWP supplemented wit .54 μM NAA (.1 mg l^{-1}) but roots formed on this medium are generally thickened and are not as long as those initiated on the lower level of the auxin. Raising NAA levels to 5.4 μM (1 mg l^{-1}) often induces rooting quickly (with 2 weeks) but the roots are thicker still and very much shortened. At 53.8 μM NAA (10 mg l^{-1}), growth of the shoots is inhibited and leaves usually turn yellow. Sometimes, abnormally thickened roots do form and grow but there is considerable darkening at the base of the shoots, and, the shoots do not grow.

BWP plus either 0.01 or 0.05 percent activated charcoal tends to initiate roots more often than BWP alone. However, when supplemented with NAA from .054, .54 to 5.4 μM (.01, .1 to 1 mg l^{-1}), the extent and quality of root formation is much improved. Normal roots appear at .54 μM NAA (.1 mg l^{-1}) when .01 percent activated charcoal is used, but in the presence of 5.4 μM NAA (1 mg l^{-1}) and .01 percent w/v activated charcoal, roots are initiated quickly - i.e. within 2 weeks, but they are abnormally thickened.

6. LIQUID CULTURES

Tests to initiate and maintain a precocious branching system of *Sapium* in liquid media have been carried out. Liquid has the advantage of exposing explants more uniformly to a given nutrient medium; it obviates the need to remove any remnants of agar during surgical manipulations and it facilitates maintenance of a high humidity environment for the shoots. Material initiated on agar but transferred into a BWP liquid medium supplemented with low levels of BAP and benzothiazole-2-oxyacetic acid (BTOA) (.09 μM to .22 μM, .02 to .05 mg l^{-1}, .01 to .1 μM, .01 to .02 mg l^{-1} respectively) disclosed that .16 μM (.04 mg l^{-1}) and .1 μM (.02 mg l^{-1}) BTOA is a suitable starting point for further development. Various levels of adenine sulfate .25 μM, 2.5 μM and 25 μM (.1, 1, 10 mg l^{-1}) and soluble polyvinylpyrolidone (PVP 360) at .7 percent have been tested as well. PVP enhances growth of cultures. In general the liquid cultures retain their green color and have a healthy appearance and retain their multiplicative or branching growth modes, at least for short periods. They do, however, tend to revert eventually to an elongated internode mode of growth and hence exhibit the shortcoming of multiplication encountered on semi-solid.

7. VARIABILITY IN RESPONSIVENESS

Experienced investigators are familiar with variable responsiveness of different species of the same genus, different subspecies of the same species, and different varieties, cultivars or even "clones" of the same species. In Sapium the variation in responsiveness has been dramatic. Figure 2 shows three examples of extreme variation according to the elite tree under investigation. Photographs were taken 57 days after initial explantation (cf. Fig. 1). Cultures had been initiated on 30 January 1985, transferred intact without surgical intervention on 22 February and subcultured by separation on 8 March. The medium used was identical in each case.

Such variation makes it difficult to envision routine and indiscriminate application of a single medium to initiate, multiply and maintain a vigorous culture of Sapium. The variation in response, therefore, serves as another example of a "no model" model so often encountered by plant tissue culturists.

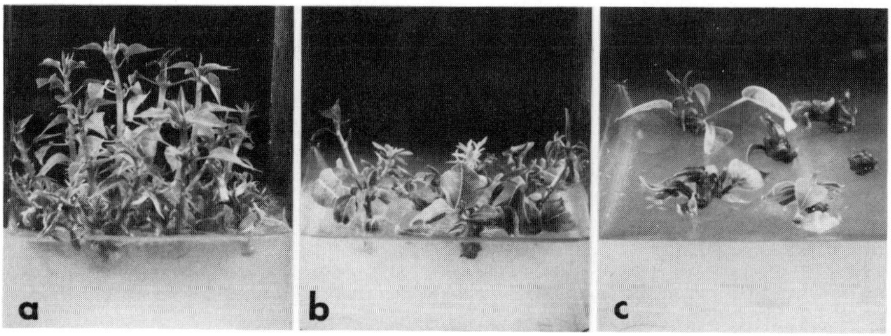

FIGURE 2. Shoot multiplication from three elite trees of Sapium. each picture shows the growth habit that is average for the clone. Note the extreme variation among the specimens. Magnification of a and b is X 0.8; c is X 1.1.

ACKNOWLEDGEMENTS

We thank Dr. H. William Scheld for supplying Sapium cuttings from select trees. Partial support for our preliminary work derived from SIMCO Inc. of Weston, Connecticut. National Science Foundation grant #PCM-8260154 'Clonal Propagation of Chinese Tallow' to one of us (A.D.K.) has funded the more recent studies. Acknowledgement is also made of general laboratory support from a National Aeronautics and Space Administration grant NSG-7270 "Plant Cells, Embryos and Development in Space."

8. REFERENCES

1. ANONYMOUS [SCHELD HW et al] 1983 Sapium-Chinese Tallow Tree. In Firewood Crops: More Shrub and Tree Crops for Energy. Nat Acad Sci, Washinton, DC, pp 46-49
2. BONGA JM, DJ DURZAN 1982 Tissue Culture in Forestry. Martinus Nijhoff/Dr W Junk Publ, The Hague, 420 pp
3. GUPTA PK, AL NADGIR, AF MASCARENHAS, V JAGANNATHAN 1980 Tissue culture of forest trees: Clonal multiplication of Tectona grandis L. (Teak) by tissue culture. Plant Sci Lett 17: 259-268
4. HENRY JF 1979 The sylvicultural energy farm in perspective. In KV Sarkanen and DA Tillman, eds, Progress in Biomass Conversion. Vol 1, Academic Press, Inc New York, pp 215-255
5. KOTWAL MRIDULA M, PK GUPTA, AF MASCARENHAS, 1983 Rapid multiplication of Sapium sebiferum Roxb. by tissue culture. Plant Cell Tissue Organ Cult 2: 133-139
6. KRIKORIAN, AD 1982 Cloning higher plants from aseptically cultured tissues and cells. Biol Rev 57: 151-218
7. KRIKORIAN AD, M SINGH, CE QUINN 1981 Aseptic micropropagaton of Cinchona: Prospects and Problems. In Rao, ed, Tissue Culture of Economically Important Plants. COSTED AND ANBS, Singapore, pp 167-174
8. LIN WC, AC CHEN, CJ TSENG, SG HUANG 1958 An investigation and study of Chinese tallow tree in Taiwan. Formosa For Res Inst Bull 57: 1-37
9. LLOYD G, B McCOWN 1980 Commercially-feasible micropropagation of Mountain Laurel, Kalmia latifolia by use of shoot-tip culture. Proc Int Plant Propagators Soc 30: 421-427
10. MORGAN RP, EB SHULTZ Jr 1981 Fuels and chemicals from novel seed oils. Chem Eng News 59: 69-77
11. MOTT RL 1981 Trees. In BV Conger, ed, Cloning Agricultural Plants via in vitro Techniques, CRC Press, Boca Raton, pp 217-254
12. MURASHIGE T 1974 Plant propagation through tissue culture. Annu Rev Plant Physiol 25: 135-166
13. MURASHIGE T, F SKOOG 1962 A revised medium for rapid growth and bioassays with tobacco tissue cultures. Physiol Plant 15: 473-497
14. SCHELD HW, JR COWLES 1981 Woody biomass potential of the Chinese Tallow tree. Econ Bot 35: 391-397
15. SCHELD HW, NB BELL, GN CAMERON, JR COWLES, CR ENGLER, AD KRIKORIAN, EB SHULTZ 1981 The Chinese Tallow tree as a cash and petroleum substitute crop. In Tree Crops for Energy Co-production on Farms. Proc Solar Energy Res Inst Sponsored Workshop, Nov 11-14, 1980, Estes Park, Colo SERI Conf Nr 801172, Solar Energy Res Inst, Golden, Colorado, pp 97-111
16. SCHELD HW, JR COWLES, EC ENGLER, R KLEIMAN, EB SHULTZ 1984 Seeds of the Chinese Tallow as a source of chemicals and fuel. In RP Morgan and EB Shultz, eds, Fuels and Chemicals from Oilseeds: Technology and Policy Options. AAAS, Washington, DC, pp 81-101
17. SCHENK RV, AC HILDERBRANDT 1972 Medium and techniques for induction and growth of monocotyledonous and dictotyledonous plant cell cultures. Can J Bot 50: 199-204

18. SHIN CL 1973 [Untitled paper in Chinese on the agronomic aspects of the Chinese Tallow tree]. Chekiang For Res Inst Bull 44 pp
19. SHULTZ EB, RP MORGAN, HM DRAPER 1981 Oil seeds for energy in rural areas of developing countries. *In* RS Ganapathy, ed, Agriculture, Rural Energy and Development, Int Assoc Adv Appropriate Tech Dev Countries (IAAATDC), Ann Arbor, pp 485-503
20. SINGH M, AD KRIKORIAN 1980 Chelated iron in culture media. Ann Bot 46: 807-809
21. VENKETESWARAN S, V GANDHI 1982 Mass propagation and genetic improvement of forest trees for biomass production by tissue culture. Biomass 2: 5-15

28. IN VITRO CULTURE OF PROSOPIS SPECIES

M. JORDAN

1. INTRODUCTION - POTENTIALITIES OF PROSOPIS SPECIES IN ARID LANDS.

The northern part of Chile (Lat. 18 - 27°South) has one of the most extreme desert climates in the world. The region is characterized by an almost complete lack of rainfall, i.e., not more than about 1 mm per year is registered. Although practically no vegetation is found in the desert called Pampa del Tamarugal, located between southern Lats. 19° to 22°, the endemic *Prosopis tamarugo* tree, (family Mimosaceae), grows naturally in some areas of this desert. Within the last 30 years about 20.000 ha of mesquite have been planted. Of these approximately 14.000 ha were planted with *P. tamarugo*(4), 3.000 ha with *P. alba* and *P. flexuosa* (24) and 400 ha with mixed populations. Besides these stands there are older ones planted with *P. tamarugo* which cover an additional 3.200 ha. Controversy exists regarding the relationship between water supply and xeromorphic adaptations (13,20,27). It has been suggested that water can be absorbed from the atmosphere and translocated to the roots under certain conditions (26,30). On the other hand, the deep-reaching root system, going sometimes down to 15 m and more, absorbs freatic water moving underground from the Altiplano regions. In some areas the trees also obtain water from seasonal floods that run down from the mountains. In addition to water stress resistance, *Prosopis* trees show a high tolerance to salts. Salts accumulate as crusts upon the soil surface or lower horizons (caliche deposits) as a result of cristalization during water evaporation.

Prosopis spp. are trees or shrubs of great importance to desertification programs in the chilean northern hyperarid region and, probably, for other arid and marginal regions of the world (4,5,7,8,22). Besides erosion control, mesquite trees provide shelter and forage for livestock. Also, the fruit is appreciated for its nutritive value (9,23,29,31), showing a high sugar and protein content.

Prosopis wood was used in the past in the salt (nitrate) mines as a fuel. It is assumed that over-exploitation of endemic tree populations contributed to the degradation of some natural forests that still exist in northern valleys.

In the absence of water, microbial activity cannot provide new organic matter by degradation of litter. Prosopis species are important in this respect, since symbiosis with Rhizobium allows nitrogen fixation, improving soil conditions.

Although P. tamarugo is mainly restricted to the Pampa del Tamarugal, P. chilensis is found farther south in desert regions where higher precipitation rates are recorded (100 - 250 mm). These regions are found as far south as Lat 33°S. and up to elevations of 2500 m above sea level.

The tropical Andean region, from Colombia to the northern Chilean desert and northwestern Argentina, is one of the natural geographic areas of development of Prosopis species. For Chile, Burkart (1) described 6 native species: P. tamarugo, P. alba, var panta, P. flexuosa, P. strombulifera, P. burkartii and P. chilensis var chilensis. Of these P. tamarugo and P. burkartii are considered endemic. Within these species there is much variation in biomass, vigor, phenological traits, growth habits, pod shapes, sizes and colors. There also is great intraspecific variability. In Prosopis spp. intra or interspecific cross pollination occurs frequently, compounding these differences.

Normally, reforestation programmes in Chile have been based on seedlings originating from seeds. Due to the great variability inherent in this method, it would be better to use selected clones. Vegetative propagation will allow multiplication of material selected for best adaptation, biomass production, fruiting characters, disease resistance and other useful traits. These parameters must be studied over long periods in dry climates before implementing major reforestation programs. Preliminary results have shown that some Prosopis, specially P. tamarugo, are rather difficult-to-root species. Experiments carried out with adult tree cuttings showed no rooting response and even cuttings derived from seedlings are hard to root. To accomplish large scale asexual propagation, micropropagation methods may be advisable especially when only small quantities of selected material are available.

2. ORGAN CULTURE

Organogenesis and growth _in vitro_ of _P. cineraria_ have been reported earlier (11,12). Our work with _P. tamarugo_, _P. chilensis_ and _P. alba_ deals primarily with the induction of rhizogenesis to obtain plantlets from single shoot-node sections and shoot-tips in culture (17,18,19). We also worked with shoot-tip callus and cell suspensions.

2.1. Nodal sections

Single shoot-nodes of _P. tamarugo_ were obtained from 9 month old seedlings raised from seed collected from 15-year old trees at the Pampa del Tamarugal. These seedlings were grown in the greenhouse under a regime of 16 h light/day and 28°C. For _P. chilensis_, seeds were chosen from trees of similar age from the Peldehue site near Santiago. Nodes were cultivated on different nutrient media and phytohormone combinations in a growth chamber with a light regime of 18 h/day (100µE $m^{-2}s^{-1}$, Philips TL 40 W/54 fluorescent tubes) at 24-25°C, unless otherwise stated.

Within 10 days the _P. tamarugo_ shoot nodes formed callus on most of the surfaces in contact with the media, on the upper part and proximal end of the nodes. Shortly after that the buds on the nodes flushed; most of the callus tissues turned brown after 2-3 weeks. Rhizogenesis began after three weeks in culture, principally in the light. Roots formed either in the callus tissue or directly on the node (Fig. 1). Under _in vitro_ conditions, _P. chilensis_ nodes initiated a fast growing green callus in all regions of the explants; i.e., at the basal and lateral sites of the nodes, and on leaflet ribs or folioles. Roots formed later either in the callus or directly on the organs.

Morphogenetic events in response to different hormonal treatments are shown in Table 1. Best hormonal concentrations leading to plantlets in _P. chilensis_ were also adequate for _P. tamarugo_, although in the latter regeneration was less frequent.

Table 1. Effect of growth regulators on callus initiation, rhizogenesis and plantlet regeneration of *Prosopis chilensis* and *P. tamarugo* shoot-nodes after 35 days in culture[c] (19).

Phytohormones			% Callus		% Roots		% Bud-break		% Plantlets	
NAA	K ($mg\ l^{-1}$)	GA3	P.c[a]	P.t[b]	P.c	P.t	P.c	P.t	P.c	P.t
1.0	0.10	0.00	76.9	55.5	30.8	11.1	61.5	66.6	23.1	11.1
1.0	1.00	0.00	92.7	12.5	12.2	0.0	68.3	87.5	9.8	0.0
3.0	0.10	0.00	98.6	81.3	30.5	0.0	29.2	18.8	20.8	0.0
3.0	0.01	0.01	85.7	85.7	38.1	14.3	23.8	42.9	33.3	14.3
3.0	1.00	0.00	100.0	62.5	12.5	0.0	31.3	50.0	12.5	0.0
5.0	0.00	0.00	75.4	57.1	50.8	14.3	6.6	57.1	31.1	14.3
10.0	0.01	0.00	97.1	55.5	57.1	0.0	37.1	55.5	42.9	0.0

[a]P.c = *Prosopis chilensis*

[b]P.t = *Prosopis tamarugo*

[c]On Murashige & Skoog (1962) nutrient medium, pH 5.7. NAA levels lower than 1 $mg\ l^{-1}$ or higher than 10 $mg\ l^{-1}$ did not generate plantlets and are omitted from the table.

FIGURE 1. Regeneration responses in nodal sections of *Prosopis* (17).
A. *Prosopis tamarugo* (left) and *P. chilensis* (right) greenhouse-grown plants.
B. Nodal explant of *P. tamarugo* showing callus tissue and roots (arrows) after 30 days. Axillary bud begins growth.
C.D. Direct rhizogenesis in *P. chilensis* (C) and *P. tamarugo* (D) showing elongation and secondary branching of the roots (arrows).
E. Multiple root initiation in callus and ribs in *P. chilensis*.
F. Rhizogenesis in a foliole after contacting nutrient solution on a paper bridge.

Naphthalenacetic acid (NAA) levels above 1 mg l^{-1} are necessary to obtain roots. Kinetin (K) sometimes stimulated bud break of preformed buds and helped to form healthier regenerated plants that survived better after potting and transfer to the greenhouse. Gibberellic acid (GA_3) was not necessary for bud break. In combination with a high level of kinetin, GA_3 was inhibitory to root formation and not conducive to the formation of plantlets.

Responses seem to be affected by differences among the seedling donor plants and by the physiological condition and age of the nodes, and the growth habit of the plantlets. *P. chilensis* plantlets became, after 9 months in the greenhouse, erect plants, 50 - 70 cm high, with leaves in the upper region. Leaves abscised from half way up the stem downwards. In contrast *P. tamarugo* was, at the same age, only 20 - 30 cm long, with a creeping habit and no loss of leaves.

The effects of temperature, light regime and different antioxidants are shown in Tables 2 and 3. In cultures of *P. alba* (Table 2) and *P. tamarugo* (Table 3) root and plantlet formation was better at 25°C and 30°C than at 20°C.

Table 2. Temperature-dependent morphogenic responses of *Prosopis alba* nodal sections grown *in vitro*[a] (19).

Phytohormones NAA (mg l^{-1})	K (mg l^{-1})	Temp. C°	% Callus Initiation	% Root Formation	%Bud break Pre-formed buds	%Plantlets
0.0	0.0	20	55.5	0.0	100.0	0.0
		25	87.5	18.8	62.5	18.8
		30	93.8	6.3	81.3	0.0
3.0	0.1	20	96.2	0.0	55.5	0.0
		25	100.0	4.0	68.0	0.0
		30	93.1	13.7	79.3	10.3
5.0	0.0	20	42.8	4.7	27.2	4.7
		25	31.8	18.8	86.3	18.8
		30	72.7	40.9	22.7	36.3
10.0	0.01	20	96.2	0.0	55.5	0.0
		25	100.0	21.4	17.8	17.8
		30	88.8	7.4	29.6	3.7

[a]On Murashige & Skoog (1962) nutrient medium pH 5.7. Growth regulator combinations that were used were those that gave the best results with *P. chilensis* (Table 1).

Table 3. Effect of antioxidants and temperature on morphogenic responses of *Prosopis tamarugo* shoot-nodes grown *in vitro*[a] (19).

Responses	Cysteine (10mg l^{-1}) light %	Cysteine (10mg l^{-1}) darkness %	P.V.P. (1g l^{-1}) light %	P.V.P. (1g l^{-1}) darkness %	Ascorbic acid (100mg l^{-1}) light %	Ascorbic acid (100mg l^{-1}) darkness %
20°C						
Roots	0	0	0	0	0	0
Callus	3	4	0	12	0	0
Bud-break	38	52	67	24	0	0
Browning	0	0	86	29	0	0
Green leaves	52	72	86	29	100	0
Shed leaves	7	44	0	0	0	0
Plantlets	0	0	0	0	0	0
25°C						
Roots	13	15	13	0	0	0
Callus	29	19	13	33	93	81
Bud-break	42	0	67	61	0	0
Browning	0	0	54	72	29	18
Green leaves	96	0	75	44	–	–
Shed leaves	29	0	0	0	0	0
Plantlets	13	7	8	0	0	0
30°C						
Roots	9	11	0	23	0	0
Callus	9	6	29	82	0	18
Bud-break	82	79	19	36	62	29
Browning	64	29	0	0	81	0
Green leaves	41	61	33	22	91	41
Shed leaves	73	43	0	9	0	0
Plantlets	9	0	0	9	0	0

[a]Nodes were obtained from 9 month-old greenhouse-grown plants on Murashige and Skoog (1962) medium supplemented with NAA 5 mg l^{-1}, light intensity 57 $\mu E\ m^{-2}s^{-1}$, photoperiod 18 hours.

There were no distinct differences in root and plantlet formation between light and dark grown cultures (Table 3). A clear beneficial effect of antioxidants in *P. tamarugo* was not evident, although cysteine reduced browning of the cultures at 20° and 25°C. A higher percentage of browning and shed leaves was found in segments cultured at 30°C. Browning was rare in cultures of *P. chilensis* and *P. alba*.

The differences in growth behaviour and browning *in vitro* between *P. tamarugo*, *P. chilensis* and *P. alba* and the growth habit of intact plants could be due to changes in concentration or some natural compounds in the tissues. We decided to determine if differences in phenolics could explain the differences in behaviour of these three species *in vitro*. Measurements of total phenols (28) showed similar levels, in terms of dry weight, in *P. tamarugo* and *P. chilensis*. In both species the decrease in concentration down the stem was similar but the types of phenolic compounds, detected by cellulose TLC were different (unpublished results). Previous work (25) has shown that there are similarities in flavonoid patterns in *P. chilensis* and *P. alba*; the presence of several other natural compounds has to be considered in further study.

2.2. *Shoot-tips*

During growth or rest periods, shoot-tips proved to be the best explants to regenerate plantlets, due to a lower rate of fungal infection. Normally, root formation occurred within 3 weeks after the tips had resumed growth (Fig. 2). This was observed in *P. chilensis* and *P. alba*, but not in *P. tamarugo* (Table 4). Root primordia were induced in the tissue beneath the endodermal layer (Fig. 2). In *P. chilensis*, primordia grew radially through the endodermis (stained red with safranin), until axillary root(s) protruded through the epidermis. In this way, regenerated plantlets were obtained without subculture. To overcome the low rooting response in the initial experiments, non rooted tips were, in a subsequent experiment, transferred to MS-medium (21) containing 1 mg l^{-1} of indolebutyric acid (IBA) as the only hormone. Thereafter, rhizogenesis was observed within 2-3 weeks. In a series of experiments including benzyladenine (BA) at concentrations higher than 0.5 mg l^{-1}, tips showed multiple shoot formation.

Decapitation enhanced the sprouting of axillary buds in selected material. However, this approach can be used with better results with resting buds.

Sprouting of new buds is important in terms of contamination control since better asepsis is obtained with young tissues.

A second method to obtain pathogen-free shoot-tips of similar age was tried. Juvenile cuttings bearing the three or four uppermost axillary buds (apical tip discarded) were cultured horizontally in liquid media (16) in Erlenmeyer flasks. Appropriate hormone levels (NAA, 0.01 mg l^{-1}; BA, 0.5 mg l^{-1}; GA_3 0.4 mg l^{-1}) enhanced simultaneous sprouting of the buds on each section. Under these conditions, a high number of buds could be obtained within

Table 4. Morphogenic responses of shoot-tips from three *Prosopis* species cultured *in vitro* with different phytohormonal combinations[a] (After 18)

Species	Phytohormones ($mg\ l^{-1}$)			N° Shoot-tips	% Root formation (30 days)	% Shoot development	N° Plantlets	% Rooted tips in subculture (60 days[b])	Total
	NAA	BA	GA3						
P. chilensis	0.01	0.5	0.40	20	0.0	100.0	0	0.0	0
	0.3	0.1	0.01	26	0.0	65.4	0	45.5	5
	1.0	0.1	0.01	29	20.7	96.6	5	25.0	8
	1.0	0.01	0.00	35	11.4	97.1	4	33.3	10
P. alba	0.01	0.5	0.40	19	5.2	94.7	1	0.0	1
	0.3	0.1	0.01	26	0.0	97.5	0	60.0	9
	1.0	0.1	0.01	27	3.7	70.4	1	0.0	1
	1.0	0.01	0.00	28	0.0	96.4	0	40.0	2
P. tamurugo	0.3	0.1	0.01	18	0.0	94.4	0	0.0	0
	1.0	0.1	0.01	18	0.0	88.8	0	0.0	0
	1.0	0.1	0.00	18	0.0	94.4	0	0.0	0

[a] NAA-levels higher than $1mg\ l^{-1}$ formed callus only and are omitted from table. (Basal medium after 16)

[b] Subculture on MS-medium, IBA ($1mg\ l^{-1}$)

two months. Tips and young nodes were then rooted using different phytohormonal combinations as described before.

3. CALLUS AND SUSPENSION CULTURE

3.1. Callus formation and organogenesis

Calli derived from nodal regions, leaflets or folioles were cut in small portions and cultivated in Petri dishes on solidified agarose MS-media with 5 mg l^{-1} NAA and 15 mg l^{-1} cysteine. Callus was also cultured on filterpaper bridges in test tubes with liquid media of the same composition as the solid ones, except that agarose was deleted. Older sections from bigger calli turned brown, especially in regions in contact with the solid media. Nevertheless, yellow or greenish new calli appeared from the central core and grew well. In two cases, small pieces of callus (less than 1 mm in diameter) began to organize, forming a multiple shoot-like structure with green pigmentation. (Fig. 3,A,B,C). The frequency of this response is still too low to determine optimal conditions for regeneration in callus tissues. On the other hand, root formation occurred often, especially in new callus derived and isolated from leaf ribs and folioles still joined to the cultivated tip. Root formation also occurred in the still green folioles on paper bridges in liquid media inside culture tubes.

3.2. Cell suspensions

Cell suspensions were obtained from calli (derived from nodal sections) in liquid medium in Erlenmeyer flasks on an Eberbach shaker (100 rpm at 15 min intervals) under culture conditions as described for nodes and shoot-tips, using MS or Gamborgs' B5-medium (10) supplemented with 5 mg l^{-1} NAA. Under these conditions, cell proliferation continued and, after 15 days, single cells of several sizes and shapes could be observed. After transferring these cells by means of micropipets to subcultures, cell aggregrates appeared after 10-15 days. Using micro-culture slides, sequences of isolated cells were followed, checking viability with bromophenol-blue (0.1%). Single rounded cells divided and raised groups of 4 or more similar cells, in contrast, long narrow cells did not divide. In B5-medium aggregates and possible proembryonic structures were observed later (Fig. 3,D,E,F).

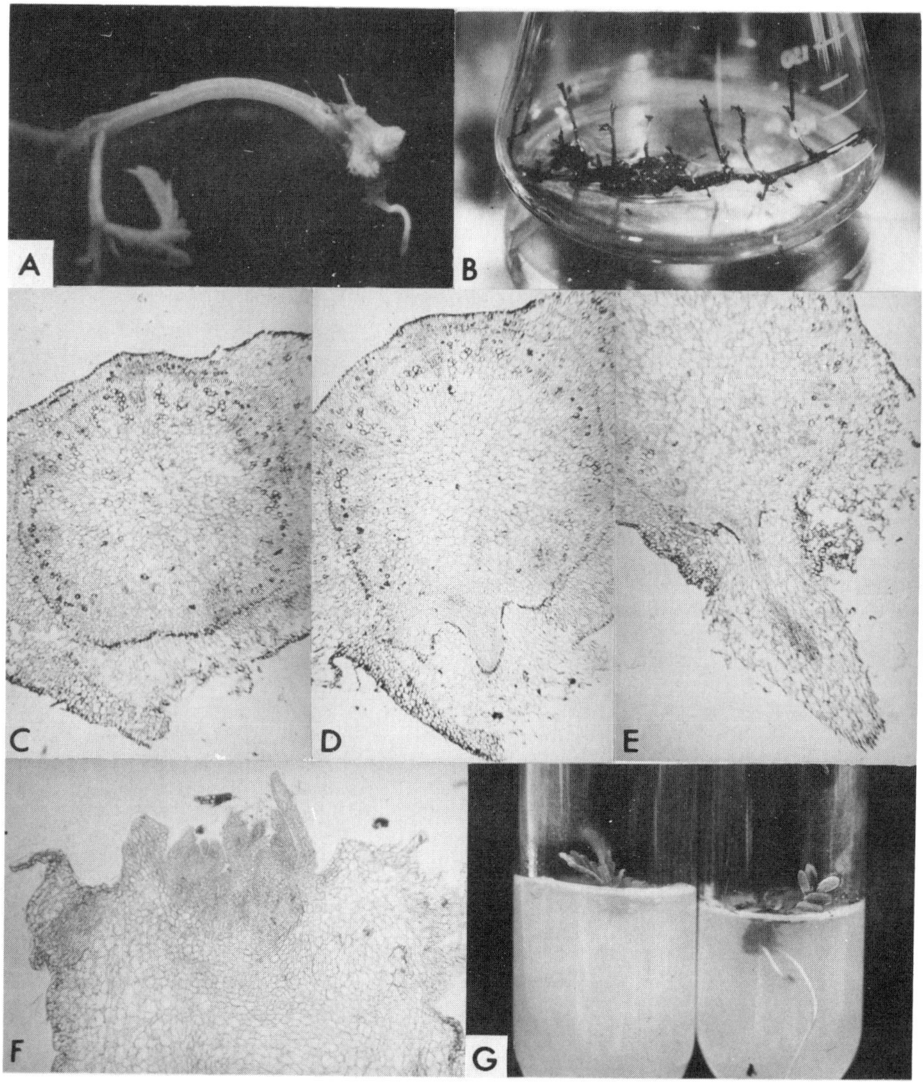

FIGURE 2. Regeneration of <u>Prosopis chilensis</u> through shoot-tips(18).
A. Elongated shoot-tip with leaves and roots after 21 days.
B. Axillary bud-sprouting and callus formation on shoot sections in liquid medium with NAA, 0.01; BA, 0.5; GA3, 0.4 mg l^{-1}.
C.D.E. Cross sections showing callus and rhizogenesis. Endodermis and vascular elements are visible.
F. Multiple-shoot formation in a cultured tip after 4 weeks in the presence of NAA, 0.3; BA, 0.1; GA3 0.01 mg l^{-1}.
G. Root formation and plantlet regeneration from single subcultured shoots grown form apical sections as seen in B.

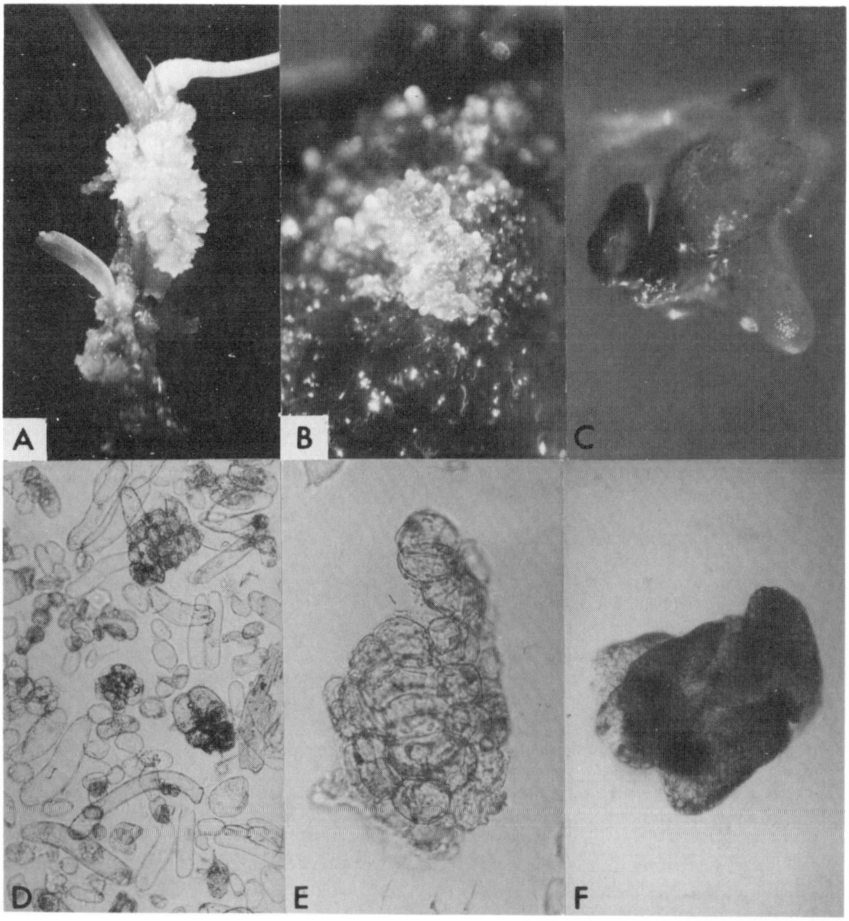

FIGURE 3. Regeneration from callus and cell suspensions.

A. Shoot-node of P. *chilensis* showing intensive callus proliferation along the segment.

B. Close-up of a P. *tamarugo* callus. New whitish callus proliferates well despite surrounding older brown tissues.

C. Close-up of a shoot-like structure developed in subculture from a whitish globular outgrowth of callus.

D. Suspension of P. *chilensis* cells. Dark spots (small cell clusters) may be the origin of proembryoids.

E.F. Proembryoid(s) isolated from a cell suspension in Gamborg's *et al*. B5 -medium after 2-3 weeks in the presence of 2,4-D. (Jordán and Cortés, unpublished observations).

4. CONCLUDING REMARKS

Several types of *Prosopis* explants have the potential for asexual multiplication *in vitro*, despite limitations such as browning in some species and contamination. At the present time, nodal sections, shoot-tips, callus and suspensions obtained from young material perform organogenesis and/or possibly the initial stages of embryogenesis. Genetic and season-dependent variation may influence responses in the tested material. Furthermore, the juvenile condition of donor plants is a major factor which triggers morphogenic events *in vitro* as well as *in vivo*. Air-layering or trench-layering assayed with 9-month old seedlings or trees or *P. chilensis* and *P. tamarugo* were not successful. Cuttings from adult trees of *P. tamarugo* soaked in various phytohormonal concentrations in dimethylsulfoxide showed no rooting response in contrast to the results reported for other *Proposis* species (6). On the other hand, juvenile cuttings that included the 4 uppermost buds or leaves rooted if treated with IBA in non sterile conditions but only if constantly supplied with air (Arce pers. comm.).

The IBA presumably promoted downward translocation of endogenous root-promoting substances from the upper leaves to the rooting zone (14). The oxygen may have favored this transport by changing conditions in the tissues from anaerobic to aerobic and/or by promoting leaching of inhibitors from tissues. It is also possible that rooting is promoted by linked IAA-aminoacids. These are stable in the presence of peroxidases and occur in legume seeds (3). *In vitro* responses of *Prosopis* explants may also be affected by such naturally occurring compounds as non protein aminoacids, e.g. pipecolic acid and 4-hydroxipipecolic acid, proline, and alkaloids (2,15,25). The function of these natural compounds in triggering morphogenesis and regeneration in mesquites will have to be studied further.

5. ACKNOWLEDGMENT

Financial support for this research was provided by the U.S. National academy of Sciences/National Research Council by means of a grant from the U.S. Agency for International Development, Grant N° FGT-CL-1-83-31 to Prof. Dr. Orlando Balboa. We wish to especially thank Dr. F.Felker for valuable information about

Prosopis growth requirements. The authors are also grateful to Miss I. Cortés, Miss A. Goreux, Mr. P. Arce, Mr. H. Moya, Mr. J. Pedraza, Mr. C. Roveraro and Mr. S. Teillier.

6. REFERENCES

1. BURKART A 1976 A monograph of the Genus *Prosopis* (Leguminosae subfam. Mimosoideae). J Arnold Arbor 57: 219-525
2. CATES RG, DF RHOADES 1977 *Prosopis* leaves as a resource for insects. *In* EB Simpson, ed, Mesquite. Its Biology in Two Desert Scrub Ecosystems. Dowden, Hutchinson and Ross, Stroudsburg, Pa, pp 61-83
3. COHEN JD, RS BANDURSKI 1978 The bound auxins: Protection of indole-3-acetic acid from peroxidase-catalyzed oxidation. Planta 139: 203-208
4. CORFO-INFOR 1981 Programa Pampa del Tamarugal-Corfo. Estudio de las especies del género *Prosopis* en la Pampa del Tamarugal, Vol II Instituto Forestal, Santiago, Chile, pp 101
5. FELKER P 1979 Mesquite: an all purpose leguminous arid land tree *In* GA Ritchie, ed, New Agricultural Crops AAAs, Westview Press, Boulder Colorado, pp 89-132
6. FELKER P, PR CLARK 1981 Rooting of mesquite (*Prosopis*) cuttings. J Range Manage 34: 466-468
7. FELKER P 1983 Production of woody biofuels from mesquite (*Prosopis* spp.). Annu Rep US Dep Energy, pp 103
8. FELKER P, GH CANELL, PR CLARK, JF OSBORN, P NASH 1983 Biomass production of *Prosopis* species (mesquite), *Leucaena*, and other leguminous trees grown under heat/drought stress. For Sci 29: 592-606
9. FIGUEREIDO AA 1975 Lebensmittelchemisch relevante Inhaltstoffe der Schoten der Algorobeira. Ph D Thesis Julius-Maximilians-Univ, Germany
10. GAMBORG OL, RA MILLER, K OJIMA 1968 Nutrient requirements of suspension cultures of soybean root cells. Exp Cell Res 50: 151-158
11. GOYAL Y, HC ARYA 1981 Differentiation in cultures of *Prosopis cineraria* Linn. Curr Sci 50: 468-469
12. GOYAL Y, HC ARYA 1984 Tissue culture of desert trees: I. Clonal multiplication of *Prosopis cineraria* by bud culture. J Plant Physiol 115: 183-189
13. HULL HM, CA BLECKMANN 1977 An unusual epicuticular wax ultrastructure on leaves of *Prosopis tamarugo* (Leguminosae). Am J Bot 64: 1083-1091
14. JARVIS BC, A BOOTH 1981 Influence of indole-butyric acid, boron, myo-inositol, vitamin D2 and seedling age on adventitious root development in cuttings of *Phaseolus aureus*. Physiol Plant 53: 213-218
15. JEWERS K, MJ NAGLER, ZIRVI KA, F AMIR 1976 Lipids, sterols and a piperidine alkaloid from *Prosopis spicigera* leaves. Phytochemistry 15: 238
16. JORDAN M, L ITURRIAGA, W FEUCHT 1982 Effects of nitrogenous bases on root formation of hypocotyls from *Prunus avium* L. "Mericier" and "Bing" grown *in vitro*. Gartenbauwissenschaft 47: 46-48

17. JORDAN M, O BALBOA 1985 *In vitro* regeneration of *Prosopis tamarugo* Phil. and *Prosopis chilensis* (Mol.) Stuntz from nodal sections. Gartenbauwissenschaft 50: 138-142
18. JORDAN M, J PEDRAZA, A GOREUX 1985 *In vitro* propagation studies of three *Prosopis* species (*P. alba*, *P. chilensis* and *P. tamarugo* through shoot-tip culture. Gartenbauwissenschaft (In press)
19. JORDAN M, G. MONTENEGRO, O. BALBOA, I CORTES 1985 Propagación de plantas económicamente importantes en zonas áridas de Chile. Medio Ambiente (In press)
20. MOONEY HA, SL GULMON, PW RUNDEL, J EHLERINGER 1980 Further observation on the water relations of *Prosopis tamarugo* of the Northern Atacama Desert. Oecologia (Berl) 44: 177-180
21. MURASHIGE T, F SKOOG 1962 A revised medium for rapid growth and bio-assays with tobacco tissue cultures. Physiol Plant 15: 473-497
22. NATIONAL ACADEMY OF SCIENCES 1979 Tropical Legumes: Resources for the Future. Washington DC, pp 153-163
23. PAK N, H ARAYA, R VILLALON, MA TAGLE 1977 Analytical study of tamarugo (*Prosopis tamarugo*) and autochtonous chilean feed. J Sci Food Agric 28: 59-62
24. SIMPSON BB, OT SOLBRIG 1977 Introduction. *In* BB Simpson, ed, Mesquite. Its Biology in Two Desert Scrub Ecosystems. Dowden, Huchinson and Ross, Stroudsburg, Pa, pp 1-25
25. SOLBRIG OT, K BAWA, NJ CARMAN, JH HUNZIKER, CA NARANJO, RA PALACIOS, L POGGIO, BB SIMPSON 1977 Patterns of variation. *In* BB Simpson, ed, Mesquite. Its biology in Two Desert Scrub Ecosystems. Dowden, Hutchinson and Ross, Stroudsburg, Pa, pp 44-60
26. SUDZUKI F 1969 Absorción foliar de humedad atmosférica en tamarugo. *Prosopis tamarugo* Phil. Univ Chile, Fac Agron, Bol Técn 30: 1-23
27. SUDZUKI F, C BOTTI, E ACEVEDO 1973 Relaciones hídricas en tamarugo (*Prosopis tamarugo* Phil.) en la localidad de Canchones. Univ Chile, Fac Agron, Bol Técn 37: 1-23
28. SWAIN T, WE HILLIS 1959 The phenolic constituents of *Prunus domestica*. I. The quantitative analysis of phenolic constituents. J Sci Food Agric 10: 63-68
29. THAYER DW, JO MURRAY 1977 Physiological, biochemical and morphological characteristics of mesquite wood-digesting bacteria. J Gen Microbiol 101: 71-77
30. WENT FW 1975 Water absorption in *Prosopis*. *In* FJ Vernberg, ed, Physiological Adaptation to the Environment. Intext Educ Publ, New York, pp 65-75
31. Yang HH, SP YANG, DW THAYER 1977 Evaluation of the protein quality of single cell protein produced from mesquite. J Food Sci 42: 1247-1250

29. EUCALYPTUS

PK GUPTA and AF MASCARENHAS

1. INTRODUCTION

Eucalyptus trees are native to Australia and also occur naturally on the islands to the north including Timor, New Guinea and the Philippines (22). Two species of *Eucalyptus*, *E. deglupta* and *E. urophylla*, have not been recorded for Australia. These two species of *Eucalyptus* tolerate lower latitudes than any found in Australia, where the northmost point is 10°41'S.

Since the early years of this century several *Eucalyptus* species have been universally identified as promising candidates for rapid production of woody biomass. Their wood is excellent for pulp and paper production and as a source of fuel. The leaves of several species of Eucalyptus yield useful essential oils. Eucalyptus grows under various agroclimates, altitudes and soil types. It is found in arid regions and in areas where the annual rainfall exceeds 100 cm. Eucalyptus trees have been widely planted in South America, Africa, Asia, Spain, Portugal, Middle Eastern countries and North America. The area of commercial *Eucalyptus* plantations today covers more than 4 million hectares in 58 countries. In India different species are planted throughout the country. Among these, *E. tereticornis* commonly known as 'Mysore germ' has been the most extensively grown in varying climatic zones in an estimated 500,000 hectares. The other commercial species include *E. citriodora*, *E. globulus*, *E. torelliana*, *E. grandis* and *E. camaldulensis* (10).

Eucalyptus is naturally regenerated by seed. The breeding of Eucalyptus is a slow and difficult process, because of long generation time and also the problem of carrying out controlled crosses in large numbers. All conventionally used methods of vegetative propagation have been tried with *Eucalyptus* but with most species have generally resulted in failure. By the time the tree has been evaluated, and quality trees have been selected on the basis of rapid growth rates, quality of wood, oil content, disease and insect resistance etc., it has already passed the stage at which it can be propagated vegetatively (31). Tissue culture technology is now making progress for micropropagation of

Eucalyptus species as shown in Table 1 and 2. Unfortunately, as with other vegetative methods, there are also problems with the development of suitable tissue culture methods for adult selected elite trees.

On several plantations of E. torreticornis, E. globulus, E. torelliana and E. camaldulensis in India, fast growing trees have been identified. Their woody biomass is over ten times higher per year than the average. In selected, fast growing E. citriodora trees the total oil content of leaves and the citronellal concentration in the oil have been found to be very high (3.5% total oil [dry wt. basis] as compared to the average of 0.5%). This chapter describes a method for micropropagation of mature, fast growing and high oil containing 'elite' trees of E. citriodora, E. camaldulensis, E. globulus, E. terreticornis, and E. torelliana. Field data of tissue culture plants show promise in re-forestation programs for increasing biomass production.

2. MATERIALS AND METHODS

2.1. Plant materials

Fresh, sprouted, lateral branches (10-20 cm length) containing preformed apical and axillary buds were collected from upper branches of mature fast growing selected 'elite' trees of E. camaldulensis, E. globulus, E. terreticornis, E. torelliana and high oil containing (3.5%) 'elite' trees of E. citriodora at a forest in Tamil Nadu, a state in the south of India. Branches were brought to the laboratory in plastic bags with wet cotton. These 'elite' trees of Eucalyptus were selected by the Tamil Nadu Forest Development Cooperation, India.

2.2. Surface sterilization

Nodal segments (15-20 cm) cut from lateral branches were surface sterilized by the following method: 1) they were washed several times with running tap water; 2) treated with detergent Tween-20 (0.01% v/v) for 2-3 minutes and washed with distilled water (2-3 times); 3) sterilized with 0.05% $HgCl_2$ for 10 minutes and washed with sterile water (6-8 times) in a sterile cabinet. For all species, except E. citriodora, nodal segments 10 mm long, containing preformed buds were inoculated on Murashige-Skoog(MS)-2 medium. E. citriodora was inoculated on (MS-1).

2.3. Culture media

The basal MS medium contained the mineral salts of Murashige and Skoog (1962) (48) and sucrose at 2%, but no edamin, kinetin (KN) and indole-3-acetic acid (IAA). Additions or omissions of supplements were made to the basal medium as given below. Concentrations of supplements are given in brackets in mg/l. In most cases semi-solid media, solidified with 0.8% Difco Bacto Agar, were used unless stated otherwise.

MS-1: MS salts with vitamins and glycine + KN (0·2) + N^6-benzylaminopurine (BAP) (0·3) + calcium pantothenate (0·1) + biotin (0·1) + agar

MS-2: MS salts with vitamins and glycine + KN (0·2) + BAP (0·5) + calcium pantothenate (0·1) + biotin (0·1) + agar

MS-3: MS salts with vitamins and glycine + KN (0·05) + BAP (0·1) + calcium pantothenate (0·1) + biotin (0·1) + agar

MS-4: MS salts with vitamins and glycine (1/2 strength) + α-naphthaleneacetic acid (NAA) (2·0), liquid

MS-5: MS salts with vitamins and glycine (1/2 strength) + IAA (2·0) + NAA (2·0), liquid

MS-6: MS salts with vitamins and glycine (1/2 strength) + IAA (2·0) + β-indolebutyric acid (IBA) (2·0) + β-indolepropionic acid (IPA) (2·0) + NAA (1·0), liquid

MS-7: MS salts with vitamins and glycine (1/2 strength) + IAA (10·0) + IBA (10·0) + IPA (10·0) + NAA (10·0), liquid

MS-8: MS salts with vitamins and glycine (1/2 strength) + activated charcoal (0.25% w/v/) (E. Merck) + agar

All media were adjusted to pH 5.8 and autoclaved (1·1 kg cm^{-2} at 121°C for 20 minutes) after addition of growth regulators.

Semi-solid media were added to Corning brand (25 x 150 mm) tubes (20 ml/tube); whereas, liquid media were added to 100 ml Erlenmyer flasks (10 ml/flask). Test tubes containing liquid medium were provided with a filter paper support.

2.4. Culture condition

Cultures in tubes were incubated for 18 hours at 25°C in light (1500 lux) and for 6 hours at 23°C in the dark, whereas flasks were incubated on a rotary shaker (100 rev/min) with continuous illumination (500 lux) at 27°C.

2.5. Cold treatment

For shoot bud induction, primary cultures of E. citriodora, E. camaldulensis, E. globulus were cold-treated at 15°C for 72 hours in continuous light (500 lux) and then transferred to 25°C growth chamber. Primary cultures of E. torelliana and E. terreticornis were directly incubated at 25°C.

3. RESULTS

3.1. Shake flasks

After 30-35 days in culture, shoot buds sprouted from nodal segments in 35-40% of the cultures in all five species. At this stage, the explants were transferred to fresh medium. Within 10-15 days, 5-8 multiple shoots developed per explant of each species.

3.2. Subculture and elongation

The multiple shoots of E. camaldulensis, E. terreticornis, E. torelliana and E. globulus, growing in shake flasks were separated and transferred to MS-2 medium. Individual shoots of E. citirodora were transferred to MS-1 medium. Within 20-25 days each shoot developed 6-10 multiple shoots. E. citriodora shoots elongated to 40-50 mm, but shoots of the other four species did not elongate more than 10 mm on MS-2 medium. At this stage, they were transferred to either solid MS-3 medium in test tubes or liquid MS-3 medium in shake flasks. Within 8-10 days, the multiple shoots of all four species elongated to 40-50 mm. For subsequent subcultures, individual excised shoots of E. camaldulensis, E. globulus, E. tereticornis, and E. torelliana were placed on MS-3 medium, those of E. citriodora were placed on MS-1 medium.

3.3. Rooting

Shoots which had elongated to 30-50 mm on the subculture medium, were excised for rooting. E. citriodora, and E. torelliana shoots were treated first with MS-4 for 48 hours, E. globulus with MS-5, E. terreticornis, with MS-6, and E. camaldulensis with MS-7 in the dark at 23°C for 72 hours. The base of each shoot (2-5 mm) was dipped in liquid rooting medium (MS-4, MS-5, MS-6 and MS-7). Following this treatment, shoots of all the species were transferred to MS-8 medium and incubated in 18h/day light at 25°C. Within 10-15 days, 50-70% of the shoots of all five species had rooted (Table 3).

3.4. Transfer to field

Generally there is a high mortality when plantlets are transferred from the sterile culture environment to the free-living condition (7). A gradual adaptation to the soil environment is usually required, involving a gradual

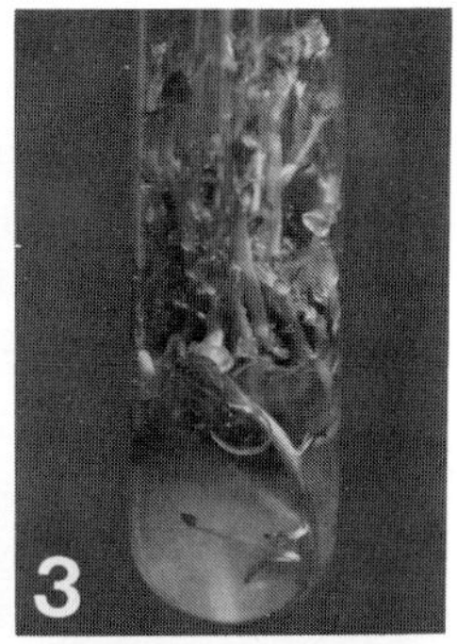
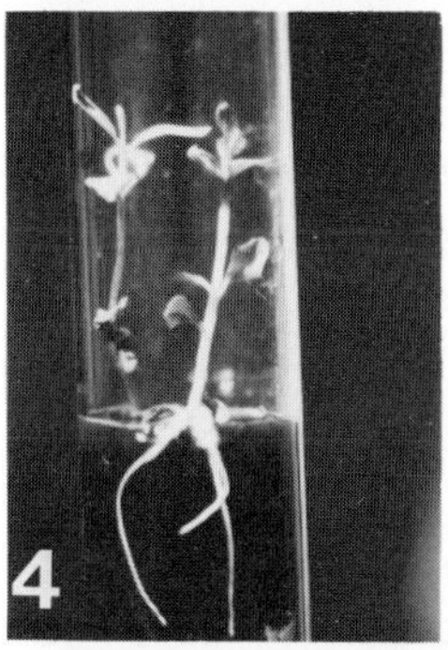

FIGURE 1 Multiple shoot development from explants of "elite" trees of *E. tereticornis* on MS-3 medium in subculture.
FIGURE 2 Multiple shoot development from explants of "elite" trees of *E. camaldulensis* on MS-3 liquid medium in subculture.
FIGURE 3 Multiple shoot development from explants of "elite" trees of *E. citriodora* on MS-1 medium in subculture.
FIGURE 4 Rooted plantlet of *E. citriodora* on MS-8 medium.
FIGURE 5 Tissue culture plants (2 1/2 years old) of *E. citriodora* growing in the field.
FIGURE 6 Flowering of tissue culture plants of *E. citriodora* after 2 1/2 years.
FIGURE 7 "Elite" trees of *E. tereticornis* at a forest in Tamil Nadu, in the south of India.
FIGURE 8 Tissue culture plants (6 months old) regenerated from fast growing 'elite' trees of *E. tereticornis* growing in the field.

simplification of the nutrient medium and increased light intensities (49). Water stress is often a problem and high atmospheric humidities have to be provided. Roots penetrating the nutrient medium often lack root hairs, which are necessary for a high percentage of survival in the field. When plantlets on MS-8 medium attained a height of about 50-60 mm they were carefully removed with a pair of forceps from the culture tubes. Roots were washed thoroughly with tap water to remove traces of nutrient and plants were transferred to small plastic pots (8 x 8 x 4 cm size) containing a mixture of sterile soil and sand in the ratio of 1:1 (v/v). The pots were incubated at $25^{\circ} \pm 2^{\circ}C$ under a 18 hour photoperiod and watered at 48 hour intervals. To prevent dessication the plants were covered with glass beakers. Only 40-60% of the plants survived (Table 3).

Within 25-30 days, new leaves emerged and plants appeared green and healthy. This indicated that the plants had become established in soil. Potted plants were then transferred to a greenhouse where they were kept for 5-6 days without the glass beaker covers for further hardening. Subsequently they were transplanted to earthen pots (30 x 30 x 15 cm size), containing soil, compost and sand and they were watered daily. After 15-20 weeks, when new sets of leaves emerged and the plants had attained a height of 30-40 cm, they were transplanted into pits (90 x 60 x 45 cm depth) filled with a mixture of good soil (river silt) containing 4 kg of farm-yard manure per pit. Before planting, the pits were drenched with 0.1% copper fungicide to avoid fungal attack. The survival in the field at this stage was near 100%. Field plants were watered at three day intervals. Each plant received 15 g of an N:P:K (15:15:15) fertilizer, "suphala" (Fertilizer Corporation of India), for the first 6-8 months at 30 day intervals, after that, once a year. Over 1000 plants of _E. terreticornis_ and _E. torelliana_, and over 200 plants of _E. camaldulensis_, _E. citriodora_ and _E. globulus_ are now growing in fields. Some of these plants are now 4-5 years old. Plants derived from seeds, germinated in pots, were used as control. These seeds were obtained from the same 'elite' clones that were used for micropropagation. Seedlings and micropropagated plants of the same age were transferred to the field at same time and given the same treatment.

4. FIELD DATA

Analysis of total oil and its main components, citronellal and citronellol was determined on a dry weight basis as described earlier (27). Analysis showed that leaves of 6-month- and one-year-old field grown tissue culture plants regenerated from mature 'elite' E. citriodora trees had identical concentrations of total oil, citronellal and citronellol as the parent trees (Table 4). Tissue culture plants, regenerated from mature trees, flowered and set seeds within two and one half years. For plants raised from seed, flowering took longer (4-5 years). Growth of tissue culture plants regenerated from elite trees was 3 times faster and more uniform than growth of the seed-raised plants of the same clones. This shows that clonally propagated tissue culture plants from 'elite' clones retained the characteristic of high oil content and fast growth, while this characteristic was lost in seed-raised plants.

5. CONCLUSION

Field data of plants produced by tissue culture show that: 1) rapid clonal propagation of adult Eucalyptus species is possible; 2) this method can be used to establish seed orchards of selected genotypes; 3) reforestation with millions of plants of superior genotype is feasible if costs of tissue culture plants were competitive to nursery-grown seedlings; 4) Eucalyptus breeding could be shortened, because tissue culture plants regenerated from mature trees flowered earlier than seed raised plants.

Work is in progress to reduce the costs of tissue culture plants by rooting of tissue culture shoots directly in vivo and mechanization of this technique.

6. REFERENCES

1. ANEJA S, CK ATAL, 1969 Plantlet formation from tissue cultures from lignotubers of Eucalyptus citriodora Hook. Curr Sci 38:69-71
2. BADIA N 1981 Eucalyptus rudia technique de micropropagation par la culture de noeuds in vitro. In Proc IUFRO sect s2 01 5. Int Workshop "In Vitro" Cultivation For Tree Species, Fontainebleau, France AFOCEL, pp 135-142
3. BARKER PK, RA DE FOSSARD, RA BOURNE, 1977 Progress toward clonal propagation of Eucalyptus species by tissue culture techniques. Int Plant Propagators Soc Comb Proc 27:546-556
4. BENNETT IJ, JA McCOMB, 1982 Propagation of jarrah (Eucalyptus marginata) by organ and tissue culture. Aust For Res 12:121-127
5. BENNETT IJ, RA McCOMB, 1986 Unpublished (personal communication)
6. BLAKE TJ, 1972 Study on lignotubers of Eucalyptus oleliqua L'Herit. New Phytol, 71, 327-334
7. BONGA JM, 1977 Application of tissue culture in Forestry. In J Reinert, YPS Bajaj, eds, Plant Cell Tissue and Organ Culture, Springer-Verlag Berlin, pp 93-108
8. BOULAY, M. 1983. Micropropagation of frost resistant Eucalyptus. In Proceeding of a Workshop on Eucalyptus in California, June 14-16th, Sacramento, California, pp 102-107
9. BOULAY M, 1986 Unpublished (personal communication)
10. CHATERVEDY AN, 1983 Eucalyptus for farming. UP For Bull, No 48, pp 6-7

11. CRESSWELL RJ, RA DE FOSSARD, 1974 Organ culture of Eucalyptus grandis. Aust For 37:55-69
12. CRESSWELL RJ, C NITSCH, 1975 Organ culture of Eucalyptus grandis L. Planta 125:87-90
13. DE FOSSARD RA, 1974 Tissue culture of Eucalyptus. Aust For 37:43-54
14. DE FOSSARD RA, 1978 Tissue culture propagation of Eucalyptus ficifolia F. Muell. In Proceedings of a Symposium on Plant Tissue Culture. Science Press, Peking, China, pp 425-438
15. DE FOSSARD RA, 1981 Plant Tissue Culture Propagation, Sydney, Film Fiche Corporation
16. DE FOSSARD RA, RA BOURNE, 1976 Vegetative propagation of Eucalyptus ficifolia F. Muell. by nodal culture in vitro. Int Plant Propagators Soc Comb Proc 26:373-378
17. DE FOSSARD RA, PK BARKER, RA BOURNE, 1977 The organ culture of nodes of four species of eucalypts. Acta Hortic 78:297-302
18. DE FOSSARD RA, MT BENNETT, JR GORST, RA BOURNE, 1978 Tissue culture propagation of Eucalyptus ficifolia. Int Plant Propagators Soc Comb Proc 28:427-435
19. DE FOSSARD RA, C NITSCH, RA CRESSWELL, ECM LEE, 1974 Tissue and organ culture of eucalypts. NZ J For Sci 4:267-278
20. DIALLO N, E DUHOUX, 1984 Organogenesis et multiplication in vitro Eucalyptus camaldulensis. J Plant Physiol 115:177-182
21. DURAND-CRESSWELL R, M BOULAY, A FRANCLET, 1982 Vegetative propagation of Eucalyptus. In JM Bonga, DJ Durzan, eds, Tissue Culture in Forestry. Martinus Nijhoff, The Hague, pp 158-181
22. FAO, 1979 Eucalyptus for Planting. FAO
23. FRANCLET A, M BOULAY, 1982 Micropropagation of frost resistant Eucalyptus clones. Aust J For Res 13:83-89
24. GORST JR, 1981 Effect of indole-3-butyric acid and riboflavin on the morphogenesis of adventitious roots of Eucalyptus ficifolia F. Muell. grown in vitro. MSc Thesis, Univ of New England, Armidale, Australia
25. GREWAL S, A AHUJA, CK ATAL, 1980 Clonal multiplication of medicinal plants by tissue culture. 5. In vitro proliferation of shoot apices of Eucalyptus citriodora Hook. Indian J Exp Biol 18:775-776
26. GONCALVES AN, 1975 The growth and developmental physiology of Eucalyptus tissue culture system. MS Thesis, Ohio State Univ, Columbus
27. GUPTA PK, RS NADGAUDA, RR HENDRE, AF MASCARENHAS, V JAGANNATHAN, 1978 Multiplication of Eucalyptus citriodora using seedling segments. In Proc All India Symp and 3rd Conf of Plant Tissue Culture, Feb 7-11th, MS Univ, Baroda, India, pp 63-64
28. GUPTA PK, AF MASCARENHAS, V JAGANNATHAN, 1981 Tissue culture of forest trees - clonal propagation of mature trees of Eucalyptus citriodora Hook by tissue culture. Plant Sci Lett 20:195-201
29. GUPTA PK, AF MASCARENHAS, 1983 Essential oil production in relation to organogenesis in tissue culture of Eucalyptus citriodora Hook. In KL Giles, SK Sen, eds, Plant Cell Culture in Crop Improvement, Plenum Press, New York, pp 299-308
30. GUPTA PK, UJ MEHTA, AF MASCARENHAS, 1983 A tissue culture method for rapid propagation of Eucalyptus camaldulensis and E. torelliana from mature trees. Plant Cell Rep 2:296-299
31. HARTNEY VJ, 1980 Vegetative propagation of the eucalypts. Aust For Res 10:191-211
32. HARTNEY VJ, 1981 Vegetative propagation of eucalypts in vitro. In Proc IUFRO Sect s2 01 5. Int Workshop "In Vitro" Cultivation For Tree Species, Fontainebleau, France, AFOCEL, pp 175-180

33. HARTNEY VJ, 1982 Tissue Culture of Eucalyptus. Int Plant Propagators Soc Comb Proc 32:98-109
34. HARTNEY VJ, PK BARKER, 1980 The vegetative propagation of eucalypts by tissue culture. In Fast Growing Trees IUFRO Congress, Sao Paulo, Brazil, pp 1-7
35. HOLDEN PG, DM PATON, 1981 Sterilization of field grown Eucalyptus for organ culture. J Aust Inst of Hortic 3:5-7
36. ISHII K, T FUKUZUMI, T YOSHIMOTO, 1982 The extractives of tissue culture cells of Eucalyptus dives. Mokuzai Gakkaishi 28:388-392
37. JACQUIOT D, 1964 Application de la technique de culture de tissus végétaux a l'étude de quelque problemes de la physiology de l'arbre. Ann Sci For (Paris) 21:311-465
38. KITAHARA EH, LS CALDAS, 1975 Shoot and root formation in hypocotyl callus cultures of Eucalyptus. For Sci 21:242-243
39. LAKSHMI SITA G, 1979 Morphogenesis and plant regeneration from cotyledonary cultures of Eucalyptus. Plant Sci Lett 14:63-65
40. LAKSHMI SITA G, CS VAIDYANATHAN, 1979 Rapid multiplication of Eucalyptus by multiple shoot production. Curr Sci 48:350-352
41. LAKSHMI SITA G, B SHOHARANI, 1985 In vitro propagation of Eucalyptus grandis L. by tissue culture. Plant Cell Rep 4:63-65
42. LEE ECM, RA DE FOSSARD, 1974 The effects of various auxins and lignotuber tissues of Eucalyptus bancroftii Maiden. New Phytol 73:707-717
43. MARCAVILLACA MC, ER MONTALDI, 1964 Culture "in vitro" de tejidos de eucalipto. IDIA supl 12:62-64
44. MASCARENHAS AF, PK GUPTA, VM KULKARNI, U MEHTA, RS IYER, SS KHUSPE, V JAGANNATHAN, 1981 Propagation of trees by tissue culture. In AN RAO, ed, Proc COSTED Symp on Tissue Culture of Economically Important Plants. Singapore, pp 175-179
45. MASCARENHAS AF, S HAZARA, U POTDAR, DK KULKARNI, PK GUPTA, 1982 Rapid clonal multiplication of mature forest trees through tissue culture. In A Fujiwara, ed, Plant Tissue Culture 1982. Proc 5th Int Cong Plant Tissue Cell Cult, Tokyo, Japan, pp 719-720
46. McCOMB JA, IJ BENNETT, 1982 Vegetative propagation of Eucalyptus using tissue culture and its application to forest improvement in Western Australia. In A Fujiwara, ed, Plant Tissue Culture 1982. Proc 5th Int Cong Plant Tissue Cell Cult, Tokyo, Japan, pp 721-722
47. MEHRA-PALTA A, 1982 Clonal propagation of Eucalyptus by tissue culture. Plant Sci Lett 26:1-11
48. MURASHIGE T, F SKOOG, 1962 A revised medium for growth and bioassay with tobacco tissue culture. Physiol Plant. 15:473-497
49. MURASHIGE T, 1974 Plant propagation through tissue culture. Annu Rev Plant Physiol 25:135-166
50. QUYANG Q, HZ PENG, QQ LI, 1981 Studies on the development of embryoids from Eucalyptus callus. Sci Silvae Sin 17:1-7
51. SAN KARA RAO K, R VENKATESWARA, 1985 Tissue culture of forest trees, clonal multiplication of Eucalyptus grandis L. Plant Sci Lett 40:51-55
52. SUSSEX IM, 1965 The origin and morphogenesis of Eucalyptus cell populations. In PR White, AR Grove, eds, Proc Int Conf Plant Tissue Culture, Berkeley, McCutchan, pp 383-391
53. YAMAGUCHI T, T FUKUZAMI, 1982 Volatile compounds from callus cells of woody plants. In A Fujiwara, ed, Plant Tissue Culture 1982. Proc 5th Int Cong Plant Tissue Cell Cult, Tokyo, Japan, pp 287-288

Table 1. Callus initiation in explants of *Eucalyptus* species.

Eucalyptus species	Explants	References
Eucalyptus alba	Hypocotyls	(38)
E. bancroftii	Stem, lignotuber	(13,19,42)
E. camaldulensis	Cambial tissue	(37,52)
E. cladocalyx	Cambial tissue	(37)
E. gomphocephala	Cambial tissue	(37)
E. grandis	Stem, leaf blades	(13)
	Hypocotyls	(26)
	Anthers	(38)
E. gunnii	Cambial tissue	(37)
E. laevopinea	Stem, leaf, lignotuber	(13)
E. melliodora	Stem, leaf, lignotuber	(19)
E. nicholii	Stem, leaf, lignotuber	(19)
E. obliqua	Lignotuber	(6)
E. robusta	Stem, leaf, lignotuber	(26,53)
E. terreticornis	Cambial tissue	(37,26)
E. x trabuti (hybrid) *botryoides* x *camaldulensis*	Stem segment	(43)
E. viminalis	Lignotuber	(6)
	Juvenile leaves	(36)

Table 2. Morphogenesis in cultues of *Eucalyptus* species

Eucalyptus species	Explants	Growth Response	References
E. alba	Hypocotyl	Plantlets	(38)
E. camaldulensis	Seedling segments	Plantlets	(32,34)
	Seedling segments	Plantlets	(20)
	Mature trees (nodal segments)	Plantlets	(30)
E. calophylla	Seedling segment	Plantlets	(5)
E. citriodora	Lignotuber	Plantlets	(1)
	Seedling segments	Plantlets	(27)
	Seedling segments	Plantlets	(40)
	Cotyledons	Callus, shoots and roots	(39)
	Shoot apices (2-year-old trees)	Multiple shoots	(25)
	Apical and axillary buds (20-year-old trees)	Plantlets	(28) (44,45)
	Callus, seedling explants, 20-year-old trees	Field data and essential oil synthesis	(29)
E. Curtisii	Seedling segments	Plantlets	(30)
E. dalrympleana	Nodes from 2-year-old trees grafted to rejuvenate	Plantlets	(21) (23) (8)
E. diversicolor	10-year-old tree	Plantlets	(5)
E. ficifolia	Seedling segments	Plantlets	(16)
	Seedling segments	Plantlets	(33,34)
	Nodes from mature trees	Plantlets	(14,17,18)
E. globulus	Seedling segments	Plantlets	(24,32, 34)
	Nodal segments (10-20-year-old trees)	Plantlets	(45)
E. grandis	Nodal segments (4-7-month old trees)	Plantlets	(11)

Table 2 continued

Eucalyptus species	Explants	Growth Response	References
	3-year-old trees	Plantlets	(13,17, 12)
	Seedling segments	Plantlets	(32,34)
	Nodes from mature trees	Plantlets	(35,41, 51)
E. gunnii	Nodal segments (2-3-year-old trees) (Frost Resistant clones)	Plantlets	(28,8)
	Nodal segements (seedling)	Embryogenesis	(9)
E. "Leichow"	Seedling segments	Embryoid formation	(50)
E. marginata	Seedling segments	Plantlets	(4)
	axillary buds from mature trees	Plantlets	(46)
E. megacarpa	Seedling segements	Shoots	(5)
E. nicholii	Seedling segments	Rooting	(13)
	Seedling segments	Shoots	(32)
E. nilens	Seedling segments	Shoots	(32,34)
E. nova-anglica	Seedling segments and nodal segments from (4-10-month-old plants)	Plantlets	(47)
E. obtusiflora	Seedling segments	Plantlets	(34)
E. oreades	Seedling segments	Shoots	(32)
E. pauciflora	Seedling segments	Shoots	(33,34)
	Nodal segments (2-year-old trees)	Plantlets	(21)
E. polybractea	Nodal segments from adult trees	Shoots	(17,3)
	Nodal segments from basal coppice of adult trees	Plantlets	(15)

Table 2 continued

Eucalyptus species	Explants	Growth Response	References
E. regnans	Seedling segments	Plantlets	(32,34)
E. rudis	Seedling segments	Plantlets	(33)
	Nodal segments from adult trees	Plantlets	(2)
E. robusta	Seedling segments	Plantlets	(26)
E. tereticornis	Nodal segments (10-20-year-old trees)	Plantlets	(45)
E. torelliana	Nodal segments (10-12-year-old trees)	Plantlets	(30)
E. viminalis	Seedling segments and nodal segments from 4-10-month-old trees	Plantlets	(47)

Table 3. Multiplication of mature elite trees of *Eucalyptus* species.

Eucalyptus Species	Media for initiation of multiple shoots	No. of shoots /primary explant[a]	Media for subculture[b]	No. of shoots /shoot in subculture	Media for rooting treatment	Media for root development	Rooting %	Survival in soil (pots)[c] %
E. citriodora	MS-1	6-8	MS-1	12-15	MS-4	MS-8	70	50
E. camaldulensis	MS-2	2-3	MS-3	6-7	MS-7	MS-8	50	65
E. globulus	MS-2	5-6	MS-3	10-12	MS-5	MS-8	60	70
E. tereticornis	MS-2	5-6	MS-3	8-10	MS-6	MS-8	80	70
E. torelliana	MS-2	6-8	MS-3	10-12	MS-4	MS-8	70	50

[a] 30-40% of nodal segments survived and developed multiple shoots after surface sterilization
[b] Subculture Period - 30 days
[c] 20-50,000 plantlets can be produced from a single nodal segment by this method

Table 4. Oil analysis[a] of tissue culture raised plants grown in field from a mature elite *E. citriodora* trees.

E. citriodora tree	Total Oil %	Citronellal %	Citronellol %
Elite parent tree	3·50	92·00	7·00
6 month old tissue culture plants	3·01 ± 0·12	89·85 ± 6·5	6·5 ± 0·46
One year old tissue culture plants	3·12 ± 0·17	90·15 ± 5·2	6·3 ± 0·38

[a]Oil analysis was carried out with leaves of 10 tissue culture plants from each age group, (6 and 12 months old). Percentage was based on dry weight of the leaves.

30. MICROPROPAGATION OF CASUARINA

M. M. ABOEL-NIL

1. INTRODUCTION

Casuarina species are members of the family Casuarinaceae, which was considered to be a primitive angiosperm, but lately was found to be closely related to Betulaceae and Fagaceae (10, 20, 25, 30). Casuarinaceae is a small family of four genera, which includes many species of evergreen trees and shrubs with a wide natural distribution in Australia, Southeast Asia, Polynesia and New Caledonia (14, 26). Casuarinas are characterized by scale-like univeined leaves arranged in whorls on green spreading branchlets (3, 25). Flowers are unisexual and plants are monocious in most species (23). The staminate flowers are borne in catkins and the pistillate flowers are spherically arranged on cylindrical inflorescent heads (cones). Seeds are formed inside samara fruits, which are enclosed by two valves (20, 21, 24, 29). Casuarinas are able to fix atmospheric nitrogen through symbiosis with an actinomycete (*Frankia*) (11, 23).

Casuarinas are important biological components in moist coastal areas or deserts in their natural habitat which spreads from temperate to tropical climate zones (24, 29). Their ability to tolerate inhospitable environments has been greatly enhanced by their high tolerance to salt and drought conditions and by their ability to fix atmospheric nitrogen. Some species are characterized by rapid foliage and root growth, which enable them to capture their sites efficiently and to play a significant ecological pioneering role (9, 12, 14, 16, 23).

Considerable attention has been given to casuarina because of its good potential for afforestation in regions with poor forest resources (3, 9, 13, 14). In large-scale plantations, casuarinas can be easily established and managed (14). They can be utilized for sand stabilization, shelter-belt, fuelwood, soil rehabilitation and animal fodder (9, 13, 14, 18, 24, 28, 29). Their fast growth and excellent fire wood qualities suggest good potential to overcome fuelwood shortage of many underdeveloped countries (17). Casuarina species that fix atmospheric nitrogen through actinomycete symbiosis have low nutrient requirements and restore soil fertility to agriculturally depleted lands (9, 23).

2. CASUARINA GENETIC TREE IMPROVEMENT

Several casuarina genetic tree improvement programs were established to explore the possibilities of improving and introducing this species into the agroforestry systems of some tropical and subtropical countries (13, 14, 28). Tree improvement starts with provenance trials of seeds from wild or improved populations, followed by cycles of selection and breeding aimed at increasing the frequency of desired genes in subsequent generations (19). Provenance trials showed high inter- and intra-specific genetic variability in a number of important characters, such as; growth rate, stem straightness, and tolerance to adverse environmental conditions (17). Selections were made for several ecotypes. Significant genetic gains were produced and found to be suitable for large-scale plantations (13, 14, 29).

Salt and drought tolerance are among the most desirable traits in breeding of casuarina (9, 14, 17). El-Lakany and Luard (16) found that survival after transplanting, and growth rate of casuarina were affected by soil salinity tolerance. Species were ranked according to their salt tolerance as follows: _C. glauca_, _C. obesa_, _C. equisetifolia_, _C. glauca_ x _C. cunninghamiana_ hybrid, _C. cristata_, _C. cunninghamiana_, _C. stricta_, _C. littoralis_, _C. torulosa_, and _C. decaiseneana_ (16). _Casuarina cunninghamiana_, _C. glauca_ and _C. equisetifolia_ are the most planted species and all are salt tolerant (9, 14, 17). _Casuarina glauca_ was reported to be also tolerant to water logging, calcareous or heavy clay soils (4, 15, 17).

In a study comparing C. cunninghamiana with C. glauca, El-Lakany and Shepherd (17) found that height, diameter, branch length and number of cladodes per branch varied significantly within and between the species. Generally, C. glauca grew faster than C. cunninghamiana. The latter had a thicker stem diameter and less total biomass (14, 17). Variations observed in this study suggested a good potential for genetic gains by selection and breeding for shelterbelt plantation or timber production.

3. CASUARINA VEGETATIVE PROPAGATION

Natural regeneration of casuarina is by seed or coppice (24). It is propagated commercially by sowing seeds in nursery beds. Vegetative propagation by air layering, suckering and rooting of cuttings is possible but the rate of multiplication is slow (14, 24, 34, 35). Application of auxin to girdled air layers stimulated rooting in young trees if treated between April and June, but rooting decreased sharply with aging of ortets (34). Treating lateral tender cuttings of C. equisetifolia with a commercial rooting hormone induced rooting in 42% of treated cuttings after 104 days of planting, while untreated cuttings did not root (35).

Techniques for vegetative propagation can be very useful tool in tree improvement program to bring together all superior and desirable clones for clonal testing and breeding. In afforestation and reforestation programs, a clonal plantation of uniform superior clones is a promising method to achieve high levels of genetic gain in the shortest possible time by capturing all additive and dominance gene effects (19). In vitro Vegetative propagation may provide better multiplication rates than conventional methods of air layering or rooting of cuttings as with many horticultural plants (32). So far, attempts to culture cells or tissues of casuarina have not been reported (5, 6, 8, 31).

4. CASUARINA TISSUE CULTURE

Micropropagation was achieved by shoot regeneration from stem segments and callus cultures of juvenile seedlings of *C. cunninghamiana*, *C. glauca*, their interspecific hybrid and *C. equisetifolia* and of mature flowering male and female trees of *C. equisetifolia* (AboEl-Nil, unpublished data).

4.1. Explants

Two types of explants were used to regenerate shoot and to induce callus formation.

4.1.1. *Seedling explants*. Seeds were germinated in peat: vermiculite: perlite mixture of 1:1:1, watered by sub-irrigation and incubated in 27°C under 16 hours photoperiod. Epicotyls were excised and used as explants when they were about 3cm long.

4.2.2. *Mature trees*. Stem tips of about 5cm long were obtained from lateral branches high in the crown of five year old flowering trees grown in the field.

4.2. Sterilization and establishment.

Explants were washed in 3% Alconox solution for 10 minutes then sterilized in 1.2% sodium hypochlorite solution for 10 minutes. Explants were then rinsed five times in sterile water in a sterile atmosphere. Segments of about 4mm were excised from stems and cultured on the surface of agar gelled media. The medium used was that of Murashige and Skoog (MS) (33), except for the growth regulators which varied from one culture stage to the other as indicated below. All cultures were incubated at 21±1°C under 16-hour photoperiod at 70 $\mu E/m^2/sec$ from cool white fluorescent lamps.

4.3. Callus initiation.

Callus was induced from juvenile and mature stem segment explants on MS medium supplemented with 5.0 μM dimethylallyl aminopurine (2iP) and naphthaleneacetic acid (NAA) (Table 1). Callus growth was induced on media supplemented with 0.005 μM NAA, while buds and callus were formed at lower concentrations of NAA (Fig. 1 A, B). Juvenile explants produced more buds than mature explants of *C.*

equisetifolia. At high NAA concentration of 0.5 μM only green friable callus was formed.

TABLE 1. Morphogenesis of *Casuarina equisetifolia*, *C. glauca*, *C. cunninghamiana*, *C. glauca* x *C. Cunninghamiana* hybrid stem segment explants and *C. equisetifolia* mature stem segments on MS medium supplemented with 5.0 μM 2iP and three concentrations of NAA

Explant	NAA concentrations in μM			
	0.5	0.05	0.005	0.0
Seedling epicotyl segments	Green callus and small roots	Few buds and callus	Buds and callus	Elongation of explants
Mature stem segments	Friable green callus	Few buds and callus	Buds	Elongation of explants

4.4. Organogenesis

Buds were regenerated from callus tissue and stem segment explants on MS media containing 6-benzylaminopurine (BA) at concentrations from 11.0 to 2.2. μM combined with indoleacetic acid (IAA) at 0.5 μM (bud induction medium) (Fig. 1 C, D, F) and (Table 2). In two consecutive subcultures at six-week intervals, buds continue to regenerate and elongate in the absence of exogenous auxin. Bud formation frequency was lower and callus growth was slower in the absence of auxin than when media was supplemented with auxin. Exogenous cytokinin was essential for bud differentiation from callus and stem segments. At the high auxin concentration of 0.5 μM, a dark green friable callus formed on cultured explants of mature trees. This callus did not regenerate buds readily. Dark green friable callus differentiated buds at low frequency when cultured three times on a medium containing 4.4 μM of BA. Callus on explants of juvenile or mature specimens, cultured on media with 5 µM of 2iP combined with 0.05 to 0.005 µM of NAA, was granular in texture, pale green and slow growing (Fig. 1 E). This callus regenerated buds readily on a wide range of auxin and cytokinin combinations.

Buds elongated into shoots upon long incubations on bud induction medium or transfer onto the same basal medium containing 2.2. µM of BA and 3 g/l of activated charcoal (Fig. 1 G).

TABLE 2. Morphogenesis of *Casuarina glauca* callus on MS medium supplemented with two concentrations of IAA and three concentrations of BA

IAA concentrations in µM	BA concentrations in µM			
	11.0	4.4	2.2	0.0
2.8	Shoots with axillary branching and callus growth	Shoots, callus, no branching	Few shoots pale green callus growth	Roots and callus growth
0.0	Shoots with axillary branching	Shoots axillary branching	Shoot with no branching	Poor callus growth

4.5. Rhizogenesis

Roots were formed frequently on stem segment explants and on callus surface in contact with the medium when cultured on media with NAA at 0.5 µM or IBA at 6 µM. This observation indicated that the genus is easy to root in culture and that rooting requirements are broad. Rooting of elongated shoots was induced at high frequency on MS medium containing 6 µM of IBA and 3g/l of activated charcoal (Fig. 1 H). Rooting and root growth was achieved after four-week incubation on rooting medium. Rooted shoots were established *extra vitrum* by transferring into a soil mixture of peat:

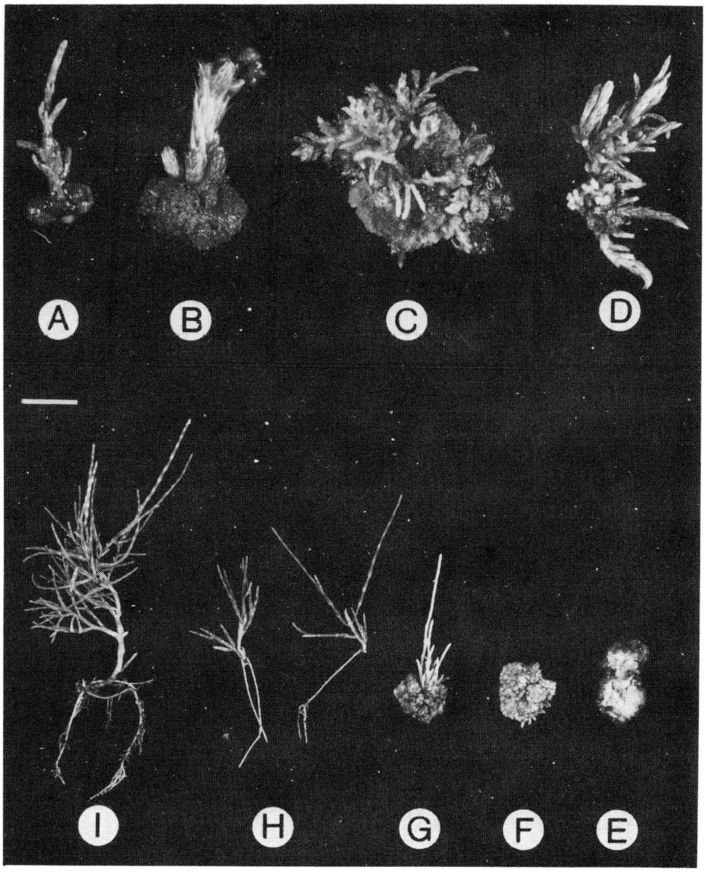

FIGURE 1

In vitro morphogenesis of Casuarina spp. Sieb. A. Callus formation at base of epicotyl explant cultured on MS medium supplemented with 5.0 µM 2iP and 0.05 µM NAA; B. Callus formation at cut ends of a mature stem segment explant cultured on the same medium as in A.; C. Adventitious bud formation from callus growing on medium containing 4.4 µM BA and 2.8 µM IAA; D. Proliferation of axillary buds from epicotyl explant cultured on MS medium containing 5.0 µM 2iP and 0.005 µM NAA; E. Callus culture derived from an epicotyl explant grown on the same medium as in D.; F. Bud differentiation (organogenesis) on callus cultured for 6 weeks on MS medium containing 11.0 µM BA and 2.8 µM IAA; G. Multiple elongated shoots growing on MS medium with 2.2 µM BA and 3 g/l activated charcoal; H. Rooted shoots after 3 weeks incubation on MS medium with 6.0 µM IBA and 3 g/l activated charcoal; I. A plant after extra vitrum establishment. (Scale Bar represents 5 mm for A and C, 2 mm for B, 4 mm for D, and 10 mm for E, F. G. H and I).

perlite: vermiculite at 1:1:1 ratio and incubation in 95% relative humidity for two weeks. Established plantlets were transferred to the greenhouse for further growth before field outplanting. Foliage grown *extra vitrum* was morphologically distinguishable from that formed *in vitro*. The former was much coarser and had less branching than the latter. In the early stage of greenhouse growth, plantlets had from one to four main stems that dominated the rest of the excessively branched crown (Fig. 1 I), while sexually produced seedlings usually only had one main stem with strong apical dominance.

5. CONCLUSIONS

The potential of *Casuarina* spp. for micropropagation by organogenesis has been established. Two approaches were used with juvenile and mature explants alike. One was callus induction followed by bud differentiation, and the other was direct bud induction on stem segment explants. Cultural requirements for regeneration of casuarina were very broad since callus and bud differentiation was achieved on a range of auxin and cytokinin combinations.

It is a prevailing dogma among tissue culture specialists that mature woody and tree species of gymnosperms and angiosperms are difficult to micropropagate (5, 6, 7, 22). A number of mature tree species required special rejuvenation treatments to facilitate their vegetative propagation by tissue culture (2, 22). Casuarina was found to be easy to micropropagating without the special rejuvenation treatments required for many mature tree species.

Application of rapid tissue culture propagation of elite casuarina trees, that are saline and drought tolerant and suitable for economic use as shelterbelt or for wood production, would have a tremendous impact on tree improvement programs and eventually on reforestation (El-Lakany, pers. comm.). Micropropagation of casuarina can be immediately utilized for clonal propagation of unique full-sib hybrids for clonal testing, for multiplication of elite trees that are scarce or endangered, and for germplasm preservation and exchange. Micropropagated plants had more basal branches than seed propagated plants. This trait may be of significance for using casuarina in shelterbelt plantations in which excessive basal branching and wide crown are highly desirable characters in contrast to timber production in which absence of basal branching with narrow crown is required (13).

Major research problems remain to be solved before large-scale micropropagation and outplanting is possible. Induction of somatic embryogenesis (1) in which production costs may be reduced is of special value to cost-effective agroforestry. Confirmation of physiological uniformity and genetic fidelity of plants produced by micropropagation is a prerequisite for any commercial application of this technology. Lastly, automation of _in vitro_ and _extra vitrum_ manipulations, and outplanting mechanization technologies are critical to achieve large-scale production of plantings for forestry applications.

6. ACKNOWLEDGEMENT

I wish to thank Dr. H. M. El-Lakany, Director of Casuarina Tree Improvement and Breeding Project, College of Agriculture, Alexandria University, Egypt, for providing casuarina seeds used in this study and for his helpful communications.

7. REFERENCES

1. ABOEL-NIL MM 1980 Embryogenesis of gymnosperm forest trees. U.S. Patent #4,217,730
2. ABOEL-NIL MM. 1982 Method for asexual reproduction of coniferous trees. U. S. Patent #4,353,412
3. BADRAN OA, MH EL-LAKANY, ML EL-OSTA, HA ABO-GAZIA 1976 Breeding and improving _Casuarina_ trees, I. Taxonomy and morphological characteristics of _Casuarina_ spp. grown in Egypt. Alexandria J Agric Res _24_:683-694
4. BADRAN OA, TA OMRAN, AB EL-FARRAJII 1979 Growth of some timber tree seedlings under different calcium carbonate percentage and irrigation treatment. Alexandria J Agric Res 27:359-375

5. BIONDI S, TA THORPE 1981 Clonal propagation of forest tree species. In AN Rao, ed, Proc COSTED Symposium on Tissue Culture of Economically Important Plants. Singapore pp 197-204

6. BONGA JM 1981 Vegetative propagation of mature trees by tissue culture, In AN Rao, ed, Proc COSTED Symposium on Tissue Culture of Economically Important Plants. Singapore, pp 191-196

7. BONGA JM 1982 Vegetative propagation in relation to juvenility, maturity and rejuvenation. In JM Bonga, DJ Durzan, ed, Tissue Culture in Forestry. Martinus Nijhoff/Junk Publ, The Hague, pp 387-412

8. BROWN CL, H SOMMERS 1982 Vegetative propagation of dicotyledonous trees. In JM Bonga, DJ Durzan, ed, Tissue Culture in Forestry, Martinus Nijhoff/Junk Publ, The Hague, pp 109-149

9. CLEMENS A, LC CAMPBELL, S NURISJAH 1983 Germination, growth and mineral ion concentrations of *Casuarina* species under saline conditions. Aust J Bot 31:1-9

10. CRONQUIST A 1968 The evolution and classification of flowering plants. Houghton Mifflin Co, Boston

11. DIEM HG, Y DOMMERGUES 1983 The isolation of *Frankia* from nodules of *Casuarina*. Amer J Bot 61:2822-2825

12. EL-LAKANY HM 1966 A study on salt tolerance of some timber tree species. M Sc Thesis, Alexandria University, Egypt

13. EL-LAKANY HM 1983 A review of breeding drought resistant *Casuarina* for shelterbelt establishment in arid region with special reference to Egypt. For Ecol Manage 6:129-137

14. EL-LAKANY H M 1983 Breeding and improving *Casuarina*; A promising multipurpose tree for arid regions of Egypt. In SJ Midgley, JW Turnbull, RD Johnson, eds, Casuarina Ecology, Management and Utilization. CSIRO Australia, pp 58-65

15. EL-LAKANY HM, OA BADRAN, SYM AHMAD, M SAAD 1982 Root distribution of young *Casuarina* as influenced by soil characteristics. Egypt J Hortic 9:31-41

16. EL-LAKANY HM, EJ Luard 1982 Comparative salt tolerance of selected *Casuarina* species. Aust for Res 13:11-20

17. EL-LAKANY HM, KR SHEPHERD 1983 Variation in seed germination, seedling growth and biomass between provenance of *Casuarina cunninghamiana* Miq. and *C glauca* Sieb. For Ecol Manage 6:201-216

18. EL-OSTA ML, HM EL-LAKANY, MM MEGAHED 1981 Anatomical characteristics of some *Casuarina* species grown in Egypt. Int Assoc Wood Anat Bull 2:95-98

19. FARNUM P, R TIMMIS, JL KULP 1983 Biotechnology of forest yield. Science 219:694-702

20. FLORES EM 1980 Shoot vascular system and phyllotaxis of *Casuarina* (Casuarinaceae). Amer J Bot 67:131-140

21. FLORES EM, MF MOSELEY JR 1982 The Anatomy of the pistillate inflorescence and flower of *Casuarina verticullata* Lamarck (Casuarinaceae). Amer J Bot 69:1673-1684

22. FRANCLET A 1979 Rajeunissement des arbres adultes en vue de leur propagation vegetative. Micropropagation D'Arbres Forestiers, Etud Res AFOCEL, Nangis, France 12:3-8
23. GAUTHIER D, HG DIEM, YR DOMMERGUES 1981 *in vitro* nitrogen fixation by two actinomycetes strains *isolated* from *Casuarina* nodules. Appl Environ Microbiol 41:306-308
24. GOOR AY, CW BARNEY 1976 Forest Trees Planting in Arid Zones. The Roland Press Co, New York
25. HUTCHINSON J 1959 The Families of Flowering Plants, 2nd ed, Oxford Univ Press, Oxford
26. JOHNSON LA 1982 Notes on Casuarinaceae II. J Adelaide Bot Gard 1:73-87
27. KARNOSKY D 1981 Potential for forest tree improvement via tissue culture. BioScience 31:114-120
28. KONDAS S 1983 *Casuarina equisetifolia* - A multipurpose tree cash crop in India. *In* SJ Midgley, YW Turnbull, RD Johnson ,eds, Casuarina Ecology, Management and Utilization. CSIRO Australia, pp 66-76
29. MIDGLEY SJ, JW TURNBULL, RD JOHNSON 1983 Casuarina Ecology, Management and Utilization. CISRO Australia
30. MOSELEY MF 1948 Comparative anatomy and phylogeny of the Casuarinaceae. Bot Gaz 110:231-280
31. MOTT RL 1981 Trees *In* BV Conger ,ed, Cloning Agricultural Plants via In *Vitro* Techniques, CRC Press, Inc., Boca Raton, Florida, pp 217-254
32. MURASHIGE T 1974 Plant propagation through tissue cultures. Ann Rev Plant Physiol 25:135-166
33. MURASHIGE T, F SKOOG 1962 A revised medium for rapid growth and bioassay with tobacco cultures. Physiol Plant 15:473-499
34. OMRAN TA, HM EL-LAKANY, AB EL-SAYED 1983 Propagation of *Casuarina* by air layers. Alex Sci Exch 22, Alexandria Univ, Egypt
35. SOMASUNDARAM TR, S STANLEY JAGADEES 1977 Propagation of *Casuarina equisetifolia* forest by planting shoots. Indian For 103:735-737

TREE SPECIES INDEX
VOLUME 3

GENERAL INDEX
VOLUME 3